David Fopp
The Youth Climate Uprising

X-Texts on Culture and Society

Editorial

The supposed "end of history" long ago revealed itself to be much more an end to certainties. More than ever, we are not only faced with the question of "Generation X". Beyond this kind of popular figures, academia is also challenged to make a contribution to a sophisticated analysis of the time. The series **X-TEXTS** takes on this task, and provides a forum for thinking *with and against time*. The essays gathered together here decipher our present moment, resisting simplifying formulas and oracles. They combine sensitive observations with incisive analysis, presenting both in a conveniently, readable form.

David Fopp (PhD in philosophy and education) is a climate justice activist and researches theories of sustainability and societal transformation at the Centre Marc Bloch in Berlin. He researched and taught at the universities of Berlin, Basel, Paris (École Normale Supérieure), and Stockholm (child and youth studies, and drama education).
Isabelle Axelsson is a youth climate justice activist and studied human geography at the University of Stockholm, Sweden.
Loukina Tille is a youth climate justice activist and studies political science at the University of Zurich, Switzerland.

David Fopp
The Youth Climate Uprising
From the School Strike Movement
to an Ecophilosophy of Democracy

Revised and expanded new edition
of "Gemeinsam für die Zukunft - Fridays For Future und Scientists For Future",
translated from German by Lucy Duggan

In cooperation with Isabelle Axelsson and Loukina Tille

Bibliographic information published by the Deutsche Nationalbibliothek
The Deutsche Nationalbibliothek lists this publication in the Deutsche Nationalbibliografie; detailed bibliographic data are available in the Internet at https://dnb.dnb.de/

This work is licensed under the Creative Commons Attribution-NonCommercial-NoDerivatives 4.0 (BY-NC-ND) which means that the text may be used for non-commercial purposes, provided credit is given to the author.
https://creativecommons.org/licenses/by-nc-nd/4.0/
To create an adaptation, translation, or derivative of the original work and for commercial use, further permission is required and can be obtained by contacting rights@transcript-publishing.com
Creative Commons license terms for re-use do not apply to any content (such as graphs, figures, photos, excerpts, etc.) not original to the Open Access publication and further permission may be required from the rights holder. The obligation to research and clear permission lies solely with the party re-using the material.

First published in 2024 by transcript Verlag, Bielefeld
© David Fopp

Cover layout: Maria Arndt, Bielefeld
Cover Illustration: Jana Eriksson
Printed by: Majuskel Medienproduktion GmbH, Wetzlar
https://doi.org/10.14361/9783839470312
Print-ISBN: 978-3-8376-7031-8
PDF-ISBN: 978-3-8394-7031-2
EPUB-ISBN: 978-3-7328-7031-8
ISSN of series: 2364-6616
eISSN of series: 2747-3775

Printed on permanent acid-free text paper.

Contents

Timeline 9

Introduction... 15

The strangers 25

Part One: The Young People's Rebellion – From Mynttorget in Stockholm to the Global Strike

Chapter 1: Swedish Beginnings
August – October 2018: An idea takes shape 33

Chapter 2: Fridays for Future and Extinction Rebellion Start to Grow
October – November 2018: Civil disobedience and the laws of humanity 61

Chapter 3: The Foundations
November – December 2018: The (climate-)scientific background 77

Chapter 4: The International Movement Develops
December – January 2019: COP meeting and climate justice 95

Chapter 5: Davos and the World Economic Forum
January – February 2019: What is valuable and what is a science of economy? 115

Chapter 6: The Prelude to the Uprising
February – March 2019: The first international meeting, and the founding
of Scientists for Future 137

Chapter 7: The Uprising
March – April 2019: The first global strike, and the London occupation...............147

Part Two: The Adults Respond

**Chapter 1: The Second Global Strike and the Preparations
for the Week for Future**
April – August 2019: The task of civil society..165

Chapter 2: Smile For Future in Lausanne and Scientists for Future
August – September 2019: The task of science...................................... 171

Chapter 3: The Week For Future
September 2019: The coordinated uprising of eight million people
across the world 193

Chapter 4: COP25 in Madrid
October – December 2019: How can we end our fossil society in a fair way? 201

Chapter 5: Corona, #BlackLivesMatter and the Climate Justice Movement
January – September 2020: The crisis and the intersectional, sustainable,
global democratic project 215

Chapter 6: Many Fights, One Heart – #UprootTheSystem
October 2020 – October 2021: Forests, agriculture, banks, courts,
and the political manifesto – all the way to a new theory of democracy 231

**Chapter 7: The Idea of Social Movements and the Journey to Glasgow –
What Is the Right Way to Live?**
November 2020 – December 2021: On a theory of democratic grassroots
movements that can change the world 261

Chapter 8: The War, Fuel, and the Global Social Contract
January – August 2022: Towards a new world order to lead us out of the crises 285

Chapter 9: Education in Times of Crisis – Learning from Young People on the Way to "Centres of Sustainability"
January – August 2022: How to change schools and universities – (eco)philosophy for a sustainable democracy 307

Chapter 10: Global democracy, the Elections, and the Future
September 2022 – June 2023: On forming a global, democratic, sustainable postgrowth society.. 347

Part Three: Facing the Future Together – A Conversation with Isabelle Axelsson and Loukina Tille

Contents 363

Chapter 1: What the Climate Strike Movement is About.......................... 365

Chapter 2: On the Relationship Between Young People and Adults
How We Can All Write History 375

Appendix and Summary

What a Global United Climate Movement Could Fight for
The Basic Principles for Social and Political Change 385

Epilogue 399

Illustrations 403

Bibliography 405

Timeline

2018

20th of August, Monday
Greta Thunberg sits down alone between the two halves of the Swedish parliament building and begins to strike

7th of September, Saturday
At the Rålambshov park, the young strikers announce the founding of #FridaysForFuture

8th of September, Sunday
Elections in Sweden. It will be months before the left-wing/green coalition government is confirmed

13th of September, Friday
First Friday strike under the name #FridaysForFuture

October
The UN special report "IPCC SR1.5" is published, explaining the difference between 1.5 and 2 degrees of global warming

31st of October, Wednesday
Founding weekend of Extinction Rebellion in London

17th of November, Saturday
The first big blockade of Stockholm by Extinction Rebellion

30th of November, Friday
Roughly 10 000 pupils go on strike in Australia, partly in response to Prime Minister Morrison's remarks about climate activists. Belgian strikes on Thursdays increase in size

9th of December, Sunday
The first internationally coordinated action by FFF and XR takes place: "Climate Alarm"

12th of December, Wednesday
Greta makes a speech at the COP24 climate summit of the UN in Katowice, Poland

21st of December, Friday
As a reaction to discussions of a carbon tax in parliament, the first largescale strike takes place in Switzerland

2019

18th of January, Friday
The first big strike in Berlin

25th of January, Friday
World Economic Forum in Davos, Switzerland. "Our house is on fire"

12th of February, Tuesday
IPBES biodiversity report by the UN is published, with alarming statistics

15th of February, Friday
Great Britain and France strike on a grand scale for the first time

13th of March, Wednesday
FFF strike at the EU offices in Strasbourg

15th of March, Friday
The first global strike with around 1.6 million participants

15th of April, Monday
The blockade of central London by Extinction Rebellion begins
In Stockholm, XR blocks the parliament bridge
Greta speaks at the British parliament after visits to Rome and Brussels

29th of April, Monday
The UK declares a climate emergency

24th of May, Friday (the weekend of the EU elections)
The second global strike

4th-9th of August
The big European FFF meeting SmileForFuture in Lausanne, Switzerland

20th to 27th of September
The Global Week For Future with global strikes together with trade unions and NGOs. Around 8 million young people and adults take part in the strike marches

23rd of September, Tuesday
Greta makes her "How dare you" speech at the UN climate summit in New York

7th of October, Monday
Extinction Rebellion blocks the city centres of London, New York, Paris, and Berlin

29th of November and 6th of December
Fourth global strike, in Stockholm in the suburb of Rinkeby
In Madrid, 500 000 people gather around Fridays for Future

2nd-13th of December
COP25 meeting in Madrid

2020

21st-24th of January
World Economic Forum

4th of March
Statement by Scientists for Future on the new EU climate law

16th and 17th of July
Young people from FFF – supported by leading scientists – write a letter to the EU with demands; and 20 activists from the Global South address the G20 states

April to August
The climate justice movement moves online due to the Corona pandemic and listens to the #BlackLivesMatter demonstrations

20th to 21st of August
Two years have passed since the first day of the strikes and the activists return – wearing masks – to Mynttorget. Young people present demands entitled #FaceTheClimateEmergency

25th of September
Global Day of Climate Action

2021

25th of March, Friday
Global strike, #PeopleNotProfit

24th of September, Friday
Global strike, #UprootTheSystem, part 1

22nd of October, Friday
Global strike, #UprootTheSystem, part 2

31st of October–13th of November
COP26 meeting in Glasgow

2022

25th of March, Friday
Global strike, #PeopleNotProfit

1st-3rd of June
Stockholm+50: UN environment conference, 50 years after the first conference when Olof Palme called for a law against ecocide

9th of September, Friday
Strike in Stockholm before the elections, exactly four years since the beginning of the strikes

11th of September
Elections in Sweden

23rd of September, Friday
Global strike, #PeopleNotProfit

2023

14th of January, Friday
Strike in Lützerath, Germany

3rd of March, Friday
Global strike, #TomorrowIsTooLate

9th of June, Friday
Last school day for many activists in Stockholm, after five years of school strike for climate

Introduction

Children are sitting on the ground in front of their national parliaments or town halls. In Stockholm. Bern. Kathmandu. New York. Kabul. Manaus. Berlin. They are striking for the climate and for their future, in city squares and online. And they are calling attention to a democratic mistake: the rainforests are burning and being cut down. The banks are investing in the fossil industry. Across the world, democracies need to be transformed, and a new way of living together globally needs to be found – and they, as children, have no say in this, even though it affects their lives the most. So now they are sitting there, refusing to accept this any longer.

How should we react to this? I ask myself this question in August 2018, when I leave the rooms where I teach at Stockholm University and go for the first time to meet Greta Thunberg and her fellow strikers, who are sitting outside the parliament. When I meet them on one of the first days of the strike, I know a lot about the climate crisis, but in fact, it becomes clear in the next weeks that I know astonishingly little. I understand the crisis, but still not really. And so I decide early on, when Fridays For Future is just being formed, that I will come back every Friday for the seven hours that the young people spend striking at the edge of Stockholm's Old Town. I don't want to play along anymore: this strike is an emergency brake, interrupting work and university life. Therefore, I follow their call for non-cooperation with the "fossil society"; at least for one day every week.

These encounters with the regulars at the school strikes in Mynttorget – the square close to the parliament – will change all our lives. Very soon, we'll be meeting the most important climate researchers in the world. I'll be getting to know the children's way of thinking and seeing grief and despair, but also great empathy and excitement about the growing global network which is a constant work in progress, and which will go on to make history as #FridaysForFuture/Climate Strike (FFF). Mynttorget in Stockholm will become the hub of the global movement. And gradually I will try to help bring together a worldwide movement in the world of adults, beyond FFF. Mynttorget is where we will cofound Scientists For Future, and from there we will organise the Week For Future, with 8 million participants.

One task is central throughout these weeks. We need to develop and carry out a plan to make the world sustainable, fair and democratic within around ten years. We shouldn't simply be patting the young people of FFF on the shoulder. As a society, we must react to what they want and give them a secure future. And in that sense, their story is also ours. It is the story of our shared future.

This story has two sides. On the one hand, it is a sad story. Probably the saddest story imaginable. It is the story of hundreds of species which are being eradicated, forests that are being razed and burnt, hundreds of thousands of people who are fleeing drought and floods (Wallace-Wells 2019). And above all, it is the story of children and young people everywhere in the world getting information from social media and worrying every day, dreading conflicts over water and food, a kind of panic which never quite disappears. What will it look like for us, they ask themselves, when we're as grown up as those two-legged creatures who belong to the same species as us – the ones who are in charge? In a world that is two, three or four degrees warmer, with the danger of irreversible tipping points and feedback loops in the climate system (Lenton 2019), life will be hell for many people, especially those in the regions known as the Global South. This is also the story of established NGOs working in parallel, which have tried many things and still not managed to change policy. And it is the story of politicians and highly specialised scientists who know all this but who barely do anything because they seem to be in a state of paralysis.

But that is not the whole story. If we look more closely, a window opens to a positive mission, maybe the biggest challenge we can picture. It has just begun, or rather, it began in an unbearably warm week in August 2018. That is when a few children and teenagers from various suburbs of Stockholm decided to join Greta – the teenager they had read about in the news; the child armed with a sign reading "School Strike for the Climate", between the huge stone blocks of the Swedish parliament, on strike. At that moment, a story starts which only develops slowly at first, with very little happening for weeks and months, but then, led by the Swedish "gang of rebels", grows to become one of the biggest ever international youth movements for the environment. Six months later, on the 15th of March, 1.6 million children leave their classrooms in protest against the world of adults. Then in September, eight million young people and adults come together across continents and to some extent across generations to go on strike together.

It's like something from a storybook, as one of the regular Stockholm strikers says. Not the widespread narrative about a single child fighting alone, but the story of different groups to which the young people belong. A story which has not yet been told; one which is also about friendships, about solutions to political and activist challenges, which brings together young people from all corners of the world. And it may be the story of a group of young people, but it is also about their attempt to wake up the adults, to work together with them, and to make their task clear to them, not least the scientists who have gradually joined together in the huge network of Scientists For Future. It is the story which we now have to tackle together and which we have to translate into reality in the next fifteen years across the world: the story of a global democratic transformation in all areas of life, in which we all help to create a life worth living with enough resources for all and without going beyond the limits of the planet: global heating, loss of biodiversity, pollution and acidification of the oceans.

But it is also a story with a long, complicated backstory, including the indigenous peoples who have long been fighting for their way of life and for the protection of nature, despite constant repression from governments and financial interests. In particular, it is also the story of emancipatory grassroots movements, the fight against colonialism, protests by the women's movement over the last century, and the struggle of workers and BIPOC communities (Black, Indigenous and People of Color), who want justice and the protection of human rights for everyone. Without these civil disobedience movements and the fight for human rights and for a democratic approach to one another (as well as to nature and its value), this adventure would not now be conceivable. This backstory is explicitly part of the young people's frame of reference when they decide to sit down and go on strike.

It is important to understand this story. Now a new chapter can begin: that of the adults coming together across the world, beyond Fridays For Future. There have been suggestions as to how this chapter of organised protest by adults might look – with the "Week For Future" and the COP meeting of the UN in Madrid, which are discussed in this book. There, young people have begun to connect with grassroots groups to become a unified global movement.

But the politics have not changed, and nor have the rules, the thinking and the economic orientation which plague and threaten nature and humanity. Despite the Paris Agreement, the UN says that the promises made by governments, with their NDCs (national plans to reduce emissions), will still lead to almost three degrees of global heating – even if these promises are kept, which is hardly likely (Chestney 2021). And there it is again, the sad story. An earth which is three degrees warmer will be unbearable for hundreds of millions of people (and billions of animals); ultimately it will be unbearable for the majority (Xu et al. 2020). And that world will become reality within the lifetimes of the main characters in this story, if we do not immediately enact "far-reaching and unprecedented changes in all aspects of society" (IPCC report, 8th of October 2018). Diffuse fears about this future world shape children's lives like nothing else. We can resolve those fears if we treat the crisis as a crisis. We humans can look ahead and plan with the help of science and the imagination. And when those two work together well, there is probably no stronger force on this planet.

On the making of this book – its structure and the people who worked on it, science and politics

The narrative form used in the first part of this book is hopefully more fitting than a purely factual text when it comes to explaining the facts and problems faced so far by the young people and us scientists, and those which now define our task.

Between the first strike day and the first global uprising of the 15th of March, with 1.6 million children, there were 26 Fridays in Mynttorget in Stockholm, a complex social fabric was woven and became the basis for the global network. This is where arguments with politicians take place, and where the most famous scientists gather. More and more cameras appear, along with people from the media. This is where placards are painted and comments by politicians are discussed. This is the place to which the young people return after their travels to the WEF in Davos and the COP in Poland, full of their adventures. And this is where friendships are forged and trust is built: a core group emerges, one that wants to change the world. Altogether, there will be more than 250 Fridays in the small square at the edge of Stockholm's old town, but also via the internet and the phone; five years with the most active young people and scientists across the world, who soon join, in Swiss towns, in Uganda, Australia, Brazil and Canada. Thousands of decisions must be made at this site of democratic experiment: what should the movement be like in the first place and how can a global network be knitted together? Which aims are the most important? What is the role of science? And above all: when might the strike end? When would the world be a place where young people could feel comfortable and safe?

This narrative insight into the history of the climate movements (as well as the sister movement Extinction Rebellion) is told from my subjective perspective as a lecturer at Stockholm University who grew up in Sweden and Switzerland and knows both cultures. In that sense, this is a book about the role of science and education in relation to political activism. The focus is not on a single discipline such as environmental science or climate science, but on the attempt to approach that subject together with other university subjects: philosophy, political science, economics, psychology, and education, for example.

During these months, I also try to incorporate my experiences in Mynttorget into my teaching at the university as senior lecturer in drama education and youth studies. I use the young people's speeches in my seminars on social and ecological sustainability, education and democracy, and together with the students I try to use role play and other research methods drawn from the arts in order to understand what is scientific, and what is true. For instance, I ask them to write plays about the legal cases brought by children across the world against their own governments, because they are doing too little to prevent the climate crisis.

What exactly is the role of the university in this shared history? In terms of method, this book might be seen as a kind of auto-ethnographic study, reflecting on the development of the new movements for climate justice and on university life. In the spirit of post-qualitative research, the central insight is that we as scientists are also entangled in problematic power relations and that we have to expose them and respond to them actively, rather than pretending to be neutral (Leavy 2009). Instead of making the young people the object of a traditional empirical sociological investigation, my approach starts by trying to work together with them to ask how society could be changed and a global sustainable democracy created.

This story is then followed by a collective view of the past and future, reflecting many conversations with Loukina Tille and Isabelle Axelsson, from Sweden and Switzerland, two of the young activists at the centre of the global climate movement. What is the movement about, and what are its demands? What does "climate justice" mean, and what about "Listen to the science?" And finally: what role do all of us have, and what is our task?

From the beginning, reflections on what had happened at high speed were an important part of the movement. The Sweden-Switzerland axis was crucial in that context, and so were the ideas and initiatives of Loukina Tille and Isabelle Axelsson. When they begin to strike, they are still school pupils who have suddenly left their classrooms. Now they are studying politics and human geography at university in Zürich and Stockholm. Loukina Tille, from Lausanne, helped build the climate movement in Switzerland. Already very early on, she was in contact with Isabelle and the other Stockholm activists, and she regularly led global meetings of all strikers, organised the first international meeting of four hundred young people at the University of Lausanne, and was one of the organisers of the global strikes. At around the same time, Isabelle Axelsson joined the Stockholm strike. Together with Loukina, she planned and carried out trips to the European Parliament in Strasbourg, built up the global organisation with her peers, and met with Loukina at the conferences in Lausanne and at the World Economic Forum in Davos. Without these two, and their interest in exchange with science and with scientists globally, the movement would look different.

Towards an ecophilosophy of democracy

This book also offers an introduction to ecophilosophy and a unified science of democracy in times of interdependent sustainability crises. This is the case not only in the narrower sense of a philosophy of climate ethics, but as an attempt to explore and present interrelated ideas about our place in nature, and the creation of a convivial, just world. It includes questions of moral and political philosophy, highlighting that the idea of justice should always be complemented by the notion of being humane, making relations and structures of domination visible and opening a world where everyone can experience the dignity of us all; thereby repairing and creating a common fabric of integrity which links us to nature as vulnerable creatures – and gives us a unique common task to create a world in which we can live together as equal and free democratic animals (see appendix for such a political program and a new framework for a convivial, global democracy).

This leads to the enterprise of a new science of sustainable democracy, integrating insights from many sciences, from neuropsychology to political science, exploring the substance of democracy and how this relates to its formal aspects: focusing on the idea of meeting on an equal footing beyond structures and relations of domination (gender, class, ethnicity etc.).

What if we could explore all of this – even in a creative way (arts and drama education) – at every school and university, in every village and town? The final chapters sketch the idea of such a prototype centre as a possible core of all educational institutions. Trying to understand the movements as well as university life, the book explores what a sustainable, democratic approach to the world could be – and what it would mean to create the social (political, economic, cultural) spaces needed to connect with each other and the environment, to produce enough resources in a regenerative way for all to live a life in dignity.

The chapters follow the chronology of the movements, but also give an introduction to the basics of earth sciences (first chapters), economics (chapter on Davos), justice, global ethics and politics (chapter on corona and Madrid), sociology (chapter on Glasgow), democracy (chapter on the "many fights, one heart"), and especially education and philosophy (chapter on education).

On this journey, one main question is: what is this endeavour of science, and scientific research and education which leads us into and maybe leads us out of the crises? What does this exploration of humans as embodied, social, imaginative beings living in problematic power relations mean for curricula, teaching methods, research ethics, and for the institutional reframing of these topics, leading to a new picture of scientific endeavour as a regenerative and transformative exploration?

The intergenerational challenge and the idea of a united, global movement

Fridays for Future was founded by Greta together with Mina, Edit, Eira, Tindra and Morrigan on the 7th of September, 2018. It is a movement initiated by young people, and that has consequences. As Roger Hart's Unicef text (1992) so wisely delineates, there is a whole "staircase" of possibilities for intergenerational cooperation: from projects initiated and carried out only by young people, to projects initiated by adults and organised for children. #FFF is a "youth-led" movement, initiated and led by young people. Young people have the right to organise themselves. Adults can help – if they are asked; or if they organise themselves as "parents", "artists", "scientists", and so on.

So adults have the task of giving young people confidence that they believe in them and in their ideas, and that they believe them capable of acting independently. A universal power imbalance between children and adults comes into play here, according to the Swedish theatre director Suzanne Osten (2009), who shaped the culture of her country. The "grown-ups" have the responsibility to make sure that the younger ones are doing well. The well-being of the younger ones has priority. The older ones can listen and help. But above all – according to the fundamental idea of this book – they must continue working on the "hand" of the united climate movements, which is made up of "fingers" such as Fridays for Future, Extinction Rebellion, the NGOs and so on. Patting the young people on the shoulder is clearly not sufficient.

The question is then: what if we joined together as "People for Future" in a global, democratic, united movement or network that anyone can simply join (see the appendix of this book) – respecting and celebrating the existing movements and their history and identity? A movement which stops the Amazon rainforest from being cut down and the German coal power plants from running, which prevents the Swiss banks from financing the fossil industry and which comes together worldwide to build a sustainable, fair society, a movement which takes care of everyone's needs and ends the injustices of colonial history – with regenerative forestry and agriculture founded on plant-based products, a fossil-free public transport system, a global network of renewable energy and a really democratic economy that helps everyone flourish and stays within the limits of the planet; this is roughly the image of the future towards which Scientists for Future are working. And strengthening a sustainable democracy beyond the nation-state, being fair to the Global South and eliminating power relations of domination based on gender, class, and ethnicity. If we look closely, this is the movement that has emerged in the last five years.

The task

"Are you happy with how things are going?" asks a reporter from the *Financial Times* in March 2019, when a million children are on strike. "We are happy when we see the young people who are standing by our side everywhere and doing the same as us. Not going along with it anymore. That makes us happy," say the young people. "But nothing has happened yet," they add. "Nothing has changed. Emissions are rising across the world." Even in the wealthiest countries, such as Switzerland, Germany and Sweden, emissions are hardly going down. Sweden has a green Deputy Prime Minister when the young people begin their strike. They strike in front of these parliaments because governments do not, in this sense, take the science seriously. If Sweden adhered to the Paris Agreement, which almost all states across the world signed in 2015, its parliament would have to pass laws reducing emissions every year by more than twelve percent, say scientists at the University of Uppsala (Anderson et al 2020). The young people are on strike because with their talk of "climate neutrality" and "net zero emissions in 2050", governments are precisely not keeping to the Paris Agreement. They are on strike because they understand that the populations of all countries would have to rise up and join together in solidarity, if a dignified life is to be made possible for all people on this planet.

We were actually all united – or rather, our governments were united – when it came to complying with the Paris Agreement. This obliges us to do everything to limit the rise in temperatures to well under two and if possible under 1.5 degrees. The IPCC-SR1.5 report, which was accepted by all UN states, shows that we will already miss this target in three or four years if we continue as we are: that we have an absurdly tiny amount of CO_2 emissions left (2020: around 350 Gt). And the Gap Report by the UN shows that the established and contractually planned fossil infrastructure (coal, oil, gas) will cause around double the permitted emissions in the next ten years (UNEP Production GAP report 2019). We are on the way to a much warmer world, already in the lifetime of the children who are currently going out into the streets, with up to three billion people fleeing from regions which have become uninhabitable and too hot (Xu et al. 2020), and with large proportions of the glaciers melting, so that water supplies across the world are under threat. That is why Scientists for Future say: We need systemic change, and all of us must take action – and take to the streets.

This book outlines how that could work. Thanks are due to Jana Eriksson, who took many of the photographs included here and who is herself part of the Stockholm climate movements.

The parts of the book which describe the first two years were written in 2020 (with the corresponding scientific data) as part of the German book *Gemeinsam für die Zukunft* (Fopp et al. 2021); the chapters about the last three years were written in Summer 2023. All activists, no matter their age, are mentioned by their first name, also as a measure of safety; the only exceptions are those who are internationally known.

There are many things to which this book is unable to do justice – with its Stockholmian, European, privileged, and restricted perspective. In that sense, this European history of Fridays for Future and Scientists for Future is not a work of journalistic research, and it should and will be told by many more different voices (see Nakate 2021). However, hopefully, this story can inspire to action and to some extent explore and explain what is important to this group of young people and scientists, and what that could mean for all of us.

Many people are concerned about the state of our world but don't know how to engage. What can each one of us do? Following the movements, we can also see the history of bottom-up initiatives which grow with the help of the grassroots movements, which are open for everyone, individuals and organisations. Together, they can be seen as a united program to change our world: including the Doughnut Economy Lab, the Fossil Fuel Treaty, the Faculty for a Future, Earth4All, bioregions, etc. But still, the central idea is that we need to come together as a humane, global grassroots movement.

The strangers

It is mid-July 2018. An ordinary day in the summer holidays. Loukina is out walking near Lausanne, stealing a few cherries and thinking about the next school year, which will be her last. Next summer she will go to university, maybe in Zurich. Images of forest fires have been flashing up in the news. It is unbearably hot. In Sweden, the trees are burning like matches. Children in hundreds of small villages all over the world hear about it too. How can this go on, they wonder. And some young people in Zurich are having similar thoughts while people jump from the high walls of the footpath to cool off in the Limmat river. The same goes for their peers in Uganda, Australia, the Philippines, in Brazil, Mexico, England, the US, Ireland and Scotland, in Italy, Finland, Japan and Germany, in Ukraine, Bangladesh, in Kenya and Argentina, all of them having these worries. They are all around seventeen, the same generation. Balder is in Holland, and he too, has a frown on his face; he will be moving to Stockholm in the next week, to spend a few months there as an Erasmus student. And there they are, spread across different parts of the Swedish capital: Tindra, Isabelle, Ell, Simon, Mina, Minna, Edit, Eira, Morrigan, Mayson, Melda, Edward, Astrid, Vega, Ebba and Greta; and so on and so forth. They don't know each other yet, any more than the other young people do, but they soon will. Some have been going to the same school for years, without noticing each other, or they have passed each other countless times in the metro stations in the city. They share the same fear and the same fury. The adults are wrecking the environment. They are systematically destroying the planet.

In Mynttorget, the square at the edge of Stockholm's Old Town, it is still quiet. If you look closely, the two huge flowerpots seem to be slightly tense. Waiting. They are watched by the oversized blocks of stone which form the royal palace wall, bordering the square. They too are waiting. Something is coming. And not just a forest fire. They are waiting for the "Fridays" to come and give them an important role.

Fridays For Future doesn't exist yet, but they all exist as individuals with their worries. They sit in their rooms, looking at pictures of floods and droughts on their phones, absurd images of forests being cut down and bleak coal mines. And of politicians not doing anything. They still feel powerless. What are they supposed to do? They don't even have the right to vote yet – how can they change anything on their own?

In the coming autumn, the evenings in Stockholm quickly get shorter, including in the square in front of the palace. Then the evening light shines through the windows of the most expensive flats in the country, which are visible from the square. They have a view of the bay which shapes central Stockholm. Standing at those windows, the opponents of these teenagers enjoy the thing that costs so much: the breath-taking experience of nature. It is here and in similar houses in Sydney, New York, Tokyo and Frankfurt that we find the bosses of BP, Exxon, and Shell, but also the financial speculators who make money from coal, gas and oil, the media moguls, and a few politicians, too. They own a large proportion of the world's wealth – in Sweden as in other countries (Cervenka 2022). They can control where investment goes and what is produced, and how. Soon the children will gather into a crowd, equal in number to the power brokers of the "fossil society". At least one or two of their own children will be among the demonstrators.

But there is a much bigger actor, one that is far less conspicuous. It is the rest of the population, who are walking past the children in the square in front of the palace, a whole variety of fellow citizens. The young people turn to them from the beginning. They want to change the situation by inserting themselves into the workings of power, using their bodies to jam the mechanism. Often, when I meet these young people in the next months, I think: they are so brave to do this every week in spite of their fears, plucking up the courage to go to Mynttorget every Friday (and to all the other squares across the world) and stay there for seven hours, accepting that they might be punished at school. And my other thought is: we need the older ones, people like you and me. Many people are still hesitating but are interested; a good number of those have to join them. Between the teenagers and those who are directly responsible for this fossil economy, ideology and politics, there are all the rest of us. This book is also about them, and especially about those who remember the climate briefly, frown worriedly, and then don't know what they are meant to do and just carry on as usual. If all of us join the children and strike or take political action, the course of history might change.

But we haven't got that far yet. There is still nothing for us to respond to. It is still July 2018, and the young people don't know anything about each other. One of them has something planned, though. On this same summer day, while Loukina eats cherries and Isabelle works in an ice cream kiosk, one of their peers is sitting on the wooden boards of the veranda in Stockholm, with a piece of wood in front of her that is supposed to become a placard. It's clear what it should say: "School strike for the climate". The "for" has to fit into the small space between the other words, which squash it from above and below. It is symmetrical, clear, and distinct.

Part One: The Young People's Rebellion – From Mynttorget in Stockholm to the Global Strike

Contents

Chapter 1: Swedish Beginnings
August – October 2018: An idea takes shape

Chapter 2: Fridays for Future and Extinction Rebellion Start to Grow
October – November 2018: Civil disobedience and the laws of humanity

Chapter 3: The Foundations
November – December 2018: The (climate-)scientific background

Chapter 4: The International Movement Develops
December – January 2019: COP meeting and climate justice

Chapter 5: Davos and the World Economic Forum
January – February 2019: What is valuable and what is a science of economy?

Chapter 6: The Prelude to the Uprising
February – March 2019: The first international meeting, and the founding of Scientists for Future

Chapter 7: The Uprising
March – April 2019: The first global strike, and the London occupation

Mynttorget framed by the Swedish Parliament on the left and the Royal Palace on the right.

Chapter 1: Swedish Beginnings
August – October 2018: An idea takes shape

Preparations

The most beautiful room in our institute at Stockholm University juts out of the façade at a height of five metres, surrounded by three glass walls, looking out over a small wood. This is where I am supposed to be planning the new semester, and especially a workshop on the topic of the climate crisis and sustainability. "Do you have any ideas?" I ask a sheep which is standing outside in the woods, looking at me through the glass. How should we explain the urgency of the climate crisis without the students switching off? So that they, the future teachers of Sweden, will dare to make space for empathy with other people and for fascination with nature? What do the school children themselves think about our way of treating nature, globally? Start there; that could work. Make a quick note. Opening my computer, I notice a news item online. A child is sitting alone in front of the parliament, in the centre of the city, less than twenty minutes away, on strike.

The strike before the strike – the first meeting

When I visit Greta and the other activists for the first time during their strike in front of the parliament, Fridays For Future does not yet exist; there is only the basic idea for a strike. They sit between the two parts of the grandiose parliament building every day for the last three weeks before the Swedish elections, not only on Fridays. It is a Tuesday in late August, and it is unbearably hot. I sit down and ask them what they have to say to us all.

After seeing the news item, I asked my university colleagues: "Should we head over there? We should at least listen." On the Monday morning before that, Greta had picked up her sign, which now read "School strike for the climate" in big letters, in black and white, and cycled to the parliament, where she found a place right in the middle of the centre of Swedish political power, unrolled her mat and then sat down alone on the ground.

Now she sits there and says: "This is a crisis." A few other youngsters and two or three adults are sitting a little way away. "A crisis?" "Yes, a crisis." There is an A4 sheet of paper in front of them, covered in scientific facts to prove it – that is, they show what we adults, or some of us, have done to the environment over the last fifty years.

> Här är några saker som alla borde känna till;
>
> › Sveriges genomsnittsutsläpp CO2 per person och år är 11 ton, enligt Naturvårdsverket. Vissa säger att vi får släppa ut max 2 ton per person och år, men det är fel. Enligt forskare på Uppsala universitet måste vi minska våra utsläpp med minst 10-15% per år med start idag (Sveriges utsläpp har inte minskat senaste 30 åren, om man räknar med konsumtion), och vi måste ner till nollutsläpp inom 6-12 år, av hänsyn till fattigare länder, som måste kunna höja sin levnadsstandard. T.ex bygga infrastruktur, som vi redan har. Allt enligt Parisavtalet.
> › Vi i Sverige lever som om vi hade ca 4,2 jordklot. Genomsnittet för världen är ca 1,7. Sverige är bland de topp 10 länder i världen som har störst ekologiskt fotavtryck. Enligt Carbon footprint.
> › Dagens utrotningstakt är upp till 1000 ggr högre än vad som, historiskt sett, anses vara normalt, enligt National Geographic.
> › År 2018 är andelen CO2 i atmosfären ca 410ppm (miljondelar), och inom 10-12 år förväntas vi nå 440. Den rekommenderade maxnivån är 350ppm. Innan den industriella revolutionen var nivån ca 280ppm. Vi passerade 350ppm år 1987, enligt Keelingkurvan, Mauna Loa-observatoriet.
> › Det går åt ca 23,5 ton biomassa för att framställa en liter bensin. Alltså 23,5 ton gamla träd och dinosaurier, och några tiotals miljoner år, för att en bil ska kunna köra en mil, enligt universitetet i Utah.
> › År 2017 konstaterade en tysk undersökning som pågått på olika platser i Tyskland under 30 år att mängden insekter hade minskat med 75-80%. Ett tag senare rapporterade franska forskare att upp till 70% av vissa fågelarter hade dött, eftersom att de inte hade några insekter att äta.
> › Koldioxidhalten i atmosfären styr temperaturen på jorden. Sist som halten CO2 i atmosfären var lika hög som den är nu så var havsytan ca 20-25 m högre än den är idag.
> › Samma skogsskövling som alla upprörs över i regnskogen pågår även i det norra skogsbältet i minst lika stor utsträckning.
> › Människan har utrotat ca 83% av alla däggdjur, 80% av alla marina däggdjur, 15% av alla fiskar och 50%

It is only much later that I really read that piece of paper from the first days. In front of the parliament in the heat of late summer, I only glance at it. I see a few familiar and a few entirely unfamiliar numbers and comments. The whole page is covered in fine print. We have to reduce emissions in richer countries by at least 10–15 percent every year, from today. We humans have eradicated 83 percent of the population of land mammals. We are now... The information is so condensed that it could make you dizzy. They are just letters and numbers, but behind that is the pain caused by humans to other humans and animals. How can we deal with it? Maybe that is why it takes me months to look at this piece of paper properly. It is simply hard to digest. But it is full of the knowledge the state ought to be disseminating and teaching to pupils at school. Weeks later, when we are meeting every Friday and a climate scientist joins us from the university, it becomes clear to me how well-read Greta and some of the other activists are; but it's not just that they know the literature, it's also that they understand the connections, and above all that they can judge their importance: what is the central point, what are the greatest risks, what is our role as adults who are wreaking destruction we could prevent. I am the one who is learning here. In particular, they are calling attention to calculations of risk: which are the numbers the politicians are working with less than ten metres away from us in the parliament; and is that really responsible or are they kidding themselves, ignoring risks such as tipping points, gambling on technologies that don't exist, closing their eyes to UN reports? Are they passing the buck to the children's generation when it comes to reshaping society?

I discuss the meaning of these facts in the next months with my colleagues at the various relevant institutes at the university. They agree with the young strikers.

But for now, it is enough for them to say that we are in a crisis and that the adults ought to present it that way too. I have brought kiwi smoothies in plastic cups; I feel ashamed of the plastic rubbish I've brought along and after talking to them for a while, I leave again, utterly confused, moved, and perturbed. Above all, I think: these young people on strike are not just there as themselves, they're also making space for an idea; the idea that no one has to accept the way the world is behaving. Even if we are small, we can step into the middle of this machinery, refuse to follow the rules of the adults, and skip school.

Half an eternity later, when a bitterly cold winter has arrived and the gang of rebels has been striking in Mynttorget for twenty Fridays, right there in front of the parliament we will all build a snow elephant, or in fact the left foot of a snow elephant, and laugh and complain about the journalists who ask their investigative questions: whether the young people are being controlled from behind the scenes, whether they earn money with their activism, and so on. Greta and the others will have travelled to Katowice, to Strasbourg and to Davos, and they will have made speeches which are broadcast around the world. But at that time, on that August day, we obviously know nothing about that. There are still just a few children sitting between the stone blocks of political power.

When might they be able to end their strike, I wonder. I definitely have to come back and hear more from the ones whose future is at stake. And even in these first days, that's not just Greta. Because one by one, others have joined her. Tindra, Mina, Edit, Eira, Morrigan, Melda, Mayson and so on. There are still not many of them, but they make all the difference. Greta's idea has taken hold. The core group of young people have found each other, and they are making plans.

Greta, 15, skolstrejkar för klimatet

The beginning of Fridays For Future – on a Saturday

And so begins the actual story of Fridays For Future – on a Saturday. That is when the children's three-week strike becomes something else, #FFF, a movement. This is because the Swedes are supposed to be electing a new parliament on the following day, a Sunday – and so, a coalition of climate activist groups has announced a demonstration. At the edge of the city centre in the notorious Rålambshov Park, there is a small stone amphitheatre. There we all gather, maybe a thousand people, listening to songs and to speeches about the climate crisis.

Suddenly the group of strikers around Greta are announced; they are starting to be well known after their three weeks of daily striking. Three other schoolchildren walk with her into the open space. "Hej." "Hej," everyone answers. "Please get your phones out," says Greta. "I will now switch to English and make an announcement." A pause. I rummage around for my phone and press "record". "Hej, I am Greta Thunberg, and this is Mina, Morrigan, and Edit. We have school-striked for the climate for the last three weeks. Yesterday was the last day. But we will go on with the school strike every Friday as from now, we will sit outside the Swedish parliament until Sweden is in line with the Paris Agreement. We urge all of you to do the same: sit outside your parliament or local government wherever you are [...]. Everyone is welcome, everyone is needed. Please join in. Thank you." Many people post the video. Some of those who see it will start to strike. And they are not in the suburbs of Stockholm, but in Brussels, Zurich, Berlin, Melbourne, and Rio. Meanwhile, Greta goes home and makes a short film of her own, which remains pinned at the top of her twitter account for months afterwards. In a small wood, she records her basic idea: sit down in front of the parliaments, every Friday; the situation is so urgent that the children have to do something. She ends this appeal with the hashtag #FridaysForFuture.

28th of August: The French environment minister, Nicolas Hulot, resigns in protest at Emmanuel Macron's climate policies.

Mynttorget

The strikers have invented or created two things, the idea of the strike – and their life on the square, Mynttorget, as a special place, a kind of democratic space. This is where the movement will come into being. The police direct them to the square. They are not to sit directly between the parliament buildings: Mynttorget is directly in front of the parliament, or just beyond it if you are coming from the seat of the government, the Rosenbad; it is crammed in between the royal palace, the Old Town, and the parliament. At the beginning of September, the strikers are still finding their way. Everything is unfamiliar. Because it is a school strike, it takes place during school hours, from eight in the morning till three in the afternoon. What should they do with those seven hours on the square? If not school, then what? Often, the ten regular Fridays For Future strikers sit quietly, leaning against the wall in front of the parliament, enjoying the autumn air. The atmosphere is serious. They are aware that they are meant to be at school and that they are taking a risk; punishment is a possibility. A few fish jump out of the water of the Mälaren lake; the royal guard marches past. Now and again, a seagull circles their heads, or even a sea eagle. Sometimes there is silence for many minutes. Then someone suggests a game or tells a story. Politicians walk past and disappear into their parliamentary offices without saying hello. Buses pass by. Sometimes a car stops, leaves them a crate of bananas, and drives off with a friendly beep of the horn.

A generation rises up

It is a whole generation that is rising up, slowly but surely. It has already been simmering for a few months. In the USA, the Sunrise Movement is growing (Holthaus 2020). In a few months, it will persuade the young Democrat representative from New York, Alexandria Ocasio-Cortez, to propose a "Green New Deal": an ambitious proposal for reaching zero emissions in the energy sector within ten years, creating green jobs and introducing social security systems (Klein 2019). The young people in Mynttorget take a curious and critical view of the idea of the "Green New Deal". Is it a serious endeavour to deal with the climate crisis, or just an attempt to kickstart the economy? Some of the Sunrise teenagers have forced the American state into a legal case during the last few months, because it is not facing up to the climate crisis and is risking the lives of future generations ("Juliana v US"; see https://www.youthvgov.or

g). In the Netherlands, too, teenagers sue the government for doing too little in the face of a catastrophic climate crisis. And they win. (On the possibility of changing the law through cases brought by young people, for a sustainable future, see Holthaus 2020.) Close to Utrecht, ten-year-old Lilly (@lillyspickup) goes out walking and becomes famous for her funny video clips, in which she urges everyone to prevent and pick up plastic litter. Lilly becomes a permanent "member" of FFF and meets the youngsters from Sweden at the EU parliament. But right now, they still don't know anything about her, or anything about each other; they haven't come to Mynttorget yet.

Still, these young people are standing on strong shoulders. Already for decades, grassroots movements in the Global South, led particularly by women and indigenous people, have been breaking a path for them. They have literally been getting in the way of the oil and coal industry in Ecuador, Canada, Australia and near Manaus in the rainforest of Brazil (see Margolin 2020).

The invention

What awaits the young people at Mynttorget? Most of all, a basic idea. From the beginning, the group has established what will become the core of the young climate movement. It will still be weeks and months before I quite understand this, so familiar am I with the old patterns of activism and political commitment. The young people have invented something new. #FFF is an invention in the best sense of the word. And it only has a few ingredients: the schoolchildren, the parliament, striking on Fridays, the A4 factsheet, the hashtag #FridaysForFuture; and the sign. That might seem obvious. But it is a very special combination of ingredients.

FFF as a movement is directed at someone; it addresses those in power: the protesters sit in front of parliament. They are not blocking petrol stations or coal power stations, they are not striking at home or in front of their schools, but in front of power, the powerful. They dare to make a direct approach to those who have responsibility. Through this, they can focus the full energy of hundreds of thousands of people, they can become the voice of a generation which is rising up. With Occupy, ten years earlier, some of us occupied squares in general, but not parliaments. The young people establish direct communication with those responsible. That gives the movement a target, not only in spatial terms, as a meeting point for people in cities, but also politically: "The rules must change," Greta says early on, in her first speech in Helsinki.

Ingredient number two: they rebel. FFF is a rebellion, because the movement chooses Friday. It is a real strike, not a demonstration. Goodbye to the giant demonstrations of the 90s and 00s, which often just stopped after a while. School attendance is a legal requirement, and by breaking that requirement, the young people demonstrate the urgency of their cause and prove their determination, refusing to cooperate with a system that makes their own future impossible. That too is new (or at least rare – during the American civil rights movement, young people used similar conflict strategies; see Chenoweth/Stephan 2012): a collective act of civil disobedience by children. But the school strike is a nonviolent rebellion, and anyone can join. People can also start by themselves; usually, no one can demonstrate on their own.

The third ingredient: Greta always places her A4 sheet of facts next to her on the ground. The children are not proposing a political manifesto in which they only argue for one stance or for specific measures, but are instead pointing out the science, the overwhelming climate research, the IPCC reports, and the goals to which all the states in the Paris Agreement have committed themselves. That is not negotiable. This means that a radical compass is available to everyone – one which the global community has already agreed on. Furthermore, FFF becomes hugely educational: the young strikers reach hundreds of thousands of people, spreading knowledge about the key facts, not only regarding the mechanisms behind global heating, but also more broadly about our relationship with living nature and about global justice. They can use their factsheet to show that they are on strike against an education system that doesn't take itself seriously. That all contrasts with earlier political movements which quickly became mired in policy disputes.

Fourthly: there is the strike sign, always easily visible. That means that the young people are not only addressing the people in power in front of the parliament, but are also turning in the other direction to all schoolchildren and the whole population, calling on everyone to join them. "Everyone is welcome, everyone is needed." Anyone passing by, the entire public, is being addressed. They literally don't need to cross any thresholds. They just have to stop walking.

Fifthly: The young people established #FridaysForFuture as a hashtag, not as an association or organisation. They will use social media like no other movement before them. They constitute a grassroots movement, not a hierarchical NGO, conscious about injustices between different parts of the planet. All children in the world can and should be part of it. A generation is rising up.

At the Swedish elections on the 8th of September, the Social Democrats win 28 percent of the vote, the centre-right Moderates 19 percent, the right-wing nationalist Sweden Democrats 17 percent, the green liberal Centre Party 8, the Left 8, the Christian Democrats 6, the Liberals 5 and the Greens 4 percent. Negotiations begin, and will continue into January, until another Green-Red coalition government is established, supported by the liberals and the green (neo-)liberals.

The first young people join – the gang of rebels comes together

During these days, a group forms which will work together closely in the next months and years – a small, but very particular group, the rebels of Mynttorget. At first, there are five, and later on ten young people who get the global movement underway. The media, which only focuses on one of them, misses the real main character: the group of young people to which Greta belongs.

It is early morning on a Friday in September. Greta comes to the square at eight, as always. Slightly later, the "regulars", as they are soon called, arrive from the old town. "I saw an article in the news about Greta and thought: She can't sit there completely alone." "Yes, I also saw it in the news. I'd known about the climate crisis for a long time, but not what I should do. There was nowhere to go. Then I thought immediately, I'm going to go there. Clicking on petitions, that won't save the world." Many of them say that they saw a child their own age sitting on the ground because the climate was getting hotter and the environment and human beings were suffering. And that they couldn't accept that. Some of them had got the tip about the strike from their grandmothers.

And so they sat down as well, hesitantly at first, gradually becoming more resolute. "We had it as a topic at school, basically a week about the end of the world. And then there was a break. And then a new topic. That felt surreal. For such a huge problem." "At first, I just wanted to sit here for three weeks. But when you understand how serious the situation is, you can't stop. So we went on. I went to the school administration and said: I'm not coming in. It's the last year of school, but I have to set priorities."

Barely anyone in the group is "only" a climate activist. They don't only come to Mynttorget because of the climate or the environment, but "also" because of the climate. Most of them are here because something is not right about society, they say, because they have the feeling that school leaves the real questions aside and because they can no longer bear how politicians look away. The social aspect, climate justice, comes up early on in the texts they write. And they want real change.

"It would be so powerful if practically the whole of Stockholm went on strike. If we got the unions on our side. If all the bankers stopped going to work. Then there would be a problem." The climate crisis, for many of them, is only one aspect of an overall picture of problems, and of structures and attitudes that have to change, in relation to animals and to other people. "Why are we not all just kind to each other?" asks someone. Hard to believe that humanity is destroying itself. "And why don't politicians do their jobs? That's meant to be their main task: to find rules and laws that give people security."

Often on these Friday mornings, a few passers-by stop and say something supportive, or something critical, before walking on. Sometimes someone expresses something that's also a prevailing mood in Mynttorget: worries, despair and a scientific interest in the future. Is it even possible to stop global heating at 1.5 or 2 degrees, or will effects come into play that keep feeding into each other? What does science say – are we on the way to an earth which is 3,

4 or 6 degrees warmer? When will that happen? What does it mean for young people across the world?

In these situations, I become aware of how important it is that I get hold of precise information and really understand the arguments for different future scenarios. I don't want to sugar-coat anything, but I also don't want to be dramatic. There ought to be a network, I think to myself, with the brightest climate scientists, philosophers and social scientists from my university, and from other universities too. And so in these first weeks, the idea of Scientists For Future emerges. I write to a whole range of professors from various different subjects, asking them to support these young people. It will take a while for the academic world to react. Sometimes I stand on the square with a sign reading #ScientistsForFuture. I want to make it clear that I represent the university. What is necessary to stop the machinery of the "fossil society"? How do we get out of the crisis?

But then the young people switch from seriousness to playfulness, as so often happens, and start imagining a way for the group to communicate with each other from their homes by stretching gigantic strings across the sky above Stockholm (and we find pictures online from a time when the sky really was almost completely full of telephone wires).

But something else emerges in the square. "This is a crisis situation," they say. "We can't just suddenly stop coming here. And so we've developed a way of getting on well, building friendships, and at the same time doing the organisational work for the strike, in a good combination." "Yes, it's like a myth. It's so strange. None of us knew the others. How we could be connected like this by our shared worries! We're all really great, clever people."

Sometimes during these days, I think about whether there is one leader of the group which is increasingly becoming the core of the climate movement. But they are so different that it is more of a cooperation on equal terms. It is difficult to make out hierarchies of status in the behaviour of the young people in Mynttorget. Maybe it is also this aspect which is so moving for us adults, the teachers, writers, and nurses who turn up regularly. It is as if these six or seven young people have decided to stick together even in the most difficult situations and during conflicts. And difficult situations will come. They are attacked in the right-wing media. Strikes with 50 000 participants have to be moved to a different location within twenty-four hours; the entire movement's strategy has to be negotiated.

In that sense, the phrase "gang of rebels" is not quite right. Gangs are structured hierarchically. "And the whole group works so well together, and has done for such a long, extremely challenging time, because we are all exactly perfectly different in age, and so we don't get in each other's way in terms of status," one of the young people explains to me. The older ones care for the younger ones, but without a hierarchy. The youngest is 13, and then each year is represented by at least one person. Two are 14, two are 15, and so on. The oldest are 18 or 19. They intuitively do what's known as "community organising".

But in the first weeks in Stockholm, none of them really has a specific responsibility. That sometimes leads to an email from the media being left unanswered for a long time, but it also helps: no one is limited to one role. This also means that no one can fail. In other countries such as Switzerland, Germany or Austria, systems are quickly built by the strikers, involving schools digitally, establishing contact people and working with NGOs. In Sweden there is no such thing. Quite the opposite: the small group of strikers are alone for months, without structural networks, without NGOs, without political parties supporting them.

They want to keep their independence. But that is also understandable and necessary: after all, at first, Fridays For Future is only an incredibly vulnerable idea, which could develop in any direction or be suppressed, and which first has to be established and anchored in public awareness. For that reason, they say:

We don't suddenly want to be mixed up with NGOs like Greenpeace or young parties like the Greens or the Left. But that means they have an entirely new, unfamiliar task. How do you start a movement?

The late morning is beginning. The young people take photos for Twitter and Instagram and send them out into the world. What is this strike exactly? This is also the question the young people are asking themselves. They are taking a big risk, I think to myself. They are not going to school, but taking the underground early in the morning to the Old Town, with all their fears and worries, and accepting the consequences, first of all in concrete terms, when it comes to the lessons they are missing and the work they'll have to catch up on. But above all, this weekly statement changes their relationships with their teachers, with their classes, with their parents, with society. They are standing very visibly in a public space. The windows of the parliament buildings on three sides of the square sometimes seem like giant eyes.

At the beginning, for months, they simply sit down. There are no marches – those come later in other places, such as Belgium, Australia, Germany, and Switzerland. They are just there. I think about that a lot, because it creates a very special atmosphere, a seriousness, something dignified. What is happening? You might think: oh, some children sat down in front of the parliament. An action. Just like going to school is an action. Or an "event" is taking place; they're "communicating" something to society. But there is something missing (or even false) in these descriptions. What they are doing is more like expressing an idea, I say to myself. So passers-by not only encounter them when they bend down and greet them, but they also meet the idea that life is important. The young people make this possible, a real encounter that gives rise to a task for all of us. Or, as the philosopher Levinas (1969) would probably say, a "demand" that emanates from a human face: to care, to give them a future. Even if no one saw them, I think then, it would be important. They are taking a stand. In some ways it is the strangest possible phenomenon, this non-cooperation on Fridays, free of violence but with such explosive potential.

The days are long and soon they are icy cold. For all of us, they represent a radical end to "business as usual". Much later, the global strikes arrive, and the dynamic in Mynttorget shifts. The regular strikers become organisers, making huge events happen. They get to know the infrastructure of the city, they find out how to book stages, locations for strikes, security measures. They spend hours organising the days which will bring together thousands of Stockholmer, they prepare speeches, they contact scientists together with me, and they write press releases. They design and order posters, hanging them all over the city and its suburbs. They often use the Greenpeace office to prepare for television interviews communicating the basic idea of FFF. And they start to travel, visiting the north of Sweden to talk to the indigenous population, with whom many of the young people remain in contact. In that region, climate change has already had disastrous effects. In this way, they bring the movement forward step by step. This is followed by trips to Europe, to Strasbourg, Madrid, and Lausanne. They make connections with rebels who are the same age as them, in different countries. And they become the people who lead strikes with megaphones in their hands, raising their voices to sing the coal and oil back into the ground, tens of thousands marching behind them. But it will be many months before Fridays For Future becomes this kind of movement. At the start, everything is uncertain. But they don't give up. It is September and what has happened so far is that the same five or six of Greta's peers have joined her and keep coming back, week in, week out.

The task and the recipe

They set themselves a task and take it on. Slowly, carefully finding their way, and then with ever more certainty and confidence. They don't just see the strike as a strike, but as a project. Now we are here, in the square, what needs to be done? How do we stop the people who are destroying the world?

In the square, real traditions develop in a short time. A guestbook is passed round and will continue to be passed round for the next 50 weeks. And the children and young people soon develop a recurring game, the improvised – absurd – cooking programme. All of them take turns to participate and they each add a new idea. On this Friday, which may be in the third or fourth week of the strike, the programme presents the recipe for a deliciously prepared threshold.

Take one threshold. As ordinary as possible. Rub the wood with your fingers to collect dust that you can sprinkle over the sauce at the end as a topping. Next find a Finnish guy with a nice big spot on his forehead that's crying out to be squeezed. Squeeze the spot carefully into the dish, add a bit of sweet and sour sauce, and spit in it vigorously, three times. Stir with a chainsaw, then spit in it again three times. Finished.

They laugh, then turn serious again and look across at the parliament.

The Swedish parliament – opponent or authority?

In the stone building opposite sit 349 representatives from 290 constituencies. Together, they have the legal means to change the conditions on which people live together. During these September days, the government is a red-green coalition. Prime minister Stefan Löfven has deputy prime minister Isabella Lövin by his side, the Green minister for the environment and climate. Following the elections, they do not have an absolute majority, and now they want to put together a new, broader coalition. What is their project?

Who should be rewarded, and how; what should be produced, and how; what should count as valuable? How should the climate crisis and the ecological crisis be combatted so that the thousands of workers in the fossil sectors of the car industry and the cement industry would find new work? What was the best way to deal with digitisation and robotisation? How to protect forests? How should food security be guaranteed, while at the same time shifting towards plant-based nutrition and regenerative agriculture and forestry? How should an international structure be established to ensure security in relation to food and infrastructure, and keep oil and coal in the ground? How could heatwaves, droughts, and floods be prevented, and how could the countries affected be helped? How could the education sector be reshaped so that knowledge of the climate crisis was at the core of all subjects and at the centre of the approach to teaching? How could care work be properly valued, the "care economy" which after all is the core of society, the caring, teaching, nourishing and healing that is still so often carried out by women and sometimes not rewarded at all? How to get to grips with all of that, with injustices, inequalities, the social and democratic crises?

The two most powerful people in the state have a unique opportunity to make a plan for Sweden and win a majority, I think to myself during these autumn days. In Mynttorget, we don't hear anything of such discussions (on political operations and topics such as lobbying work, see Kemfert 2020, "Mondays For Future"). Greenhouse gas emissions have not really been reduced during the four years with a red-green government (Urisman Otto 2022), any more than the ecological footprint. The members of parliament who pass the young women in Mynttorget and enter the massive building cause about 10 tonnes of CO_2 emissions per year, like most Europeans (including emissions in foreign countries through consumption of goods; see Thunberg 2022). In around twelve years' time, if a fair system were introduced, they would only be able to emit 1.5 tonnes per year (Anderson et al 2020). How is that supposed to happen?

What is the democratic plan for such a change for all of us? And most importantly: how can the rules change in favour of the people who are most affected by the crises and have often contributed the least, in Sweden and globally; including the indigenous people in the North?

Facing the large, silent windows of the parliament, I often think: an abyss is opening within democracy. In these seemingly modern democracies, young people have no real place. They may receive a certain amount of sound teaching at many schools on the subject of "citizenship", but they are not included in political decisions. I walk up and down and look up at the gigantic windows of the members of parliament. How can that be? And it feels more wrong than practically anything else, when at some point during the next Fridays I'm not able to come back to the square at eight in the morning. For that reason, I decide not to go to the University on Fridays anymore, but to take Greta's call seriously. If a few hundred or even a thousand people do the same and gather in front of their parliaments, a political force could develop – that's the idea – which could lead to a change in our structural conditions.

Strangely enough, hardly any of the members of parliament pause as they pass the strikers. The contrast between the ministers' suits and the young women sitting on the ground could hardly be greater. And it is above all young women who are joining together. Many newspaper articles describe the whole generation of strikers as the responsible girls or young women of our time, and that is true; often they have active grandmothers supporting them. The people on the square are more female than male, but there are also quite a few – as the movement grows – who don't see themselves fitting into the gender binary. This whole situation reflects a deeper problem, according to my colleagues in the social sciences at the university. After all, the structure of the society against which these young women are rebelling is still shaped by men (Brown 2019). The men on the boards of the banks, the oil companies and the coal industry, and the world of men at the economic forum in Davos. These are complex structures which make the task of the five strikers on the square difficult, and which are also soon expressed threateningly, with pure hatred and misogyny. It can't go on like this, we think to ourselves then, the adults who return regularly to the square. We have to organise too. In London, so we hear, a new climate movement for adults is being formed, Extinction Rebellion; maybe we can build something similar here?

At the university

After my first meeting with the activists, I take the smoothies with me, open them with a bad conscience and drink them on the way to the metro station "Old Town", to get rid of the plastic rubbish. Travelling on the red line, it's only five stops to the university.

A university is a strange thing, I often think as I pass through the tunnels to the campus. In actual fact, ideas and thoughts swirl through the air wherever we are, including the centre of the city or Mynttorget. Ideas about what is normal, reasonable, what is seen as valuable or politically possible. Wherever we are, we're surrounded by them: what counts as healthy or sick, as fair, as democratic, what counts as freedom, as an appropriate teaching method, as reason, as science. Sometimes we take our time and look at one of these concepts more closely. Maybe we call it into question if we're being harmed by the way one of them has taken shape in society: the women's movement, the workers' movement, and the civil rights movement questioned and changed the prevailing worldview and what was seen as "normal" practice. The university is the place where these ideas are supposed to be made visible, you might think, and converted into new ones, more democratic, fairer, more sustainable. But precisely because they are the stuff of which all the lessons are made, even when the focus is on specific medical procedures or legal considerations, which are in turn shaped by ideas, some of them also remain unquestioned, at least so it seems in this case (see the chapter about education and the transformation of the universities towards the end of the book; and see the work of Judith Butler, Stuart Hall and others).

And so I walk from the station named "The University" past all the wonderful institutions, past Law, and then past Earth Sciences, past Climate Research, Drama and Literature; I leave Languages, History and Philosophy to my right and head to my Department of Teaching with the strange feeling that many of the things that are seen as normal here are actually highly problematic. Solemnly presented by people who mainly come from a particular part of the white upper middle classes, who ultimately don't take the climate crisis seriously. Not really. Those who teach at the university really ought to apologise to the young people in Mynttorget, I think. They should apologise for the fact that for the last forty years, the academy hasn't come up with an adequate answer.

What would my colleagues think if I joined the strike? How would my students react? Will I lose my job? Shouldn't all teaching and all research respond to the current situation of society, and therefore focus on the fact that we are facing a crisis we've never seen before (see Raffoul 2022; McGeown/Barry 2023)? What would that mean for each subject – economics, architecture, education, philosophy? For interactions between the subjects and the institutional organisation; for the method of teaching; for integrating us as imaginative, social, embodied beings living in problematic power structures? Like a strange, dark omen, the climate crisis hangs in the air.

But these thoughts are layered on top of my more concrete worries during these August days. Together with a few colleagues, I have created a completely new three-year bachelor course in the last few semesters, and not only am I responsible for many of the five-week sections of the course, some of which don't even exist yet and which I still have to develop, but I also teach many of the lectures and workshops myself. They focus on democracy, citizenship, and justice, on ecological, economic, and social sustainability and teaching methods, and they are taught via art education, especially drama, music, images and role play (on drama education: see for an overview McAvoy/ O'Connor 2022; Haseman/ O'Toole 2017; on applied theatre: Prenkti/Abraham 2021).

The drama method (in contrast with theatre) often means involving everyone all the time as part of an improvisation. Only occasionally do we come up with text-based pieces for an audience. Rather than only addressing people's minds, we are exploring what school can be, what the economy can be and what global democracy can be, using our whole personalities and our imaginations, working together with others in the theatre and dance spaces; we're testing ideas. Similar courses even exist in a few other places, such as New York University and the Central School in London. But more and more "applied theatre" or drama education courses (such as at the ZHDK in Zurich and the UDK in Berlin) are developing in this direction focusing not only on role play, devising, empathy and compassion, but on exploring sustainable democracy and transforming society (see Fopp 2016).

Now, at precisely the time of the first Friday strikes, a new year is beginning at our institution. A new class, a new group of students is arriving. I scan the twenty faces. It is the Monday after the first Friday strike. "Hello." We play a few games to introduce ourselves. Then: "How do you want to treat each other?" Twenty completely different people look at each other. "You can interrupt at any time and say 'stop' if something doesn't feel right," I say. "In improvisation, any idea is allowed. But not any action. Be respectful. And finally. All of you should look out for whether everyone is okay; to ensure that no one has to give up the connectedness or contact to themself and others. Again. Every one of you should develop a helicopter perspective; take account of your own needs and other people's needs, express them and respond to them. Support each other. It is not enough that everyone is allowed to talk and be here; inclusivity is good, but it is not sufficient. In the long term, this will only work if everyone makes sure that everyone else's needs come into play; if you look after each other. Is that okay?" They discuss it. Decide that it makes sense.

Science and games – Mynttorget in September and October

In Mynttorget, too, science is everywhere. "Listen to the science" soon becomes the young people's slogan. And a scientific worldview also comes up in their conversations. During these weeks, the young people who brought the strike into the world often sit together, before the others join them, and talk about everything under the sun. About chemical elements, for example. How strange it is that the whole world is made of so few elements, and that these only differ from each other because of some electrons or protons etc. As if a stone and a carrot were closely related. How strange the scientific worldview is anyway, I think to myself then.

As soon as a sense of morning hustle and bustle arrives, with the shops selling Viking hats and cinnamon buns opening in the narrow streets of the Old Town, most of the young people form small groups and begin to discuss the situation and talk about all sorts of things. It is still warm. No longer as hot as it was in summer, when the heat was quite unnatural. We start playing "city country river". We go through the letters of the alphabet silently, A, B, C and so on, and someone says stop. Then we have to find a city, a country, a river starting with that letter. But we could also choose something else. Dogs, drinks, dishes. And so we begin to play. In the next weeks, it will become a tradition on these Fridays. Sometimes a few other people join. I have no chance. Somehow, the young ones know all the cities in the world, many, many cities I've never heard of. With dogs it's even worse, and with drinks it's funnier, because we just make them up: anything can be a drink. Kiwi juice.

Six months later, on the evening of the first global strike, the 15th of March, I ask myself how many of the cities which we listed randomly on those Fridays, from Buenos Aires to Berlin, now have strikes going on. In how many of those cities does someone leave their classroom on Fridays – whether it's one person or tens of thousands, around the globe? It is as if we had been feeling our way towards this global dimension in these initial weeks, towards a sense of all of us as inhabitants of Earth, in our dwellings on this living planet.

Then, it is already the week of October in which Greta travels to London for the founding of the new movement of environmental activists, Extinction Rebellion, and the group is in the middle of the alphabet game: a dog beginning with L? Bye, off she goes, to the Rebellion in London. Two weeks pass. She makes a speech which goes halfway round the world, at least the activist world. She stands there in London in front of the crowds, but the microphone doesn't work. Then the thousands of environmentalists do what they always do at moments like that. They become a living megaphone. The people at the front repeat loudly what they hear, passing it on to the people behind them. Greta's speech takes ten minutes instead of three, because everyone repeats everything: "When I was eight years old," she says. "When I was eight years old," roars the square in front of Westminster Abbey. But now she is back on the square and the young people go on seamlessly with the alphabet game. The strike day begins.

Emil and Alfred – what is nature and what is healing

On one of those first Fridays, in the evening after "strike time", I have to teach, so I hurry away from the square and go back through the narrow streets of the old town and take the underground to my office. I pack my papers for teaching, most importantly Astrid Lindgren's book *Emil of Lönneberga*, and go over to the building known as the "stables", because that is what it once was. It now houses our theatre spaces. This is where I teach. Not just the new course, but hundreds of future teachers.

In room 420 they jump around, the future kindergarten teachers who have landed in Stockholm from every possible country and every part of society; it's a small hall with a polished wooden floor, without chairs, with a view of the woods that come before the sea, apple trees in between. Soon they have created – as in Lindgren's book – an imaginary waiting room at a doctor's surgery, with rows of chairs. One after another, they shuffle into the room, limping or coughing, and find a place to sit. We are in a small Swedish town, we imagine, Mariannehof, one hundred and fifty years ago. I go around asking: "Who are you?" It is all improvised. Someone says: a maid. I was kicked by the cow. A farmer, I'm pregnant. A carpenter, I hammered a nail into my finger. The pastor, I'm not feeling well. And so on. Twenty groaning characters. We discuss: how did they live in that time, not very far from us, but in a different era, with a different technology, a different relationship to nature?

We act out the scene from Astrid Lindgren's *Emil of Lönneberga* in which the ten-year-old boy saves the grown-up farmhand Alfred from blood poisoning which he got from a cut in his thumb. The drama method works like this: someone reads a page, then we jump into the text, head over heels, together. We become Alfred, who is lying on the sledge in the snow, and we become Emil, who is pulling the sledge. Others become the snowstorm which delays them. And then we all become the thumb, twenty people become a single thumb, and I cut it with a wooden stick. We talk about what really happens when someone cuts their finger. What kind of "things" are in a finger, and so on, the way they can talk about it with their five-year-olds. Later we will heal again, as a thumb. Drama is not theatre, which means that no one is standing on a stage, there is no audience, everyone is always doing something, we switch roles, make backdrops and costumes ourselves.

Alfred is critically ill. We have to go to the doctor, in the town. But that's not possible, it's snowing, everyone says. It snows and snows, and the road is impassable, says the father. Everyone has to accept that. You have to accept fate, Emil. But Emil is overwhelmed by despair. It's his friend which is sick, they grew up together. What are the others doing. They're just looking on while Alfred might be dying. Who are these people who don't really take anything seriously. Who see the feverish farmhand and don't look for a solution, for a cure. Not really. That is the question we ask ourselves in the room with the polished floor; why do some people take life seriously, while others don't? What goes on in their heads? There, in the middle of the night, at four in the morning, Emil has stayed with Alfred the whole time, and he gets up, leads his horse out of the stables and lays the farmhand on the sledge. In a fit of determination and defiance. I cannot accept that no one is helping. Gustaf Gredebäck, a researcher at the University of Uppsala, examined how even some of the smallest children can show this form of empathy (Gredebäck et al. 2015). His experiments have been confirmed many times across the globe: they suggest that even six-month-old infants often have a strong impulse not to accept it when somebody is in pain, and to be annoyed by those who inflict it. At the end of the story, the exhausted Emil reaches the doctor's waiting room, and the farmhand Alfred is healed.

How strange that we can heal on our own, or that a thumb is able to heal on its own. Or rather, not quite on its own. Some wounds are too deep. But when we help people, with a plaster or an operation, a process of self-healing begins. The extent to which we are a part of nature – and the extent to which nature itself is miraculous – is probably never clearer than in a process of healing. But we can also damage nature so badly that it cannot heal on its own anymore. Ecosystems lose their resilience, as specialists call it, my colleagues from the Stockholm Resilience Centre, among them the world-famous researcher Johan Rockström (2009). Then we heat up the oceans so much through greenhouse gas emissions that the coral reefs are permanently destroyed, and the Arctic ice melts entirely. Tipping points occur, and feedback loops, and what used to be whole can no longer repair itself. The planet, as our habitat, is then irrevocably broken, because it is too hot.

The worldview must change, at the universities, too, I think during those days. We have to rethink and reshape our fundamental relation to each other and nature. This is what a reasonable reaction to the crisis would look like; and a reaction to the young people's. What I feel after these first weeks is that they have given us a task: not to study them as an object, but to study together with them how to get out of the crises. For us as scientists, this is now the project for the next few years. Do my colleagues understand the desperation and grief of the young people in Mynttorget? Do they see the existential challenge for all of us?

In Childhood/Youth Studies research, this is referred to as the perspective of children and young people themselves and is juxtaposed with the children's perspective (see Sommer 2010). The children's perspective can and should be adopted by adults when describing the world and taking account of children's concerns. Whereas making sure that the perspective of children themselves is recognised means giving them space to speak and shape things for themselves. In accordance with this, I will keep insisting to my colleagues that we should study how we get out of the crises together with the young people.

The first Swedish strike groups are formed

In Sweden, the Fridays For Future movement begins to spread. While the young people sit leaning against the wall and enjoy the still-warm autumn sun, a few of them begin – together with some adults, such as Andrea, Janine, Torbjörn, Anders and Ivan – to write to all the Swedish communities, or more specifically to any contacts they have within the grassroots movements and climate activism, and also keep a very close look out for anything happening online. "KlimatSverige" is a great help with this; "ClimateSweden" is an alliance of all the environmental organisations, but because of a strange, happy coincidence in world history – in contrast, for instance, with the equivalent "climate alliances" in Germany and Switzerland – it is almost a band of amateurs. It's just ten people who spend their free time in the evenings collecting information about all the other groups, writing a newsletter, and creating a network of contacts across the country. Both organisations and private individuals can be part of KlimatSverige. The amazing thing about it: the big three, Greenpeace, WWF and the Nature Conservation Union just let this little group of heroes do what they like, they don't get in the way, they don't control anything, they just provide a bit of money. That is the only reason that the young people can build on the work of this network; it has enough of a grassroots character.

Whoever comes to Mynttorget in those first weeks meets the young people first of all, and then the older activists such as the teacher Jonas, the pastor Lena, the carer Cilla, and especially Janine. Janine is from Australia and sometimes comes to the square accompanied by a good proportion of her family, but most often by one of her sons. Janine herself has worked as an engineer at Ericsson, she is very curious and wants to know exactly what is going on within the growing network of Fridays For Future. Having paid attention to climate research for a while, she is active in various climate organisations, and already in the first days she joined the strike. She networks continuously through her iPad and the emerging Facebook groups. On all topics and in all possible constellations, she creates various chats and forums, and adds all new children (and adults) from Mynttorget to these groups.

Young people from other cities appear in the chats. They, too, have had enough of their governments. The Paris Agreement is not being followed. They leave their schools: Andreas in Falun, later Tyra in Norrtälje, Alde in Gävle, the Sundsvall group; and so on. And so the first lists of strike locations emerge. Andrea, a yoga teacher, takes on the task of finding contact details together with the young people and putting them into a document. This is an embryonic version of the global map created a bit later by Jan and Jens, and which is valuable for coordinating the global strike actions. The list includes the place where someone is planning to strike, the time and possibly their contact information. At the beginning, it is just a few points on the map of Sweden, but gradually more are added. And everyone posts the successes almost daily on the newly created Facebook group page, #FridaysForFuture. This will be the first source of information for strikers in Sweden and then across the globe, and will remain the key source for a long time. Week by week, new locations pop up. And the media start to report that there are thirty and then fifty groups striking in Sweden. #FFF has become a reality. And not only in Sweden. In Berlin, too, people are striking in front of the Reichstag, and soon we discover more and more cities and countries on social media. All of them want political change and a safer and fairer world for everyone. As strikers, they are the same everywhere. And yet: we have such different starting points and privileges in this struggle; coming from different generations, social classes, and so on.

The big task and the small one – status and privileges

Back at university. What would a more humane world be like? What role do we have in it, as adults or young people, as privileged or disadvantaged people?

We experiment a lot with the question of what is humane in the drama spaces. In particular, we come up with so-called status exercises (see Johnstone 1987): someone plays a king or queen who takes on a higher status in relation to their subjects and can ask for whatever they want. Then we continue improvising so that the subjects increase their status and the monarch's status is reduced, until they hardly dare to rule over their realm. As an exercise, both can be equally difficult, playing someone with a high or a low status – for Johnstone (1987), status is not the same as power, but a quality of relating to others, of taking up space and time, for example, often expressed by how confidently we move and speak. The students want me to join in. That reduces my status significantly, since it requires me as a teacher to get involved in bizarre situations; but it also increases my status if I'm open to taking part. I often ask myself on the way to the university: What exactly is my role as a male, white, middle-aged person from the middle classes (or the role more generally of such people)? With some of my colleagues I discuss this again and again. How should we deal with our own privileges; how should we give others space; how should we dismantle power structures?

"Give me that stool there!" a student yells at me. "Yes but..." "Give it to me. Now!" We are standing in the middle of a group of improvising students and testing a scene. It is about understanding how we might meet each other at eye level. For that to happen, we try to understand status, power and domination. "But the stool is rather wobbly," I say, testing out, as a servant, the high status of my king. What will he do now? If he has a fit of rage, he will ostensibly raise his status, but in fact he will lower it all the more: what kind of king flies into a fit of rage over a broken stool? But doing nothing would also undermine his status. What would be a truly self-assured move? He could make me try sitting on the stool and thus display and protect his sovereign position. And so we test, day in, day out, micro-transactions of domination, with body and soul, in order to get our own sense of how and when we place ourselves above others or discriminate against them – and how we can instead work together without blocking each other (the exercises can be found in Johnstone 1987).

In this way, we practise what we can call "substantial" democracy. It is important to see that the concept and phenomenon of "status" is different from "power". Power belongs to an individual or group, whether taken by itself or given by others, in informal or formal ways, and can be exercised or not; while status is the specific relational transaction which we perform towards someone. Status can be established on the same level – low, middle, or high – or on different ones: lower or higher than the level of the other; we can call it domination if we violently lower the status of someone (or of ourselves; as submission). It is the core ingredient of oppression, in all dimensions of privilege and discrimination (including gender, ethnicity, class, etc.). The opposite of domination is not indifference, but either adopting the same status, or moving "beyond" the status struggle, which we often call "humane" behaviour. "Meeting on an equal footing" can mean either adopting the same status or creating a relationship with the other beyond status transactions – which we often call friendship.

Many children specialise in this, even very small children, according to research (Fopp 2016). If they are not valued by adults or treated with dignity, they will do anything to topple their giant counterparts from the throne. With words, with facial expressions, with disobedience, or whatever method they can use. And they are very happy if you communicate with them at two levels: if you act like you're going to be a lion and eat them up, and then don't do it. To meet each other beyond squabbles over status, we all have to practise; that is what my colleagues argue, and what I argue at our institute. It is useless to offer ethics seminars on empathy, or come up with long-winded theories about a new ethical society, or rejoice in a new "compassionate" attitude. We have also have to put these things into practice, we have to let ourselves be observed, we have to see through longstanding patterns of domination and discrimination (Ogette 2018).

Most of us don't even notice what we are doing, which patterns we have become used to and what is inscribed in our bodies and our self-image (as white and male, for instance). In our seminars, we study these microtransactions of power in world politics, too: Trump's handshake is already well-known as an attempt to put people off their stride and humiliate them; the same patterns are reflected by (climate) policies which explicitly call for "dominance" over other people and nature. "White supremacy", as represented by many of Trump and Bolsonaro's followers, explicitly refers to that dominance (see Saad 2020). Another dramatic example is the first interview given by the director of the oil company BP after the explosion of the oil platform in the Mexican Gulf, when he talks about the thousands who relied on fishing for their livelihood and who have lost that livelihood as "small people": people he looks down on.

When it comes to analysing these power relations, it is also useful to look at the results of psychosocial research, from Theodor Adorno's *The Authoritarian Personality* to modern intersectional theories (Meyer 2017), which show how various dimensions of discrimination (racism, patriarchy etc.) influence each other (see the chapter on corona and intersectionality). A white man has a different place in society from the man of colour, and a Black working-class woman has a different position from a rich white woman (see Collins 2019).

In our exercises, we often observe that most people use high-status positions to keep other people down; but high status and positions of power don't have to result in relations of domination. There is after all the possibility that the king could discharge me from my duties as a servant, and abdicate power, leading to a democracy. The question underlying the whole course is how all of that hangs together: how do we arrange and structure our small-scale encounters in terms of domination or being humane, in the family, in school, in university lecture halls – in relation to the ways in which we organise our shared life in the larger political spaces, when designing our economic system, for instance, but also when setting the political conditions for environmental protection, the health system and so on.

What is the ideal, the compass? Domination – we often agree – ought to be replaced with more democratic spaces, in which we can reinforce each other's strength and make sure everyone is okay and keep the caring contact to oneself and others; in which everyone's needs are seen and met, so that we all have enough to lead a dignified life and we take care of the environment (see Raworth 2018).

But there is so much getting in the way of that, I then think; there are the existing power relations and privileges. It is not enough just to claim that everyone is free and equal. That's a point we first have to reach, by dismantling the very concrete relations of domination (Saad 2020): at a micro-level in the university classroom, it is about giving everyone space for discussion, not just the three or four who are the loudest. But it can't stop there; and that is what I worry about the most after the seminars. The macro-level of power relations can be felt on a small scale in our interactions: for instance, there are barely any Black students, and no Black tutors. An invisible gulf opens up between theory and practice. Because more or less all Swedish students in the humanities, in sociology, education and political science – that is, far more than half of all students – attend compulsory courses on "norm criticism" and intersectionality, where they learn things that the universities themselves hardly implement. Something is fundamentally wrong. There needs to be a discussion with the people in charge. We can't just talk about changes; we have to carry them out as well – that is the basic idea of the course we are offering our students. That means we also have to put it into action. That's why we don't just read texts but also work with our bodies in the room and try to bring about real, substantially democratic encounters.

At the end of October, the UN's climate researchers publish a report known as the IPCC SR1.5, a special scientific report demonstrating what the differences would be for the development of the planet if the earth became 1.5 degrees warmer rather than 2 degrees, compared with the pre-industrial era. The differences are dramatic, particularly when it comes to reaching potential tipping points. The report receives a great deal of attention, and the worries in Mynttorget grow even more serious.

Chapter 2: Fridays for Future and Extinction Rebellion Start to Grow
October – November 2018: Civil disobedience and the laws of humanity

How to organise a rebellion

Week after week passes. Fridays for Future develops very slowly. But towards the end of October and the beginning of November, things start to move. The adults take up the young people's call for an emergency brake and create the movement Extinction Rebellion. But what kind of actions are justified? What are the limits of legitimate democratic action? Do we have to obey the letter of the law?

It is Saturday afternoon when I find myself standing next to Greta and a hundred other protesters of all ages in the middle of the central north-south axis of Stockholm's centre. We have to make a decision in the next few seconds. Either we refuse to leave, or we will be dragged away and possibly arrested. But will they really do that, the police officers who are coming straight for us? Will they risk a photo of the young people, who are blocking the central street of the capital of Sweden, in front of the Royal Palace – and police carrying them away because they are doing what they can to oppose the climate crisis and the mass destruction of living species? That would go right round the world, even now, although most people still don't know them. It is only a couple of months since they sat down with the strike sign in front of the Parliament and initiated the FFF movement. And now we are standing together on the street, young people and adults. And along with our despair over the climate crisis and our fear of the consequences of the blockade comes the question of responsibility. Can fifteen-year-olds be put in prison? For how long? Or kept in a cold police station? Do I have to intervene? Are we prepared to end up in prison, like many people right now in London? What does it mean to take responsibility? How do we bring about real political change?

Extinction Rebellion is called that because this movement of rebels, freshly founded in Summer 2018, is fighting against the human-caused extinction of thousands of species and their habitats, as well as the obliteration of the human species via the climate crisis (Extinction Rebellion 2019). It is about showing how the picture fits together. On the sheet of paper which Greta hands out in the first weeks is the following sentence: "According to the Living Planet Report, humans have eradicated 83 % of all wild mammals, 15 % of all fish and 50 % of all plants." Anyone faced with this statistic (see Carrington 2018) will realise that because of humans' destruction of natural habitats in the last few centuries, there are barely any wild animals left on the planet, only about 4 percent; that together with our "livestock" of cows, sheep and chickens, we represent the absolute majority of all living beings, and that eating animals is one of the ways in which we most contribute to the climate crisis: forests have to be cleared, methane is emitted (Foer 2019) – the homes of thousands of species are destroyed and with them the web of life, the interconnected "earth system." That is maybe a reason why Greta was present for the founding weekend of XR (Extinction Rebellion) and encouraged us adults in Stockholm to form our own group. As so often happens, the initiative comes from young people; we react. How can we stop all of that?

Evening after evening, we meet on the web platform Zoom, long before it becomes well-known through the pandemic, and prepare for the big blockade. Now it is Saturday afternoon. The police officer comes closer. We've been preparing this protest for weeks, together with ten other people I didn't know beforehand: teachers, nurses, students. A group of adult rebels is emerging. A lawyer for Greenpeace is available for advice. What has the most effect; what draws attention and reaches the "silent masses" of the public, rather than just causing a disturbance?

The basic idea of XR is to mobilise a mass movement and to occupy the centres of society and create a "disruption," using a form of civil society uprising to hold up "business as usual," ideally by blockading central sites in front of parliament buildings (see the discussion of tactics in Extinction Rebellion 2019). It makes too little difference to paralyse individual parts of the fossil industry for a few days, as some of us had done before with "Ende Gelände" in the previous summer, when we occupied the Vattenfall coal power station south of Berlin.

The system itself must be transformed democratically in a few years; otherwise, the world will be a nightmarish place, three or four degrees warmer, by the time the children are growing old, the scientists tell us (Rahmstorf/ Schellnhuber 2019). That is why the action focuses on centres of power – and why it must draw a different kind of attention from protest actions focusing locally on a coal power station, a bank, or another part of the cogs and wheels of fossil society. At the beginning, governments don't know how to categorise XR; precisely because the uprising is so clearly formulated as an idea. This makes the principle of nonviolence all the more important for us. We train ourselves in the history of civil disobedience (refusing to obey certain laws and thereby showing that they are not adequate because they don't respect everyone's dignity and have to be changed in order to deepen democracy; see Chenoweth 2012) and the risks associated with it; and in the tradition behind it, from Rosa Parks to Gandhi and Martin Luther King.

And now we find ourselves in the street, and I have no idea what we should do. The state, as we have agreed in our discussions, ought to protect us as citizens; it ought to ensure that in fifty years' time there will still be a dignified life for everyone. That's why this blockade is legitimate. This is not about the interests of one group in society, but about the dignity of all. I look across at the police officer, who represents the state. He seems determined, and he looks at the young people. How did they manage it this time? They have woven their way between queues of cars and through the hundred people blocking the street, and despite this confusing situation I always knew roughly where they were. But then they suddenly sat down in the middle of the street, in front of the palace, with a flash of their own instinct for political action and media communication. As if there were nothing unusual about the situation, they now turn to me: "We have to let the busses through." "We should let the busses through?" I ask. "Yes, there are old people on them." On the other side of the junction, there really are three or four busses lined up with older people on them, waiting to see what will happen. In front of them are at least ten people with cameras.

Before this, the young people spent an hour handing out *pepparkakor*, Swedish cinnamon biscuits, to the drivers stuck in their cars, and talking to them. Those in the biggest SUVs often seem to be the least friendly. The *pepparkakor* they had got hold of are the only ones without palm oil, the production of which destroys the rainforests of Indonesia, and they are vegan, which is just a detail, but an important one.

Suddenly someone yells. We are all startled, including the police officer. From the corner of my eye, I see a man punching a car, again and again. A driver who had been held up for twenty minutes had suddenly driven into the demonstrators and almost injured a child. It is the father who loses control.

We all stand up reflexively, go over there and try to diffuse the situation. This puts an end to our protest without a decision being made. What would I have done if the police had threatened to arrest us? How far do we have to go as individuals? How can we help each other build a truly strong global movement and change the political rules?

XR was formed by Gail Badbrooks, Roger Hallam, Clare Farrell, Nils Agger and others to make sure that people would draw attention to the climate crisis, with quite a different approach from the earlier generations of environmental activists. One of the first protests is an occupation of Greenpeace, to point out that their form of protest does not work, not sufficiently at least; the political framework can't be changed by NGO work, only specific policies. Normal people have to come together, in a democratic bottom-up movement. And it is about displaying civil disobedience, not covering it up. That is the only way social change can take place, as shown by history; that's their theory, anyway. If a few thousand people would come and stand next to us, I think, and in a city with millions of people that is not so many, it would change the world. History shows that the key locations have to be held for several days if society is really to be transformed (see also Chenoweth/Stephan 2012).

Later on, encouraging people to get arrested will also draw criticism within the movement: not everyone has the privilege of being able to get arrested, without putting their basic existence in danger.

I look across at the car. My reaction is to be furious with the furious father. Good sir, I think to myself, don't smash up that car. Don't let there be photos of violent activists. No violence. Civil disobedience, nonviolence; the two belong together. And then my anger turns in a different direction: why do we have to stand here on the streets of Stockholm, Bo, Jörg and I, and with us a hundred other ordinary citizens? Why is the XR movement spreading like wildfire, much faster in these first months than Fridays for Future, which still consists of the fifteen regular strikers? Why doesn't everyone react when a child sits on the street because our shared life on a living planet could be threatened during that child's lifetime – and already now people in the Global South are much more severely affected by it? Tipping points and feedback loops could mean that the next seven or eight years are historically crucial; if we don't change fundamentally now, it could be too late (see Schellnhuber 2015).

The newspapers report on us, and soon the plan emerges to expand our activities in the course of the coming April and block the capitals of western industrial states: London, Berlin, Amsterdam, Stockholm. But what are we trying to achieve? We spend hours standing, sitting, dancing, and blocking the street in front of the parliament; what are those inside the building supposed to change?

The three demands of the global environmental rebels

What many people of all ages say after similar FFF and XR protests: in these hours together in front of the parliament at the road junction, they experience a kind of non-instrumental relationship to each other, and a kind of peaceful recovery of influence and social participation, which feels strong and liberating, even when the fears – of the crisis and of the police – and the grief may return very soon.

After the protest action, we all gather in the Franziskaner. The oldest German-speaking restaurant in the Stockholm old town is right by the sea. The tension only ebbs slowly, the stress of the situation in the street can still be felt. "Maybe we should have provoked an arrest," says Bo. "That's the basic idea of XR; that's the only way to get the attention which creates change in society." I have ended up between the yoga teacher Andrea and the doctor, Kasper, and opposite Bo. Almost all the hundred participants are between 30 and 55. They are teachers, unemployed people, doctors, immigrants, locals, relatively diverse. "We can't just organise blockades," I reply. "We need a political alternative, a global political movement which everyone can join easily, at any time, in all countries; and which formulates a clear alternative, names the things all governments have to change: first of all, staying within the tiny emission budgets; secondly, keeping coal, oil and gas in the ground; and thirdly, seeing this as a shared project, in which we all help each other and abolish unfair structures." Bo disagrees. "Why? No. First we disrupt, then we build; otherwise, we're just like another party and we don't change anything. Why should anything change if we don't stop this four-degree machinery?"

The basic question is this: previous generations have left the world in a desolate state. Now a program must be developed to counter this. Within ten or fifteen years, we have to have a society in which we treat each other and the treasures of this planet more peacefully, hardly eating animals anymore, producing electricity from sun, wind and water, dismantling the power relations of domination which exist in the economic system worldwide and nationally (racism, gender inequality, class), and equipping everyone with enough resources to lead a dignified life (see Göpel 2016). How can we achieve that?

Extinction Rebellion has three demands which are then formulated in a similar way by Fridays for Future. First: name the crisis as a crisis; inform the population at long last about the state of our planet and our treatment of it, from the climate crisis to ecological destruction and loss of biodiversity. Secondly: the government must put measures in place to make zero emissions of greenhouse gases possible by 2025, in all sectors, from transport, energy, finance, and agriculture to largescale industry. In Switzerland, FFF will agree on 2030 as the goal for zero emissions in the region known as the Global North, but will always point out that it is not about which dates we aim for, but the absolute level of greenhouse emissions, the global "budget," which is minuscule (Rahmstorf 2019). During these days, we get in contact with the world's leading researchers and discuss these aspects in detail – and we will not stop doing that. On the contrary, we continue to expand this cooperation in the university context as well; Scientists for Future emerges. But already early on, as we spend every free moment studying research papers, it becomes clear to us that the Paris Agreement itself actually compels governments, the governments of wealthy countries, to stop all emissions by 2030 (Anderson 2020); otherwise the goal of keeping the rise in temperature "well under two degrees" cannot be reached. The world must have eliminated well over half of all emissions within the next ten years, says even the conservative UN; emissions must go down by seven percent each year worldwide, says the UN GAP report. No country is even anywhere close to such a transformation.

In comparison with that, societies seem to have slipped into a kind of anaesthetised state, or else they have been put in this state by those who don't want the status quo to change, as explained by Harvard Professor Naomi Oreskes (2010), who has researched the lobbying work of the fossil industry sectors. More and more messages appear every day in our Twitter, Facebook and Instagram feeds: about the Arctic ice around the North Pole, which will probably melt completely in the summer within five years; about the melting permafrost ice in Russia, which was only supposed to begin to melt in seventy years; about the lethal droughts and the unimaginable suffering caused by floods in India and Mozambique, and about the forest fires in California (see Wallace-Wells 2019). It is only when we make it clear that we have a few years to turn everything around that we come to see how much work that would take.

We also become aware of how drastic that societal change would be in our conversations in the Zoom chats during these October weeks, as we go through the individual sectors. What needs to happen so that coal power plants can be decommissioned without disrupting the whole of society? So that the transport system (cars, planes) can be shifted away from oil-based fuel options? Not only will many people have to re-train and change jobs. Our whole system relies completely on fossil fuels. If we were to introduce zero emissions tomorrow, within the shortest possible time our food supply chains would collapse, as would our transport system, our electricity supply, and so on. Partly because our "goods", from clothes to food, rely directly on production methods which emit greenhouse gases, and also because they often have to be transported right around the world before they reach us. We have to stop that, though, and replace it with a sustainable circular economy. This replacement could be humanity's best and most important project, a project for all of us (see the appendix of this book). So we have to start there, I think, as we reflect on our street blockade actions.

XR's third demand is supposed to ensure that this transformation of our societies takes place democratically and in a socially just manner. The basic idea is this: a societal change as drastic as the one required by an imminent zero-emissions society needs democratic oversight in the form of citizens' assemblies, which should advise and help decide which laws would guarantee a fair transformation for everyone. Otherwise, there is an enormous danger that only privileged groups will redistribute power to their own advantage. Our argument is that we need more democracy, not less.

Does humanity have unwritten laws?

At the centre of all these discussions is a fundamental question: how do we legitimise our civil disobedience to other adults and convince a large enough proportion to join us (Chenoweth's research says that only three or four percent would be necessary for fundamental change; Chenoweth 2012)? Why is this not simply a violation of rights? Why does history show that societies can change in this way – and above all, that this is almost the only way that they can change?

During this time, my work at the university helps me, as does my work on an article for the culture section of the biggest Swedish daily newspaper, DN, which is soon published. Already two thousand years ago, Antigone made one of the famous speeches in literary history, challenging the ruler, Creon, on a stage in ancient Greece: in her speech, she says that there are "unwritten laws" of humanity, of reason, which are more important than the written laws. If we do not follow them, we are no longer human, and life makes no sense. And for those laws, it is worth risking everything. Just as Rosa Parks insisted on her seat in the part of the bus which was reserved for white people (having prepared as carefully as possible, as part of a hard-working grassroots human-rights organisation). The question then arises: what is this compass against which we can measure a law in a democracy and declare it illegitimate? What is it that is important to the strikers and which gives these unwritten laws of humanity legitimacy and allows them to turn against the existing laws in order to fight for other, more important ones, which are still unwritten? Why is this not just a riot by a mob (on the theory of civil disobedience: Braun 2017)?

In the last thirty years, including in the newer tradition of critical theory (from Habermas and Honneth to Menke and Forst – unlike Adorno and Marcuse, see Biro 2011 for the relation between critical theory and the sustainability crises), most humanities institutions at universities across the world have answered this question as follows: what is legal is not fixed, but results from negotiations, at the end of which equality and freedom often emerge as the compass, although they are often defined in different ways and can also be augmented with other norms. But in the time of "new materialism", and "convivialism" (Vetter 2021; Hickel 2020), this partly liberal, partly postmodern thinking (in popularised versions) now seems to be reaching its limits. Because you can't negotiate with planetary boundaries and tipping points. The oceans cannot be cooled by the arguments of postmodernity; suffering does not disappear that way, either.

So what is the standard by which all laws must be measured (see also the chapter on democracy)? It is found in the declaration of human rights and in the constitutions of most countries: the equally inviolable dignity of all people (Menke/Pollmann 2017; Bieri 2016). (Extending this thought, during XR actions we draw attention to Swedish emergency law; our actions are legal because they protect people's lives and health.)

And at the same time, in my lectures at Stockholm University I test out the idea that this dignity can be viewed from two perspectives. Seeing people as having dignity means, on the one hand, not dominating others, and that means eliminating relations of domination, whether in terms of gender (patriarchy), ethnicity (racism), class (exploitation) or other realms of inequality (Collins 2019); that is familiar from all humanities and social science subjects. The second perspective is that of care and attention. It is not enough just to avoid dominating others. Seeing people as having dignity also means taking their needs seriously, not letting anyone starve or abandoning them to illness, and so on. Or with a metaphor: the standard measuring civil disobedience is the maintenance and repair of what we could call our shared "fabric of integrity", the material of our dignity; what breaks when someone starves, when someone is hit, when children are left alone with their fear (see Fopp 2016). This metaphor also aims to make clear how close we are to nature and to the material from which we are made, the living vulnerable physicality which connects us with the whole network of the plant and animal world. It is this dimension of the fabric of integrity which is most important for many of us activists; it is what many of the children's and YA books I discuss with my students are about, and it is what theories such as feminism, ecologism, postsocialism and postcolonialism aim to protect (Emmett/Nye 2017; Hickel 2020; Fraser 2022). Isn't it about upholding the laws of humanity, which Antigone was already talking about? In order to protect them, we adults have to join together just as the young people of FFF are showing us, when they refuse to go to school even though they are legally obliged to do so. That is what I think during these days. The children with their alarming future have no voice in our democratic constitutional state. That is why we have to stand up.

On the 28th of October, Jair Bolsonaro is elected as the new president of Brazil, with 55 percent of the vote.

Magic in Mynttorget

It is raining again, and it is terribly cold, mid-November, a perfectly normal strike day. The young people are still sitting on the paving stones, on their yoga mats and rucksacks. Some stand to one side, with their raincoats and umbrellas. I look up from my place next to the flowerpot, lost in thought after a conversation with a journalist from one of the big international newspapers like *Le Monde*, *Der Spiegel* or *El Pais*, which now come to Mynttorget every week. It's not really raining properly, so it doesn't matter if you let an umbrella spin around. We look across at the parliament building. "Can't you float up like Mary Poppins?" I ask. "You could fly over there and shake everything up." That would be something. Magic. We are the creatures that can imagine. That might be what most distinguishes us from other animals. We can come up with fantasies of both good and evil. If the technology of the umbrella works together with the natural power of the wind and with the human will, then we're able to imagine that magic might come about.

That is the central formula of our lectures on the theory of imagination and fantasy: the magic of the imagination is somehow connected with the way in which humans elevate nature and technology to a freedom combining vulnerability and power. It becomes a problem if people set up nature or technology to use as forces against each other, rather than being united. We use this perspective to analyse children's books and films, from *Momo* to *Harry Potter* and *Titanic*, that megalomaniac masculine project of technology which collides with raw nature.

At some point months later, I begin to notice that there's something in Mynttorget that has become the centre of everything, something the children have "inherited", as they say. They are so obvious; they stand there so elegantly and naturally that they are practically hidden. The street lamps. Electricity is here in person. Maybe we as a society have still not really examined or understood what the mysterious scientist Tesla landed us all in when he succeeded in making alternating current travel long distance and thus made modern life and the modern city possible. They bring warmth, promise security, vivacity, community. And somewhere at the other end of the line is the coal, the oil, the gas, the atomic power plant. Where there ought to be wind, water, sun. Sustainability means creating structures in which nature and technology don't appear as opposites, you could argue. A wind turbine follows this better cooperative logic; not the logic of burning up resources. We ought to be expanding our use of a healthy form of electricity, just like the electricity in our nerves.

The young people stand there under the streetlamps, which always switch off just when they sit down early on Friday mornings, and use their brains. Little electric impulses jump across the synapses, connecting cell with cell, idea with idea, and out comes a plan.

The global network is born – the digital heart of Fridays for Future emerges, along with the idea of the global strike

It is an inhospitable November evening, ice is falling from the sky outside, and Fridays for Future is a good two months old. Every Friday, small groups of strikers gather in many locations, most of them still in Sweden, but also in Berlin: in front of the Brandenburg Gate, Barbara can be found happily tweeting out pictures of her six or seven companions into the world. And often that is around the number of people that turn up, six or seven, in some places they are a varied mix of young people and pensioners, gathering week after week behind a cardboard sign. Everything is registered and documented in Mynttorget and spread across the wide world.

 I walk up and down too, trying to understand our role as scientists; there are about three or four of us now who turn up every week. Janine waves to me. It's about the preparations for an event as part of "Climate Alarm," which is organised every year by all the climate movements. She explains that Climate Alarm is an internationally coordinated demonstration with speeches. The whole thing is supposed to take place in December on a Saturday, not a Friday. I nod, slightly sceptical. While the young people discuss whether they should start appearing in public under their real names and accepting the many requests for interviews, Janine tells me: "I've added you to the chat."

It is a small Facebook chat called "Climate Alarm FFF." Initially, there are only seven or eight of us taking part, a motley collection of people from different generations and different European countries, Brice from Luxembourg, Marta from Portugal, Benjamin from France, Andreas from Denmark. Young people from Sweden as well as their peers from Ireland, Scotland, Holland and so on. It is one of hundreds of chats which are started in these months. But it is a special one. All those who gather are volunteers who organise demonstrations and other activities in their local communities. It is precisely not about NGOs, but about real grassroots movements.

But chats or WhatsApp groups cannot replace actual meetings. The group soon plans a first Zoom meeting for this Climate Alarm project. Eight pm on a Sunday. So begins an important part of the history of the international Fridays for Future (online) network. Week by week, month by month, the pulse emerges which becomes the digital centre of FFF. In this chat and in Zoom meetings, the idea for a global strike is born, the date of the 15th of March is discussed and decided, the strike is coordinated and organised. In this chat, and soon also in the WhatsApp groups, the crisis over the Strasbourg trip to the EU parliament is dealt with and the numbers of participants are counted up after the global strikes. As an old relic, it will survive all technical innovations – the move to Discord, then Telegram – and it represents the information channel used by everyone when they want to reach the other central figures of the global movement quickly.

By January at the latest, there is a competing WhatsApp channel, but it is semi-anonymous, and everything quickly becomes a conflict. The same goes for the Telegram channels. This old Facebook chat keeps its name, and so the digital headquarters of FFF continues to be called "Climate Alarm". Much later, after the second global strike in May and the fights with the NGOs who join, the chat loses its function and abruptly goes quiet from one day to the next, after almost a year. It did a good job.

Early in December, like little seedlings from the original plant, other small chats sprout from it, as do regular Zoom meetings which blossom into the central communication channels of FFF: the strike dates chat, the chat for welcoming new countries, the logo/arts chat, and so on. There is no principle of meritocracy or representation here: someone may only have been to one strike, and there could be ten participants from the same country. In these early times, this principle rules over all others. FFF is a network initiated and run by young people, not an organisation.

The movement has found its digital home, and even if these meetings still only have a few participants, most of whom come from Europe, the young activists know where the pulse beats, where people can find out what's happening across the world: the indigenous population of South America is soon represented, along with the rebels from Australia, the young graphic designers, led by Yacine from France, as well as the Canadians who are working on getting their government to declare a climate emergency. New groups form in India, Ukraine, Bangladesh. The agenda is usually put together on an ad hoc basis. Early on, "FFF's demands" are always on the list, a point which will put the whole movement to the test.

And suddenly someone has the idea of a global strike. Unlike Extinction Rebellion, Fridays for Future was always, from the beginning, focused on global social justice. "Equity" and "social justice" are terms which are sent out into the world early on by Greta and the Mynttorget group, as well as by the Swiss-wide meeting of activists. Not that this point was not important for XR, but XR's focus is on the capital of each country, on specific blockades and the idea of democratic citizens' assemblies. Meanwhile, the young people of Fridays for Future have seen themselves from the first moment onward as part of a whole generation, as a global movement: "There is no planet B," say those who live everywhere on this "planet A." So the idea of a globally coordinated strike is not so unlikely; on the contrary. The question keeps coming up: when should this huge strike take place? What can it achieve? As happens so often in these weeks, once an idea is in the air, it cannot be taken back. There are no hierarchies, and no one can insist on bans or priorities. Among those of us who are in Mynttorget, there is a certain level of scepticism: isn't it better just to keep on striking once a week? A huge strike could disrupt that tradition and tie up too many resources. But then someone suggests a date, the 15th of March, and from then on the ball is rolling.

"Why should we choose the 15th of March specifically for the strike?" "Yes, we know that XR is starting their big blockade of London on the 15th of April; shouldn't we choose the 7th of March instead? We need enough distance." They go through the holidays of the most active countries, as well as the bank holidays, making sure that there won't be many people who can't strike on the chosen day because they have a day off anyway. The 8th of March is International Women's Day, and FFF wants to honour that and not compete with it.

But the fact that the world will strike on the 15th of March is ultimately down to the environment representative of Stockholm, Katharina Luhr. The young people of Mynttorget want to approach her this autumn with their demands: they are calling on her to present a plan of how it could be possible to keep to the budget of greenhouse gas emissions drawn up for Stockholm by Professor Kevin Anderson and other researchers in Uppsala. The 15th of March seems to be an excellent date for a deadline; four months should be enough to make a plan. On the 15th of March, the city of Stockholm will not have come up with any plan to reduce emissions, but that will be forgotten amid the global uprising.

Sometimes, when a few countries suddenly lose their way, I post a reminder of FFF's basic numbers on the climate budget in this main chat. Otherwise, I back off as much as possible, and only intervene if I think everything is falling apart. Often – almost always – those moments come when a small group or an individual suddenly wants to speak for the whole movement or tries to change the direction of the network fundamentally, such as by changing its name. Or when adults push to the front. "We can support the young people, but we are not FFF. We have to respect that. The young people are the movement," I often say. I insist on that, and keep pointing to Roger Hart's text (Hart 1992), which he wrote for UNICEF, and which describes how adults can help young people with the projects they come up with, without suddenly taking over leadership of those projects. As adults, we have to organise ourselves, I think.

While Extinction Rebellion does not develop so well in the next months, Fridays for Future blossoms. The fact that the original Facebook chat and the online meetings have such success, at least at the beginning, is also because they are focused on action, organised to plan concrete events on concrete days, starting with the Climate Alarm on the 9th of December, and then the 15th of March. It was never purely an organisational chat or a crowd of all the people who want to work with FFF, but a quite pragmatic chat in which everyone is preparing something together. This means that it has a clear focus. All the people who are at the Zoom meetings are both more and less than themselves: they don't have to show how clever they are, but can also just listen to what ideas there are and what the others can agree on. People can also say that they are completely new and have no idea about anything. It is really a place without prestige, a village square in the best sense, a meeting place, a digital Mynttorget, where people can solve smaller problems directly, as well as concocting bigger plans and dividing them out into other chats and Zoom meetings.

Chapter 2: Fridays for Future and Extinction Rebellion Start to Grow

Without this centre, even rebel movements probably don't work, we theorise as we shake our heads over those who are turning XR in Sweden into more and more of an NGO or organisation. XR Sweden also had a channel like that which guaranteed its success. When the channel dissolved in December, XR Sweden almost completely disappeared for months. There may be an organisation with clearly defined working groups and roles with tasks (based on the models of Holacracy and Self Organising Systems), but the centre falls away along with the focus on protest actions (see the chapter about grassroots movements later in the book for a new theoretical approach and the research literature). "Decentralised", desirable as it may be, then just means that those with the most resources and informal power do something without really involving the rest: the law of the strongest often prevails when people talk about "decentralisation".

FFF remains, in these crucial months, free of the influence of organised, paid climate professionals, from adults in NGOs and institutions. Among the adults who help out, none are employed to do so, and almost none belong to an NGO, let alone being paid by one; this is an organically growing grassroots movement. Many right-wing and some left-wing journalists and politicians don't want to believe that. But it is a youth strike. And in some countries, NGOs such as BUND, WWF, 350, Avaaz and Greenpeace try – at best – to support the strikers, which is appreciated, and at worst, to influence their methods of organising and their thematic focus. The young people themselves react critically to this. The basis of Fridays for Future consists of Tindra, Isabelle, Ell, Simon and Greta, and everyone else in the square; later, in Switzerland, it is Loukina, Lena, Fanny, Jonas, Matthias and Paula; in Uganda it is Hilda and Vanessa. Not paid, organised adults. But NGO-isation and the accompanying weakening of the actual youth uprising and their grassroots movement is a danger of which the young people themselves are aware; and it will never disappear.

The Climate Alarm day itself, for which the original FFF chat was initiated, is almost forgotten at the beginning of December, amid the excitement of FFF and XR events. It has fulfilled its historic role: it meant that a digital centre was created for the climate justice movements.

Chapter 3: The Foundations
November – December 2018:
The (climate-)scientific background

The first speech in Sweden – the evening at Oscarsteatern

And so, at the end of November, two of the most important weeks in the history of Fridays for Future begin.

It is a mild Monday evening, the 26th of November, and Greta stands on the stage of the theatre, in the centre of Stockholm. The room is full; more than 800 people have come to "An evening for the climate" with music and talks. Before and after Greta's appearance, there are speeches by Archbishop Antje Ackelen, former president of the Club of Rome Anders Wijkman, and Greta's father. In between, her mother sings Swedish songs, accompanied by her own chamber orchestra. This is the only time when the family can be seen all together in all these months. The project of the strike movement belongs to Greta, and she is careful to ensure that it does not become mixed up with her parents' activism; for years, they have been active in the cause of refugees, for example, and in Sweden they are well-known people.

"Hej," says Greta. "Hej," says the whole audience. The speech is about eight minutes long and packed with information, biographical descriptions of her Asperger syndrome, the depression she has overcome, and the connection of that with the climate crisis. For many children across the world, talking openly about an Asperger's diagnosis is liberating; it is not only an obstacle but can also be a superpower. And so Greta comes to the most important facts: the average Swede causes emissions of ten to eleven tons of CO_2 per year; that should be less than two. Emissions must sink by more than ten percent per year, because we can only emit a small CO_2 budget, otherwise the earth will become hotter by more than 1.5 degrees, and in just a few years this budget will be used up (Thunberg 2019). Humans have exterminated about 80 percent of mammals on land and water. Every year, billions of euros are still invested in fossil fuels. The atmosphere in the theatre is concentrated. Gradually, images and numbers transform themselves into a clearer awareness. Ah, that's how things look and that's where our planet is now. Even for those who knew about the climate emergency and the ecological crisis, and after all that is the majority of these grey-haired theatregoers, something peculiar happens during these minutes. Knowledge changes from something abstract to something clearly seen and felt. It becomes so clear that it's as if there's no way back. Greta gave the same speech in English the previous day, and she posts it online as a Ted Talk a few weeks later.

The basic principles

Fridays for Future could never have grown so quickly if there had not been a centre from which, in these first months, no one deviates. This includes: no specific demands beyond the reference to the Paris Agreement and the 1.5-degree reference point of the IPCC report SR1.5; taking account of Anderson and Rahmstorf's calculations of the emissions budget, meaning zero emissions within the next twelve or so years in European countries; nonviolence; a holistic approach: we need a systemic transformation, a new way of thinking, including social justice and equality.

In Mynttorget, the only question which is really controversial, especially among the students, is whether they can and should present more concrete political demands and suggestions too. This will soon become the great (productive) dispute within the movement. But the cluster of principles is clear and radical enough to hold the whole movement together for months – discussions follow it consistently, as well as all decisions, placards, interviews. The young people of Mynttorget do everything to emphasise it and protect it.

The scientists' task

And so, climate and environmental sciences take centre stage. The young people refer to them in all their interviews. And soon, the first researchers can be found who take the side of the children and young people.

Whether we have to remake our society sustainably without emissions within twelve or within thirty years is an important question, these researchers say. Either we are forced to pull ourselves together for a joint effort, or we're not. If there were no tipping points – such as the loss of the Arctic ice – or negative self-reinforcing effects such as the melting of the permafrost, probably nothing would be half as bad. There would be the certainty of linear developments. But that is not the case. There is a real danger that the whole system collapses and the planet heats up by several degrees, becoming uninhabitable for us (Lynas 2020; Rockström et al. 2009). The course we set in the next years decides what will happen in the next century.

On the one hand, we know that stopping CO_2 emissions has direct consequences and will slow down the rise in temperatures. We are not simply at the mercy of natural processes. We can take responsibility, say the scientists, and stop digging up and burning fossil fuels.

At the beginning, it is only Maria Johansson who joins the group in the square. She goes there and takes a stand, as an environmental scientist at Stockholm University. Otherwise, things are quiet for a long time. For months. Half a year. Maria works on her colleagues in the climate sciences, but they generally don't want to appear in the square as university researchers; at most, they are prepared to go there as private individuals. They agree with the movement. In private conversations, they confirm Greta's numbers, one by one. But it's not possible to take a position openly. Scientists are supposed to be neutral. Is that true? What is the task of those who are employed by the state to conduct research, and who see that this state is not acting adequately, to the disadvantage of the children? Should they just watch? Intervene?

What Maria Johansson does with her daily conversations with colleagues will show results over the months. At first, two or three will join, including Douglas Nilsson, who will play a central role in the founding of Scientists for Future. And finally, from March onward, all of us in Mynttorget will be joined by almost the whole of the Bolin Centre, which brings together climate researchers from universities across the city. They will come to the square with great enthusiasm and a giant placard: "Questions about the climate? Ask us." Among them are some of the most well-known researchers in the world. They feel that they must act and support the strikers, officially, because politicians are operating with "incorrect", misleading numbers and parameters, or ignoring the real facts entirely: the Swedish goal of zero emissions by 2045 does not tally with the Paris Agreement, they say, and above all, Sweden is not even on the way to this distorted goal (see Urisman Otto 2022).

The strikers have suddenly opened up a space for them. Some of researchers have been cursing privately in their offices for decades, but haven't dared to do anything. Now they can, with the protection of the ten young people. They say: what the politicians are doing, and in fact all the parties, is risking a rise in temperatures of two or three degrees within the next seventy years which will make nightmares come true; and the floods and droughts are already happening now, everywhere. Food and water supplies for all people are threatened (Wallace-Wells 2019).

They see the whole picture, what researchers call the "great acceleration" (see Raworth 2018): how all the curves develop in parallel and continue upwards ever more steeply, like hockey sticks. That's the acceleration: CO_2 emissions are accelerating, as is the rise in temperatures through the fossil society, the production of waste, the consumption of water, the eradication of animal species; the increase in gross national product through the fossil economy; the acidification and over-fishing of the oceans – this is how the "earth system" is reacting to socio-economic factors and becoming feverish. This "acceleration" must be stopped; that is our task.

Particularly in the early months, I am not quite sure of myself during these discussions of scientific details, and that is why I gradually build up a network. From late September, I introduce the concept of #ScientistsForFuture (much later, S4F will be "properly" founded in Germany), and sometimes I stand in the square with a Scientists for Future sign. We form study groups and read research by Schellnhuber (2015), Rahmstorf (2019), Anderson (2019), and Rockström (2019), but also work which is more critical of the emerging movement. We talk to colleagues at Stockholm University such as Frida Bender and Douglas Nilsson, and with Line Gordon at the Stockholm Resilience Centre, one of the most renowned institutions in sustainability research worldwide, and they also stand together with the young people on the stage during the second and third global strikes in May and September. In the night before a strike, they phone us: we have made a giant snowball and we need a freezer behind the stage – where can we get one? They appear with the 1000-page IPCC report and wave it at the 10 000 young people in the audience. All of them agree: most politicians distort the picture of the state of the earth and the reaction that would be needed.

Figure 1: "The great acceleration" – the background of the climate crisis

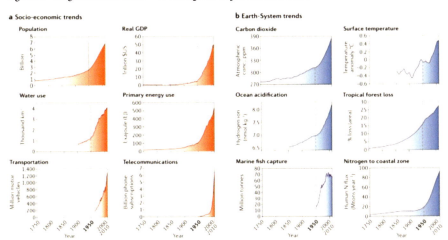

From: Steffen, W. et al. (2015)

So what is at stake? What is certain is the connection between the emission of greenhouse gases and the rise in temperatures. This correlation can be estimated very well by the scientific community, and already has been for 40 years. That means we can calculate how much CO_2 (and equivalents) can be emitted if the earth is not to become more than 0.4 degrees warmer than the 1.1 degrees that have already been caused. In 2018, this budget is around 420 Gt, if we follow the scenario in the IPCC 1.5 Special Report (as explained in a more profound way in the chapter on the Smile meeting).

Why is the talk of "net zero emissions in 2050" so unfortunate, then? Zero emissions of greenhouse gases would mean that we can no longer breathe out. That is obviously a nonsensical goal. The zero emissions which we need would mean that the carbon dioxide and methane we emit are so small in quantity that they can be reabsorbed by forests and other natural processes. Then the goal would be met, meaning that no extra CO_2 would reach the atmosphere and trap warmth. The Keeling curve, which shows the rise in CO_2 concentration in the air, would no longer rise, for the first time in decades. This concentration must be pushed down below 350 ppm again; it is currently higher than 417 ppm, probably higher than it has ever been in the history of humanity. In the last millennia, CO_2 concentration was always at around 280 ppm. In a century, we have changed the whole system radically, the entire composition of the air. The heat is being trapped. So far, so uncontroversial, so disastrous.

The problem is that it is no longer about not looking for more oil, coal and gas, as the GAP report by the UN shows (GAP 2019). Already with the existing infrastructure (coal power plants, oil refineries etc.), more greenhouse gases are emitted than is possible if we want to avoid heating the earth by more than 1.5 to 2 degrees in comparison with pre-industrial times, adhering to the Paris Agreement. And that is where the politicians – including those who otherwise avoid risks – start their Russian roulette. Many of them say that we should reach "net zero" by 2050, and that afterwards we will have to make use of significant negative emissions through which carbon dioxide will be sucked out of the air and stored in the ground (even though it is entirely unclear how that is possible with such quantities). In principle, all the scenarios imagined by European governments assume that there will be enormous technological negative emissions (see Thunberg 2022). Gigantic quantities of CO_2 will have to be stored underground. Most people have no idea about that. A few researchers describe this as a declaration of war by the generation of fifty-year-old politicians against their own children (Stiegler 2020). No scalable technology exists, and plans for "solar geoengineering" sound disastrous to most of us, since they could put the whole "earth system" in danger. It would be sensible not to chop down any more forests (on all these problems, see the chapter on "many fights").

Apart from that, governments which are supposedly some of the most progressive in the world want to take a fifty-percent risk that the goal will not be reached. And they ignore the fact that through air pollution (aerosols), up to 0.6 degrees of global heating may already be built into the system (see Rahmstorf 2020).

What does that mean for policies? By around 2030 or a few years later, there shouldn't really be any larger sources of emissions anymore, according to Stefan Rahmstorf and Kevin Anderson, who are also two of the most important scientists for the Fridays For Future movement; not in Europe or the richer countries, for reasons of justice. Schellnhuber (2015) says in his monumental work that by 2040 the whole world must leave fossil energy behind. No quantities of coal, oil or gas should be burned anymore. Not for heating, for cars, for planes or for the steel or cement industries, and hardly any methane from the animals we eat.

That is – as I see it – why the young people are sitting in front of the parliament. That parliament decided on 2045 as its zero emissions goal. Such a goal takes no account of tipping points, feedback loops, or social justice and fairness, equity; it relies on problematic technologies and takes an enormous risk that everything will go wrong. The wealthier western countries must, according to the Paris Agreement, switch more quickly to a sustainable economy than the poorer ones. They must also help poorer countries massively to finance their transformation, the Agreement says (Thanki 2019). Even some environmental and climate scientists hardly take that on board; and so many politicians ignore it. A public debate must begin, looking at what justice could mean, I say to myself in these days of autumn 2018.

Global perspectives and the unjust classroom

At this time, young people in the so-called Global South are also starting to take notice of the movement. There are messages on Twitter from Vanessa, Leah and Hilda in Uganda, a few children in Nigeria and Kenya, and several groups of students from Bangladesh, Mexico, Brazil and Pakistan.

The global structure of the economic system, which many researchers regard as unjust (see e.g. Hickel 2018) is often the subject of the sustainability lectures which my colleagues and I give at the university. You could say that the struggle of the politicians in Sweden, Switzerland and the US to preserve the "fossil order" is directed not just against the generation of young people in those politicians' own countries, but above all against the children in the most affected areas (Margolin 2020). They have made the smallest contribution by far to CO_2 emissions, and they are already feeling the consequences of global warming much more severely, the droughts, the floods, but also the results of deforestation.

For that reason, the young people chant at all the strikes everywhere in the world – and it is perhaps the only universal concept in these months – "We want climate justice." What goes for the situation within the individual countries also goes for the global situation: a small proportion of people – the richest ten percent – are responsible for emitting more than 50 percent of greenhouse gasses (Anderson 2019), and those same people own more than 80 percent of wealth, even in seemingly democratic countries such as Sweden and Switzerland (Cervenka 2022). That would be, I say to my twenty students in the university classroom, as if two of you were to own more or less everything, an enormous concentration of power would emerge – and those two simultaneously blighted or even destroyed everyone else's lives with their emissions. The question is then – why should anyone accept that? This injustice is also the reason why many young people take to the streets. That makes sense to most of the students; they already react strongly to the smallest sign of injustice when we work together. What is still missing is the determination to stand up and put an end to it; what researchers call "agency".

Historically, the piling up of enormous wealth by the richest section of western society is directly connected with extracting fossil fuels such as coal from countries in the Global South, I go on to explain in my lectures on sustainability – and with the exploitation of the people who live there (Malm 2017; Bellamy Foster 2010; Fraser 2022). The uprising which is starting to develop through Greta's work in these months also becomes an uprising by the children of those countries which are most disadvantaged by the fossil society. At the university, we agree that it is about thinking of democracy in a new way and making this new way a reality, not just within individual nations or within local economic systems, but also globally.

The climate scientist who emphasises this perspective on social justice and global fairness, again and again, is Kevin Anderson.

The idea of the emissions budget

Of all the scientists who were important for these first six months of FFF, Kevin Anderson stands out. A professor in Manchester, he is employed at the University of Uppsala in the crucial years for Fridays for Future. That is where I visit him, as do other climate activists, including Greta and her parents. The university centre where he works, CEMUS (founded by two students, Niclas Hällström and Magnus Tuvendal, and student-led as well as transdisciplinary), generally plays a central role in the spread of the climate movements. Kevin has the personality of a character in a play; wiry and equipped with a warm sense of justice, not prepared to make bad compromises just to impress colleagues, with a dry British sense of humour, and not one to avoid productive conflict, even on Twitter, with a concise, clear style. His speciality is calculating emissions budgets, meaning the piece of the pie of carbon dioxide emissions which we can still have if we want to hold up the process of global heating. Without him, the movement would have been missing an important piece of the puzzle.

The crucial insight is this: it is not even about setting goals like "We want net zero in 2050," as we hear from the EU, Switzerland and Sweden (in Sweden's case: 2045). It is only the absolute figures that count, the levels of gases being emitted. That's what it's about. The fact that governments pay no attention to these "budgets", but only talk about abstract goals, comes across as a deliberate piece of deception.

What our governments ought to decide would be to make real emissions transparent and show whether we are keeping to our budget. This is what all ministers explicitly refuse to do (including the German minister Svenja Schulze, who refused to answer several times when asked by ZDF). Probably because then most people would realise that the policy currently established in Europe does not in any way reduce emissions by more than ten percent per year. We really need plans for how that reduction could happen, sector by sector, and in a systemic, just way. That is the main job of the ministries right now, according to many researchers (Anderson et al. 2020). But even at universities, barely anyone dares to say that. Anderson also points that out, criticising the academy at least as much as he criticises policies (on possible transformations of schools and universities: see the chapter on education).

Figure 2: The required reduction in emissions (with a budget of 420 Gt)

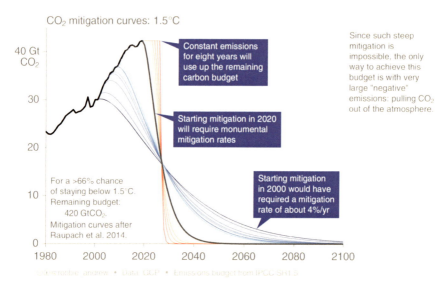

Graph: Robbie Andrew

Anderson's companion in this struggle to push through emissions budgets rather than abstract net zero goals is Stefan Rahmstorf, who visits the children in Mynttorget, a central figure at the Potsdam Institute for Climate Impact Research together with Joachim Schellnhuber, who officially advises the government, and Johan Rockström, the Swede who previously drew attention to the climate crisis here in Stockholm. Like Anderson, Rahmstorf often uses Twitter as his political megaphone and is not shy about approaching the German government directly. He reacts vehemently to Angela Merkel's first longer comment on Fridays for Future. At the world security conference in Munich, the German prime minister seems to imply that Fridays for Future is actually an initiative controlled by Russian internet trolls. But it is not Russian trolls, but Greta, Isabelle and Loukina who are working on this.

Gradually, a network of the most well-known and important climate scientists emerges, supporting the group in Mynttorget: Douglas Nilsson from Stockholm, Michael Mann from Philadelphia, Julia Steinberger from Leeds, Kathryn Hayhoe and Reto Knutti from Zurich, and so on. Twenty of the most renowned institutes worldwide are represented. Already early on, Rahmstorf tweets a graph which Greta retweets several times, showing how emissions have to "peak" in 2020, before drastically declining.

As we can all see, because the temperature difference between the last ice age and our current Holocene is only four degrees, rises in temperature by two degrees or even by four are associated with far-reaching changes to the conditions in which so many of us creatures live. If nature has a thousand years to adapt, it can do so. We are provoking such a change within a period of one hundred years. This means that the danger is very real that we will pass – or have already passed – tipping points in the climate system from which we can no longer come back.

Figure 3: Tipping points in the climate system

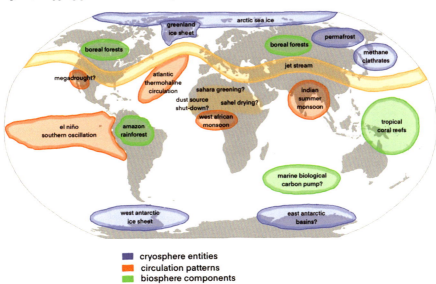

Graph: According to PIK-Potsdam

But, as argued by many in the FFF and XR movement, this approach and the whole concept of emission budgets is really almost cynical in itself. Already in the last hundred years, burning coal and oil has caused a drastic rise in the global temperature of around one degree. Hundreds of thousands of people have lost and are losing their homes due to drought and floods (Wallace-Wells 2019).

Natural scientists are not the only kind

But in these winter days at the universities, it is not only natural scientists who are waking up. Mynttorget is regularly visited by the educationalists and sustainability researchers Isabelle Letellier, Leif Dahlberg and Kristin Persson. And from November, a small circle of researchers gather almost every Wednesday evening in a dark room at our small institute at Stockholm University: "Blackbox humanities" is the name of the project.

Outside it is already dark, early now, the snowflakes whirling through the air. All different departments of Stockholm University are represented, including education and psychology, but also the Karolinska Institute for Medicine, specifically neurology and psychophysiology. Now we want to develop a new centre where Stockholm children would be able to play not just with science experiments but also with the social sciences and the humanities. The topic: how would a sustainable democracy look, one which would pay attention to their perspective and get children involved. That reminds me of quite a similar small group of researchers who pursued this question a century earlier, looking intensively for a more adequate image of humanity. From this group came parts of modern psychology (Köhler), biology (Goldstein), education theory (Wertheimer), and philosophy (Merleau-Ponty), gestalt theorists, as they are called, because they focus on holistic connections, on the whole picture, on systemic changes (Fopp 2016). And if they had not been scattered across the world due to the political contempt for humanity that prevailed before the Second World War, many of them being Jewish citizens in Europe, perhaps the university as an institution would have developed differently. The central theme of all this research: what does it mean to be truly in democratic contact with oneself and with others? How can we make this non-alienated exchange possible?

We are searching for a new concept of humanity – humans have bodies, social interactions, compassion and imagination, and are connected to their environments, including at the university and as researchers (see the chapter about education towards the end of the book). If we do not let ourselves be touched existentially by our research, and take action in society accordingly, we are not taking our research seriously; this is one of the claims made by the "transformative" tradition of research (Leavy 2009). It is about connecting research and teaching with the experiences of those affected; but also with being able to act and wanting to act. That is what we are trying out as we develop the new courses. How can this ability called "agency" be taught? A possible answer would be: through a combination of explaining facts, being playful so as to awaken the desire to act, and creating concrete possibilities. And by making sure that we teachers are setting an example. After all, students do notice if we don't take our own research seriously and behave inconsistently outside of the lecture hall.

Thinking about being in contact on equal terms – that is my actual field of research, a mixture of neurophysiology, psychology, education theory and political science. It is about understanding how we can shape our everyday world, our work, play and life in such a way – globally, too – that we don't cut ourselves and others off from ourselves unnecessarily by relations and structures of domination, becoming "alienated", but instead enter into a good form of democratic exchange. If there is a centre or a core of the future zero emissions society, a compass guiding the general talk about "respect for the limits of the planet and the basic needs of all people", then it is precisely that, I argue in my lectures: creating circumstances beyond domination in which this doubly good humane contact to ourselves and to others is possible, for everyone.

This practical knowledge is often lacking at universities (see McGeown/Barry 2023) – for example, knowledge of how easily children can become tense and cut off contact to themselves and to others, as analysed by the Alexander technique, for instance, as well as Bowlby's (2010) attachment theory and the developmental psychology of Winnicott (2005) and Daniel Stern (Fopp 2016). This means that we lack the practical, democratic knowledge to create humane social spaces, and that there is no basis for the foundations of many subjects with regard to their content (from architecture and economics to history, law, philosophy and education), or for the approach to teaching methods. These could – in sustainability studies, too – take their cue from the fact that we are social, creative, interactive creatures who can dominate each other or enter into a caring exchange (see the chapter about education).

This is also noticed by many of the students who go on strike: even courses focusing on environmental sciences seem not to have any existential anchoring in the current political situation or in the natural surroundings. But this makes it very difficult to gain a real understanding of the context. That is why the courses at the advanced CEMUS institute in Uppsala combine aspects of climate research with philosophy (see Raffoul 2022).

If we share this idea, it becomes clear that we would need a new wave of enlightenment, a really new way of thinking and of seeing ourselves as living creatures on a living planet. As long as tutors are neither able to analyse their own behaviour in the spaces of the university in terms of intersectional power relations (see e.g. Carbin/Edenheim 2013), nor knowledgeable in practical terms about how they can see through these power relations and reshape them by creating democratic spaces (Johnstone 1987), nor able to understand when real contact breaks off and how it is created physiologically, barely any real education can take place – the social sciences and the humanities, including economics, are training people up who do not defend themselves against the climate crisis, because they lack this fundamental democratic dimension in their education. Such is the claim of this tradition of research. They only possess abstract knowledge.

That could change quickly, according to my plan. The disciplines could be joined together in new sustainability centres, by a grassroots movement such as #ScientistsForFuture, with teachers from all fields. This is not just about introducing ethics, empathy and compassion in schools and universities, but about making forms of domination visible, playing with the circumstances in which they occur, and creating the opposite: making the dignity of everyone visible so that we can feel and see it. Being at the centre of the concept of human rights, this "dignity" belongs to everyone and is not linked to (good or bad) actions and deeds – as philosophers stress (Menke/Pollmann 2017); no one has to deserve or earn their dignity, and no one can lose their right to it, even if they go against ethics and damage the common fabric of integrity.

These are some themes we talk about long into night at our Wednesday evening group in the university. We've even been joined by artists such as Mats Bigert, who sketches ideas about how this "democratic materiality" can be explored. One day during these weeks, he creates (with Lars, Åsa, Michael, and many more) Artists For Future, who join the young ones in front of the parliament. And soon, their ranks are swelled by psychologists, teachers, nurses, parents, grandparents, doctors and writers, a whole society... for future.

Winter sets in and the Mynttorget group doubles in size

Then something radical happens. Mynttorget changes. The ice-cream stall has disappeared. And early on a Friday morning, a truck comes rolling up. It is the almond man. In the Swedish winter, small stalls appear everywhere in the cities, where passers-by can buy mulled wine and roasted almonds – if they think they are capable of stuffing the little sweets into their mouths without getting them stuck to their thick gloves. The almond truck stops at precisely the place in the square (this part is called Tage Erlander Square) where Greta and the other young people have police clearance for their strike. The band of rebels positions itself distrustfully. They are not looking for competition. Half an hour later they are all sitting on their yoga mats and stuffing tons of roasted almonds into their mouths.

But someone else appears in the square during these days. And this will have much bigger consequences. Isabelle and Ell, both of them seventeen years old, join the core group, and soon they are also joined by others who are the same age – Simon, Vega, Ebba, Astrid, Anton, Linna, Minna, Edward, Sophia, Johanna, and many more. With this arrival of the slightly older second half of the rebel group, Mynttorget definitively becomes the hub of the global FFF network. Many of the newcomers quickly make contact with the international movement which they themselves are helping to invent.

Isabelle and Ell often come at the same time to the strike place. They have been worrying about the climate crisis for a long time and privately following the strike from home. "I've been keeping updated since September, via Facebook." "Same here." "But why did you choose this moment?", I ask the two of them during an interview for my research, who have known each other for a long time, since they met four years ago at an animal club and environment centre.

They gained an awareness of the idea of a general strike from the context of the EarthStrike. At that time, they saw the climate crisis as a threat to the animal world, "and I watched countless documentaries about the environment." "In December, I thought: that's enough." "Enough is enough. I can't just sit at home, and I feel bad because the world is ending, something like that. And when we came, the people here were really friendly. We had really good conversations and met great people. That's why I carried on."

At first, striking felt very strange, particularly for Isabelle. "I'm always on time for every lesson at school, and then suddenly... I don't go. My teacher was confused. My parents were confused." But then everyone sees the seriousness behind her commitment.

Together with Vega, Simon and the others, they immerse themselves in the emerging international networks. They make connections with the Swiss, Jonas, Lena and Loukina, with David in Italy, Saoi in Ireland, later with Dylan in Scotland, Alejandro in Madrid, Mitzi in the Philippines, the Belgians, the Brazilians in Manaus, and on and on. Ell often runs global meetings together with Loukina and Saoi. And together with Vega and Edward, they work on building a new international Discord platform and continue the debate in the main Facebook chat. Simon increasingly becomes the internet specialist for the global movement, helping to build the new website and keeping an overview of all the different social media channels.

"We came along at the moment just before FFF exploded. Before Katowice and Davos." "Before that it was a bit strange. No one knew exactly what was happening. There were still just ten of us here every week. And then suddenly we were part of an internationally established network – we were part of it and we were the centre of it. A global sensation. From the outside." In Mynttorget, the atmosphere changes mainly because of the new intensity of the connections. Like a brain in which the nerve cells suddenly build connections and exchange information with each other at a furious pace, the square is transformed into a busy centre.

They stay up to date and join forces with the old group: who is on strike where in the world. Who is taking which position. "Oh no, that's just greenwashing. And we don't even know if that's a real person or a fake account." Italy suddenly has all its passwords stolen for the social media accounts; India, too. "Oh, those people are coming up with whole catalogues of demands, that's rubbish." "Suddenly it's exploding everywhere." "In Belgium, Switzerland, Germany – and how we celebrated when 15 000 went on strike in Australia!" Such numbers are still unthinkable in Stockholm or Sweden. "And the media. At the beginning, no one was here." The atmosphere is changing. And making people start to dream. "I would like whole cities to join us." "It's enough. We're sick of it."

Tindra can only agree. She represents a part of the connection to the group of those regular strikers who sat down next to Greta in the first days. On Fridays she often comes to the square with a homemade cake or giant muffins for all the others, vegan, of course. She says that she knew that the environment and the climate were in a bad state, and then she saw: there is something the young people could do. As time goes by, she is also the one who appears at eight with a few others and begins the day; and she helps hold together the Swedish movement as a whole.

Many of the young people have a very broad interest in society; they are well read and are also committed to other causes. They learned to look at things in intersectional terms, seeing the connections between injustice and discrimination in relation to gender, class, ethnicity and so on.

Some of them soon take care of the Instagram account for the Swedish Fridays group and get involved in planning the larger strikes, but they are just as capable of leading the masses of tens of thousands of young people their age at a march, or answering questions in television studios. And Tindra joins Isabelle and Andreas and travels to the first international meeting in Strasbourg at the EU Parliament, where they meet 60 likeminded FFF activists.

The group in Mynttorget has become so varied that it is possible to form quite different constellations for different projects.

Some work with others to make contact with trade unions. Others go to schools and give talks. And someone else joins them. Isabelle suddenly has a double in Mynttorget when Sophia turns up, her twin sister, who had been striking on the west coast in the first months, near to the biggest oil refinery in Sweden. She quickly takes part of the responsibility for the FFF email accounts and for social media.

Another person who is really part of the Mynttorget group, but who lives and strikes in Falun, is Andreas. He is so active and so frequently part of digital exchanges that he is drawn into discussions. He reflects on what happens in poems and songs, and he often makes the three-hour train journey to Stockholm for the meeting with all the other strikers.

I see how they come back, every week, despite their anxiety and sadness. It is difficult to imagine how they feel when they see how their peers and friends all over the world are already suffering now. Some of them have nightmares about their lives in thirty years, the fights for water and food. This is not a game. It is reality, it is their life. And again: I feel I have to wake up my colleagues. How can we let them fight alone? Where is everyone? Where is everyone?

And still, nothing changes politically. Yet? The negotiations for a new government are still going on, and have been since the young people first developed the idea of Fridays for Future in September.

But then the young ones get a completely different kind of support.

Chapter 4: The International Movement Develops
December – January 2019: COP meeting
and climate justice

A storm is coming

In these December days, if you were to get into a hot air balloon in Mynttorget and slowly lift off – without burning too much gas, of course – and gradually float upward further and further through air containing over 410 ppm carbon dioxide (the villainous molecules holding in the warmth of the earth), in the next weeks and months you'd see events following a pattern on the blue planet down there. Inspired by Greta's speeches or just by a picture or the rumour that there is a group of children in Sweden who are fighting back against all the irresponsible actions of the grownups' world, suddenly tens of thousands of young people are leaving their classrooms at the same time, with similar cardboard signs, gathering in the most important squares and demanding that their parents' generation finally pull the emergency brake. And even if they generally refuse to do so, they have to face the question of how they should react, as parents, as teachers, as ministers. The time of debates about compulsory education begins, about the authorities' duty to step in, keep the children in their classrooms and fine the parents. Teachers and headteachers in Berlin, Melbourne and St. Gallen are suddenly facing a dilemma. Some of them punish children so that the strike will clearly be seen as disobedient (Wetzel 2019). What is the point if the schools support it? It becomes clear in these weeks: the children have found a source of power which is stronger than all others, stronger than demonstrating, writing petitions, and occupying banks. But how do they create a united movement?

30th of November – the first largescale strike in Australia

At this time, it's obvious that the living planet is sick, especially in Australia. One heatwave after another makes people, animals, and plants collapse. On the 30th of November, ten thousand children and young people suddenly stand up and leave their classrooms in unison. It is the biggest and the first largescale strike by the movement. And it becomes the beginning of countless others, which will follow each other week after week, first in Belgium, then in Switzerland and gradually in more than a hundred countries, until on the 15th of March 1.5 million people join together for a global strike. But what came as a sensation for most people, the young people in Mynttorget had suspected already weeks in advance.

"Where is she, then?" I call. It is ten o'clock. The pictures travel round the world, showing children walking through the streets of Melbourne, Sidney, and Perth with their placards. Australia is half a day ahead of us, so they have been up for hours, and Janine is nowhere to be seen with her two teenagers. She prepared herself so well for this day, Australian that she is, together with the other children and mothers across Australia. She arrives towards midday, having had hardly any sleep. Today is not just a big day for Australia. Today Greta is going to Katowice and at the same time the other young people of Mynttorget are to hand over a demand to the city councillor of Stockholm: there must be a carbon dioxide budget and a political plan showing how emissions could decrease every year by more than ten percent. Everyone is nervous.

How is it possible that tens of thousands of young people have left their classrooms at the same time across Australia and taken to the streets? Greta's interviews and speeches were picked up by the pupils there in the past weeks. They were just the last spark in a situation that had been tense for a long time. The firm Adani had announced that it would be building a new, gigantic coal mine which would warm up the climate for decades. This led activist networks to form, and they were already used by school pupils. The weather had been out of control for a while, and temperatures of 45 degrees are no longer rare. Young people were increasingly politically aware, like their counterparts in America with their Sunrise movement. A few of them begin to get organised. They quickly build a website and reach the whole nation (see Mao 2018). Jean soon becomes one of the best-known young people and can be seen beaming into the television cameras; she finds words for all the anger at a government which is deliberately risking their future. Parents are there too – mainly mothers, in fact, as will be the case across the world in the coming months.

Along with a few NGO workers, they chat with Stockholm almost every day in the weeks before the strike and support the young people however they can. In Castlemaine, Victoria, some pupils have already started to strike on some days of the week, thus becoming a model for the others.

And then, on a cold November day, weeks before the strike, the young people in Mynttorget are sitting on their mats as usual with their eyes fixed on an iPad: there are the faces of the young people in Australia, asking questions about how to strike. How should they react to stupid passers-by, what do they need to bring, what should they sit on... The children on the other side of the world at least get the feeling that a strike is not something impossible.

The plan for the huge strike on the 30th of November becomes more solid, and then it becomes a reality. Into an overall situation that was already politically explosive – in the context of the building of the Adani Mine – the Swedish call for a school strike has come at exactly the right moment. And now, unwittingly, the activists are also helped by the conservative prime minister Scott Morrison, who declares in parliament that he will not tolerate a school strike and that pupils must attend school: "Schools are not a place for activism!" This is the best thing that could have happened. In parliament, a debate is sparked. Even more young people are mobilised and dare to join the strike plans. At the same time, an article by Greta is published in the Australian edition of the Guardian (26. 11. 2018), explaining the scientific background and also presenting information and facts to the parents' generation.

At exactly ten in the morning on the 30th of November, more than ten thousand children leave their schools, while dawn has not yet broken in Stockholm. The pictures go round the world. The movement has become international, and we are all incredibly happy. And then comes noon and chaos breaks out.

The COP24 climate conference in Katowice

We are all busy with something else when Greta quickly grabs her sign, holds it up and bounds off along the façade of the palace, away from Mynttorget. I see an unbelievably dirty car. It is right there where the sign was a minute ago. Then there's a screech, the car speeds up and turns and Greta's father sticks his head out of the window. Greta is going to Poland today, to the UN climate conference in Katowice.

The climate conference begins, and some of us in Mynttorget follow every detail on the internet. COP ought to be a meeting of representatives of the world population, who should be able to use the best research in the world to agree fair, global political change. That is why the UNFCCC and the IPCC were founded as "global" (research) institutions of the UN. How can we live together well, fairly, sustainably, on a living planet? That was what it was meant to be about.

Climate conferences are a strange world unto themselves, at least in my experience as an activist in Copenhagen in 2009 and in Paris in 2015. For us climate activists, and for nature itself, the events of 2009 were a nightmare which seemed to paralyse us for years afterwards. For us it felt as if the non-decision in Copenhagen destroyed something irrevocably, and not only the possibility of decreasing emissions. Something broke inside us, something like belief in humanity, if you want to use big words for it. Belief in ourselves, too, perhaps. Added to this came the brutal blows of the police, who beat us up at the final demonstration. For the first time, we dared to use methods of civil disobedience, rather than just watching. We knew that Copenhagen ought to have been a turning point. The heads of government ought to have pulled themselves together, freed themselves from their nationalist sympathies, put an end to the fossil fuel regime and the lethal geopolitical games they were playing, and set a new course (see Linnér/Wibeck2019).

As representatives of "civil society", we sat around in a depressing "alternative congress centre", knowing what could happen. That the next ten years would be the warmest since records began. And that the emissions would continue to rise. Naomi Klein and George Monbiot were already heroes of the alternative scene. But they irritated some of us. They proudly announced that a unified protest movement was not necessary; that our strength lay in our plurality. Why, I wondered at the time – why? We must unite, combine all our energies. Otherwise, we're powerless. I was much more concerned with the question of how to create unity. How can we create the feeling and organisation of one shared humanity on one living planet? How can we build a united movement? Through values? Demands? Structures? Places? Back in the present, in Stockholm, I ask myself: how will the young people act now, ten years later? How do they hold their movement together? How do they express the fact that they are one connected humanity?

On science, universities, and activism

In the rooms of the university, my students follow what is happening in Poland and discuss the function of science in relation to politics as well as their own societal role as students.

When I become co-responsible for the B.A. program, creating new university courses, I define for myself the task of these studies (of drama education and applied theatre) as opening the imagination to understand and practise how to interact with each other and the environment as part of a shared, democratic, sustainable humanity. How can we as embodied, social, imaginative beings living in problematic power relations playfully explore and create humane relations? And this is what the COP should be about, I think: a place which cares for the one humanity living on one planet.

But what kind of project is that, the scientific project on which we have been working for two hundred years, at universities but also through technological research and development outside the academy? One lecture is about the relationship between science, imagination, and magic, and thus about the emergence of our modern concept of the industrialised world of nation states, which divides body from soul and separates scientists from their environment, presenting them as "neutral" observers. On the A4 sheet which Greta brings with her in the first weeks of the strike, we can read the outcome: our modern worldview may have brought with it an incredible productivity and desire for discovery, but it has also ensured that most wild animals have been exterminated, and that our lifestyle emits so much carbon dioxide that we as a species are changing the shape of the whole planet. Some people claim that the Anthropocene has begun (Hardt 2018), referring to the epoch in which nature is shaped by humans.

Can a new change take place, two hundred years after the dawning of modern science, a new paradigm shift enabling us to dissolve power relations? How politically committed can science and research be (Raffoul 2022)? Luckily, most textbooks on empirical sociology, humanities, politics and education now have chapters (Leavy 2009; Cohen et al 2017) emphasising that we researchers are also part of a natural and political environment and that we can take an "objective" view of this environment, but also act transformatively (McGeown/Barry 2023).

"Objectivity" then does not mean neutrality but includes using comprehensible methods to trace what is going on in the world; being open rather than taking a single position from the beginning; using terms that have been defined; interacting critically with the field; being transparent; remaining in permanent self-critical exchange with others – but also: questioning our own positions of power, entering this environment with an existential responsibility for our privileges, in search of a more democratic world, changing our environment through research – as flesh-and-blood researchers, not as disembodied brains that don't reflect on their context (see the chapter on education).

We remain sitting for a long time in the room after the seminars and look out at the winter landscape of Stockholm. With another cohort of the course, we are performing a new English play set in a refugee camp in Calais (Robertson/Murphy 2017). It articulates the nightmarish experiences of young (climate) refugees from Syria, Afghanistan, and Congo, but also the possibility of human cooperation. The course is really about showing the students how they themselves can work through plays with children and young people. But also: what are our political roles and what does it mean to encounter people as someone who practically has no rights – or as a fellow citizen with equal rights? How can we create a global shared life with dignity for all, a worldwide convivialism?

The speech

The corridors of the COP conference centre in Katowice fill with more and more politicians and scientists, who are supposed to discuss how the Paris Agreement can be implemented in the next few years.

In October, a revolutionary report by the IPCC was published (IPCC SR1.5): what is the difference between a world that is warmer by one and a half degrees and one which is warmer by two degrees? The differences are enormous and dangerous, especially when it comes to the potentially disastrous tipping points which could push forward the warming of the earth by several degrees by the end of the century, forcing billions of people to flee (Xu et al 2020). Still, the most powerful people in the world hardly talk about leaving fossil fuels in the ground. Instead, already in the first days of the conference a scandal begins to develop: Brazil, Russia and the US do not want to "welcome" this IPCC special report, but only "take note of it"; this would weaken political decisions in the future. The Swedish environment minister intervenes at the last moment and the report is accepted by all countries – thus becoming a fundamental document for the international Fridays for Future movement. It states that to reach the goal of 1.5 degrees, the world in 2018 only has about 420 Gt of CO_2 emissions left. That is nothing; if they carry on in the same way for eight years, it will all be used up. A sense of world citizenship would have to emerge, we think to ourselves in Stockholm. The delegates would have to wake up suddenly and see themselves as joint representatives of all people and of nature.

Thursday afternoon becomes evening. Greta is supposed to be speaking at the end of a three-hour meeting. I tune into the UN live broadcast. Some prime minister of one of the almost 200 countries is holding forth, with lots of flowery words. And everyone is going over their allotted time. Greta's turn is supposed to come at eight, but it is postponed to half past, and then to half past ten, and she has to shorten the speech and adapt it. The fact that she is even allowed to speak is thanks to "Climate Justice Now," an alliance of grassroots movements and NGOs which she is representing. So it's not thanks to the Swedish government that she's been asked to speak, or because of some powerful people, but indirectly because of the little Swedish Fridays for Future groups, which have been striking in solidarity for several weeks in around fifty different places.

The prime ministers talk and talk. Via WhatsApp, I receive from a friend a leaked version of the draft agreement between the world governments. The language is bureaucratic, half political, half scientific, and they argue over every comma. As usual, it is about the main chapter: "mitigation", meaning the prevention of global warming; "adaption", meaning adapting to the damage caused by global warming; and "finance", the financing of transformation. I stare at the document and then at the screen. We ought to be able to share natural resources, or not just see them as resources but as a shared basis for life, I think. Otherwise we won't have a future. We need a global contract right now to settle this fairly, so that fossil fuels are kept in the ground for ever.

Then Greta appears. "Hey," she says. I film the speech directly from the screen. How will it go? The speech is incredibly moving, for many people it is deeply shocking, for some perhaps annoying, and it is compact, much shorter than the speech she made at the Stockholm theatre. The speech addresses all of us with quiet, furious force, and gives people the feeling that we really can change something. Right away, the media invent some historical parallels with Jeanne d'Arc and other historical figures. And from that evening on, I am constantly worried. Is it good for a young person, being exposed like that, being famous? How can we handle it in an ethically defensible way? The climate researcher Kevin Anderson tries to shield Greta by tweeting that the celebrity cult is just one part of a corrupt society.

In Mynttorget, we try to understand the issue and to act. I'm already having trouble understanding the adults who cheer the kids on and find their strike great, but don't do anything. This shouldn't actually be the children's job, I say to myself again and again. I'm there, not because I think it's great that they're on strike, but because for me the whole situation is wrong. They shouldn't have a reason to be so deeply frightened about their future.

Many young people across the world see the speech and are not just hit by the force but also feel supported in something that has lain dormant in them for a long time; namely the idea that they, the children, don't just have to watch. That it's not an absurd idea for them to rebel. A young person their age is speaking not just from the heart and with knowledge of the impending climate crisis, but about their lives. "I care about climate justice and the living planet." It is more like a fundamental statement by one generation to the world. It is about justice and about the perspective of children. "You say you love your children above all else, and yet you are stealing their future in front of their very eyes." "Our biosphere is being sacrificed so that rich people in countries like mine can live in luxury. It is the sufferings of the many which pay for the luxuries of the few" (Thunberg, 2019, p. 38).

Since the evening when the speech is broadcast around the world, Fridays for Future as an idea and as a global movement becomes unstoppable and changes the situation dramatically, not just in Stockholm. Thousands and thousands of children and young people have seen it. Three young people are there, and they talk to Greta in the COP corridors. Jonas and Marie-Claire from Switzerland, and Luisa from Germany meet at this time, all of them very well informed about the climate crisis and about the political background, and they sit down together and strike with the Swedish sign. They travel back to their countries and organise their peers. They force those around them to take a stand.

In Mynttorget, the young rebels are proud of the one among them who has dared to shake up the world and make everyone see why they have been gathering every week in the rain and the icy cold. They throw snowballs and stand together in the winter square between the parliament and the palace, and see in the global media what their courage and work have achieved. But hardly anyone understands that they are the ones who keep the movement going week after a week. On Friday, when Greta is back, they form a circle with her and improvise a new absurd recipe. The contrast between this circle of young people and the rows of COP delegates in their expensive, dark suits could not be greater.

The formation of "local groups" – the Swiss take to the barricades and demand climate justice

In the weeks afterwards, something happens that changes the whole movement. In a certain sense, you could say that this is the moment when Fridays for Future emerges as a structured network, not as an organisation in a way that an official association would be, but in the sense that FFF becomes more than a series of individual large-scale strikes like the one in Australia or later in Belgium. Now there are epicentres everywhere, in many places which organise strikes over months: on the 14th and the 21st of December, Swiss pupils leave their schools en masse, but coordinated, with core groups which not only organise a strike but do continual work and political education, like the "local groups" that emerge later in Germany and France.

A kind of stable organisation comes about, for the first time, worldwide; in every country in a different way, strengthening the movement as a whole: some people use WhatsApp, others Telegram; some don't work together with NGOs and youth political parties, others do; some strike every Friday, others once a month, and all of them feel part of a shared movement. In Switzerland they call themselves "Klimastreik", in France "youth for climate", in England "SchoolStrikeForClimate", but all of them can unite behind the hashtags #FridaysForFuture and #ClimateStrike.

In the square, the young people are glad to see Swiss schools going on strike. In many cities, children accept the fact that they will come into conflict with school boards, with their teachers and headteachers. At first, there are only a few people in St. Gallen who can no longer accept the inaction. Miri and a few others had seen Greta's speech in Poland, as she will explain six months later. She could almost be a twin of Isabelle and Tindra in Stockholm, a Swiss variant with American roots. She has a similar breadth of political interests – it is not just about protecting the climate, but just as much about social and global justice; and about the core group of pupils who organise. They are a real community, an "asocial network", as they call themselves. Soon, other schools in eastern Switzerland will join the St. Gallen strike. And as has been seen in Australia and later in Germany, it is a concrete political decision that helps the Swiss activists. Parliament will be discussing a carbon tax.

Already on the Wednesday after Greta's speech, a group of four or five pupils meet, some of whom don't even go to the St. Gallen *Kantonsschule*, and form a "strike organisation." WhatsApp groups are formed and within two days they succeed in mobilising four hundred of the roughly 1000 pupils in the school. A letter goes to the headteacher, who has certain sympathies with the seriousness and the cause of his pupils and implies that he will tolerate the strike so long as it takes place near the school rather than within school grounds.

Then things move very fast: the first large-scale strike is followed by a general meeting of Swiss activists at a church in Bern during the winter holidays. The demands of the new climate strike movement are agreed; a website and a flame logo are created. The 300 participants from all corners of Switzerland follow the communication methods and values of Extinction Rebellion and YOUNGO (the youth organisation of the UN), which are brought in by Jonas, Marie-Claire and all the others: based strictly on grassroots democracy, on consensus, with silent handwaving to show agreement, and so on.

The first demand is: all greenhouse gas emissions in Switzerland have to be brought down to "net zero" by 2030. Then: a climate emergency must be declared, meaning that the population needs to be informed of the crisis and laws have to be shaped by it. And: social justice should define everything. On the controversial demand for "system change," a compromise is formulated. And similar processes happen soon everywhere in the world: in Italy, the Czech Republic, Ireland, Japan, Brazil, and so on. The worldwide movement takes shape.

Climate justice

The concept of climate justice is important for the whole global FFF movement from the very beginning – an "empty signifier," as some researchers call it, not because it is actually empty, but because many slightly different positions can be united behind it (see for different distributive, procedural, intergenerational, and regenerative or restorative aspects: Jafry 2020). It is something like a compass showing the direction. In this, the young people are joining a long tradition (Thanki 2019). In particular, at the climate summit in 2009 in Copenhagen, the phrase became an inclusive term.

The basic idea is that the environmental and climate crisis is not limited to the technical problem of global warming (Müller 2020). The crises themselves and the way we attempt to deal with them must be seen in an ethical and political context; not only in relation to a "just transition" for the workers in the fossil sectors (energy, transport etc) but also for all other people, especially in terms of the human rights of the most vulnerable.

In the core of this tradition, we can find the following analysis. A few people, especially in the Global South, are much more severely affected by the crises; and often, they have barely contributed to CO_2 emissions (Margolin 2020). A few states (and people) are much wealthier and have better resources to avoid emissions and also to compensate for the damage. In the Kyoto Protocol, and also in the Paris Agreement, justice and equality are explicitly named as a binding compass. All states have in fact committed themselves to this.

What justice means exactly – that can be interpreted in different ways; for example, the intersectional tradition would relate it to the different dimensions of injustice and discrimination and how they affect each other (gender, class, ethnicity etc; see the chapter on corona and intersectionality). But what it is generally about, in the tradition of the climate justice text of Bali 2002 and the "climate justice" networks, is a broad view of the situation, and a stance with political consequences. It means, for example, letting those people take charge who are most affected, such as indigenous people in Brazil and in vulnerable populations in the Global South, such as Mozambique, Bangladesh and the island states which will soon be submerged.

In concrete terms, it is also about the fact that those who have already emitted more CO_2 and are also wealthier overall should contribute significantly more to the reduction and compensation of damage. That has consequences: a Swiss person causes around 10 to 14 tons of CO_2 emissions every year, but to prevent global warming, they should actually only be allowed to emit about 1.5 tons if we distributed the budget fairly. And this justice does not even yet include the broader concerns of historical debts and current wealth; in that case, richer countries would have to pay a "fair share" (cf.: Civil Society Equity Review Group 2018) into a green fund which would go far beyond cutting our own emissions.

In Mynttorget, too, we have lively discussions about the implications of "climate justice": does it mean something deeper, including looking at the system which shapes our fossil societies, in which states in the Global South deliver cheap raw materials through companies such as the Swiss Glencore, which are then sold with the support of banks such as Credit Suisse and UBS – making a profit which goes mainly to a few white male shareholders in the North? From this perspective, it is not just about who pays how much (to affected workers etc.) and who can emit CO_2 through their lifestyle, such as by eating meat or flying (through individual or national quotas, for instance); the social and economic structures would have to be changed so that these privileges and power relations disappear (Hickel 2018; and see the chapter about Economics in this book).

And in my lectures, I point out that the concept of "justice" should be supplemented by that of "being humane". We need to care, to provide enough resources for everyone, not only to get rid of problematic power relations (Fopp 2016). If whole populations in the Global South are treated as if their welfare is secondary to ours in the North – that is unjust, but it's also more than that; it is inhumane. Most of us adults in the square are plagued by a bad conscience and the knowledge of our privileges. All the more important that we change the situation. But how is that possible; global politics would have to change. Where do we start?

Making international contacts

The live stream of the Switzerland-wide meeting is as exciting as a good film. The very fact that the young people work so transparently is remarkable for many of us and ensures that those who cannot be there are still included; travel to the meeting is also paid collectively for those who don't have any resources. And everyone can follow the first hour of the meeting live on YouTube.

Can they agree, or will the movement fall apart right at the beginning? There they sit, Fanny, Paula, Matthias, Lena, Linus, Eslem and all the others, from towns and villages across Switzerland. Two who lead the meeting are Loukina and Jonas; both are about 17. Jonas comes from the Zürich area and Loukina from near Lausanne. In the next weeks and months, they quickly become two of the leading figures in the global network. The Swiss activists decide early on that they don't want to encourage "stars" to emerge (which makes the movement stronger than the movements in most other countries), and the faces making statements in the media are frequently rotated.

Soon, these two take on the planning of the first international FFF meeting, when some fractions of the EU parliament invite the young people to Strasbourg. Jonas holds together the threads of the global movement for the writing of the first Guardian article, and takes part in the Lausanne SMILE meeting. Loukina chairs together with Saoi from Ireland many discussions at the international FFF Zoom meetings, which take place every Sunday evening from now on. There, the global group works through all upcoming questions, from the planning of new large-scale strike days up to the "welcoming" of new countries which are going on strike for the first time.

Early on, Loukina also tackles a project of her own together with other activists in Canton Waadt: because politicians are failing to act, they draw up a detailed plan of their own, a "Climate Action Plan" showing what a zero emissions society in 2030 could look like. A long, comprehensive paper emerges. Sector by sector, political actions are described: the economy, energy, agriculture, changing people's diets, developing public transport, and reforming the financial system.

And Loukina and Jonas quickly set about following the path of this climate action plan at a national level together with other Swiss people, and with scientists (www.climateactionplan.ch). In Germany, there are instead eight specific demands for the government: from switching off all coal power plants by 2030, to a carbon tax of 180 euros per ton. In Sweden, the young people insist that it is the rulers themselves who must come up with a plan; they refuse to present concrete demands. The movement should be an emergency brake and not a youth party.

One Friday after Christmas, in Mynttorget, I excitedly describe the Switzerland-wide meeting. The following week, which is stormy and snowy, the Swiss activists talk to their Swedish peers on the phone. Another piece of the puzzle has found its place. The global group of rebels is growing; a kind of community feeling is becoming ever stronger. And in Belgium, Anuna and Adélaïde have had enough of the adults' passivity and are leading thousands, and then tens of thousands through the centre of Brussels every Thursday. Already weeks earlier, they had begun to interrupt lessons at school using the alarm function on their phones. More and more often, and for a longer and longer time. Now the anger and the desire for change have become unstoppable.

Every new strike, and every new strike location, is picked up by the group in Mynttorget on their daily trawls through social media, and retweeted once it has been checked thoroughly: is that definitely a real photo? Are there newspaper reports about the number of people on strike in Brussels? In Kiel? In Zürich? In Bangladesh? Pakistan? Are there independent sources? The movement has become international. And it is getting a reaction.

The adults' reaction

How do the adults react? Some are baffled by the force of the protests and the courage of the young people who are ready to leave their classrooms – because that's what often happens. In Lausanne, for example, Loukina walks through the corridors with her fellow pupils and calls on others to strike; in front of their teachers, they get up and go out into the streets to rebel.

Many people adopt a "wait and see" approach. Headteachers are put in difficult situations. And in the media (see Voss 2019), there is plenty of interest, but also plenty of criticism. "Do you travel by plane?" – that's often their first question. "Do you eat meat?" But the young people – and that is part of FFF from the beginning – do not allow themselves to be pushed aside as small, not quite grown-up individuals. They hold onto their perspective as a generation who are being betrayed by indifferent adults. They answer: yes, individual lifestyles play a role, a significant one. Many of them are vegans. But they also insist that it is just as much a structural problem. How are emissions supposed to go down by ten or twelve percent just by a few children flying less? They expose the critical attitude of the media as a political stance which is hostile to their generation and reduces them to consumers. And the structures – that's exactly what they can't change. They are not allowed to vote, and the political system is not built to take account of their future, the world in 40 years' time.

I am confused, sad, and angry. Why don't the older generation stand up when they see the desperation of their children? Many of my colleagues at universities worldwide are researching the causes of "passivity" in the face of the crisis; or rather the activity in the direction of the status quo (for an overview: Stoddard et al 2021). They systematise different dimensions: the structural ones like the late capitalistic class society, including its "hegemonial" narrative of justifying exploitation; the economic power structures which influence the financial system, media, and education; the sometimes criminal lobbying and the enormous interests behind all of this for people who own the fossil fuel, agricultural and forestry industry, unimaginable wealth. But they also mention psychological reasons: the ability and need to look away and deny (Birnbacher 2022 refers to the often spatially and temporally distanced effects of actions with climate impact); and ideological ones: racism, white supremacy, patriarchy, and so on (for alternatives, see the chapter about the WEF in Davos). I would add: what is lacking is an understanding of the substance of democracy, the caring meeting on equal footing, even at universities; the reduction of democracy to formal processes, for example of elections (see the chapter about democracy). But still, I think, most people in high-income countries could act. They could stand up and organise. And I see the young people doing exactly that.

The Christmas celebration in Mynttorget

Mynttorget is one of the only places where the young people strike every week for seven hours. In other countries, they march through the city once every three weeks, but the Swedish activists keep going with this huge effort, at personal risk. They come along week after week, miss lessons, and have to catch up over the weekend. Without them, there would be no strike movement in Stockholm. But that also ties them together, not least on public holidays. Obviously, they celebrate Christmas together, Easter, Midsommar, and all their birthdays. And now, they hope that the new year will finally bring a change. What if thousands of pupils were also to leave their schools in Sweden? They are still the "outlaws", they sit on their mats and fight for their future.

On Friday evenings and on Saturdays, some of them read hundreds of notifications on Twitter. Where are strikes taking place? Who can they draw attention to? As soon as a group appears with strike placards in any town, or even better with a video clip, they tweet back. Regardless of whether it's a single child or tens of thousands. A huge collection emerges, – and so they gradually build up a global network: Bangladesh gets in touch, Japan, Uganda, the Czech Republic, France, and all the German-speaking places. They also get support from Céline in Germany, who is responsible for the German XR Twitter account.

This network stretches across the world, bringing together sad, angry, and spirited teenagers and young adults in thousands of cities, who are watching in panic as the hesitation of politicians makes their future ever darker. As soon as they use the hashtag #FridaysForFuture and mention their city or town in the tweet, they become part of the biggest ever environmental youth movement and help to spur on the other strikers.

Early on Friday, some of the Mynttorget activists turn up with advent decorations. In the middle of the square, there are two huge flowerpots with a few straggly shrubs, just strong enough to hold Christmas decorations. As always, it is absurdly cold, the wind is blowing from Lake Mälaren and trying to drag us out to sea. Despite this, they sit down between the flowerpots as they do every week, and come up with a new recipe for their fictional cooking program, taking it in turns to improvise an ingredient. This time it's a Christmas recipe:

> Take one Christmas bauble,
> made of plastic,
> fill it with pine needles,
> cut it in half
> and put a single grain of rice inside it
> which has to be painted green
> with acrylic paint
> and then add plenty of snot
> and melt it over a flame.
> Then it's time for a sauce made of melted snow
> And spit in it one more time
> Et voilà.

Almost every week, new countries are now joining the chats with hundreds of young people; the band of rebels is being formed. In the days towards the end of December, the idea emerges of organising "deep" strikes in some countries which would take place roughly every month rather than each week. That's how the Swiss are doing it. Anna and her friends from England decide on the 15th of February, and are joined by France. Luisa plans something for the 24th of January in Berlin. And the discussions about the global strike on the 15th of March continue. I'm amazed at the ability of these young people to think ahead. How brave, I think to myself, that they are agreeing on a date in the future and risking that it won't be possible to mobilise and that it will all fizzle out. But nothing fizzles out.

How to teach commitment to sustainability – playing animals

In these days before Christmas, it's finally time to teach the seminars which I prepared for the trainee primary school teachers when Greta was beginning her strike. The basic idea for the improvisation was co-created with my colleagues: the students travel to a climate conference. They can choose whether they are politicians, researchers, civil society, media people or activists. We work out the whole scientific background and they have to prepare five-minute speeches in small groups and then present those speeches – while the "media", the "activists" and so on come up with their reactions. This form of creative work seems like a wiser idea than the typical games in which people are supposed to represent the interests of a specific country, "egoistically".

At the end of the conference, I claim that everything has been broadcast live to a northern Swedish forest, the whole conference. The students become animals who have gathered in the woods and are commenting on what the humans are up to, and how quickly they've been changing the animals' habitat. There they sit, elk, bears, ants, and birds of every kind, reflecting on what they've seen. I join them as a journalist who is going to report on their conversation in a big newspaper, and thus lower my status among the students, since I'm representing a rather irresponsible species. The north of Sweden is getting warmer much more quickly than the world on average: two degrees of warming elsewhere means 4 to 6 degrees for the animals and the indigenous Sami people, whose economic existence is completely dependent on nature.

The permafrost in Russia, which is in exactly the direction of the window in our theatre space, is melting during these winters (Welch 2019). Enormous amounts of methane are released. Of all the tipping points of the earth system, the one which makes me panic the most is that such tremendous amounts of methane could potentially contribute to a three-degree increase in global temperatures – and this point may already have been passed.

In these sessions, sometimes grief takes hold, but it is a liberating kind of grief; it can be mixed adequately with anger or it can be quietly expressed. The students, or the elk, can connect with the valuable dimension within themselves, which connects us with others and with nature. Are they being led to take action? With these "multimodal" forms of learning (theatre, painting backdrops, rhetoric etc), am I teaching them "agency", "resilience" and "empowerment", as the jargon has it, the will to change the conditions of their lives and the belief in their own power to act and potential to shape the world? It seems to me that something more fundamental comes about; they are "in contact" with themselves and others, and with an overarching humane dimension – one which is often funny – from which everything possible can then emerge (see Fopp 2016 for the difference between "being humane" and "being human"). That might also be at the root of what many people call "spirituality", which could be the core of many religions.

In these sessions, we don't just prepare ourselves, we don't just strengthen our skills, we don't just increase our knowledge, but we also just spend time together in a good way. If education is just preparation, and not life, many things go wrong, I think to myself then and look out at the winter landscape. And so we all leave for the hibernation of the Christmas holidays.

The new year begins

From then on, their outlines become clear, the European and then the global core group of strikers. Through Greta's speeches and their own national documents, they have come up with basic principles. They meet on Sunday evenings in the virtual Zoom space and for 24 hours a day in the chats, often linked together by people from Mynttorget, which becomes a kind of global hub.

During the holidays, the movement explodes across Europe and starts to become confusingly complex. What kind of structures are needed, they ask themselves. Are any structures needed at all? But most importantly: what kind of strategy? The real struggle is only just beginning. Giant companies in the fossil industry are puffing out millions of tons of CO_2 into the air, as we read every day on Twitter, supported by the banks and even by Swedish pension funds.

How is a small number of 16-year-olds supposed to tackle that? It's easy to become afraid just by thinking of the powerful people who have so much to lose. Then Greta decides to travel to Davos to meet that very elite.

The Swiss activists talk about the coming strike. They have chosen a different strategy from most other countries and alternate Fridays with Saturdays so that apprentices and parents can also join. And a week later, the Swiss pupils leave their classrooms. There are thousands and thousands of them; in Lausanne, Geneva, and Zurich alone there are 10 000 in each city mounting the barricades. The 18th of January becomes the biggest strike so far in the German-speaking region, then on the 2nd of February it is even bigger, with parents and grandparents showing solidarity and support.

The whole route along the Limmat in Zurich, which connects the train station area with Bellevue, is a mass of dots, heads, people holding their cardboard placards. The citizens of Zurich will soon be voting in a new parliament in the city and in the canton, and what the media are calling the "Greta hype" will lead to a shift towards left-wing and green candidates. The new parliament implies that it will adopt into law the basic demand of the climate strike, net zero emissions by 2030.

Hope is growing, but with it the challenges: in these weeks, people keep appearing who want to transform the Fridays for Future plant, which is only just beginning to grow, into a hierarchical association, or who want to water down the demands. Evening after evening, the young people in Mynttorget discuss these suggestions, some of which are quite bizarre. They all refer to Kevin Anderson's research, trying to defend Fridays for Future and the ideas which come from the square against these "hostile takeovers". And then, finally, German names appear in the international chats. Behind the scenes, after Luisa's meeting with Greta in Katowice there is so much going on that it's hard to imagine: WhatsApp groups are formed, schools and youth associations join together; NGOs help in some places with infrastructure and know how. One local group after another springs up. And they plan a meeting in Berlin. In Stockholm, new helpers appear in the square and make a big difference: Helena has been dedicated to the climate for a long time and helps Greta with media relations, free of charge. She has a small PR office in the realm of popular culture, and almost from the start she is in the square during the lunchtime hours with her husband Erik, who runs an online conservation magazine.

But those who pass the young people and go into the parliament building, the ones who have the formal power to make rules for a sustainable society – they literally do almost nothing. Sweden in turn gets a red-green government in these January days, supported by the two liberal centre parties. But this government still says nothing about the call for an emissions budget, and barely comes up with any plans showing how society could become renewable and sustainable sector by sector in around ten years on a socially just basis, as set out in the Paris Agreement (Anderson et al 2020). By what means could the rules actually be changed?

Chapter 5: Davos and the World Economic Forum
January – February 2019: What is valuable
and what is a science of economy?

Davos

The crisis must first be established as a crisis in people's minds; this is still the young people's basic idea – in the minds of ordinary people, and of the people in power.

Davos is located high up in a wide Alpine valley. For several years, it has been the home of the World Economic Forum: the meeting of the "masters of the world", as described in a tweet by Naomi Klein, climate activist and architect of the American plan for a "green new deal" (Klein 2019). Here, the world elite in finance, politics and the economy meet annually for a conference at New Year. Behind closed doors, deals are done (or at least arranged), new networks emerge, and ideas are exchanged.

Meanwhile, with my students, I'm sitting in a seminar room at the university, watching the official livestream of the World Economic Forum in Davos, often in the breaks between lectures, but also as part of my seminars on sustainability, democracy, and theory of science. Between the ever more frequent news reports on Greta's arrival from the Swiss media, the monitors in the seminar room show images of the various conference rooms in Davos. Rich people present their latest ideas for how the world could be a better place. Among them are those who run the hundred joint-stock companies that produce more than 60 percent of greenhouse gas emissions (Riley 2017). Aren't they the ones who should instantly be changing their business model? Or is it actually about the politicians who are legitimising that business model? The people who have gathered here could set the course for a real transformation. Could they change our democracy fundamentally, we ask ourselves in Stockholm. "Ideas about the

21st century economy." "The future of education." "AI: opportunities and risks." Those are the names of the panels.

If we want to understand the real reasons for the climate crisis, and also how we can get out of it, I think as I watch the screen in the university, then perhaps the most important step is understanding this. The logic behind this little group of people, the workings of the financial sector and how it is intertwined with the rest of the economic system, including its context of political regulations. How nature and the people of the Global South are often either excluded from this or exploited. And how we can change that, where we can start. Children and young people are nowhere to be seen in these spaces. They don't fit into this worldview. Nor do their dreams, voices, ideas, or their future.

But then things get hectic. On Thursday and Friday in the third week of January, so much happens that by late afternoon in Mynttorget, all our phones have died. It is not only that Greta is in Davos – in Berlin, the first ever really big strike in Germany is taking place. Suddenly, from one moment to the next, young people who were previously unknown, such as Jakob and Luisa, are being interviewed on TV, where Fridays For Future is being presented as if it were obviously an established youth movement.

And over all that hangs the question of what a sustainable economy would mean, and what we scientists have to say about that. In many conversations everywhere in the world, the young people of Fridays for Future are asked: "What do you want instead? Show us a plan for transformation into a sustainable society. What do you even want? What, if not growth?"

It is Thursday afternoon. Greta makes a short speech to the gathered elite in the congress centre. She goes looking for a direct confrontation, and demands democracy. Greta makes no secret of the fact that she is on the territory of the people she is fighting. She says that they, the economic and political elite, are the ones, with their mania for wealth, who are destroying the world. And those who get themselves photographed with her for magazines, the Trudeaus, Merkels and Macrons, don't truly realise that her criticism is aimed at them. Why don't they listen to their advisors? Greta and fourteen other young people will soon prosecute them at the UN and start legal proceedings against them in Germany, Brazil, France, and Canada (Gonzalez 2020).

The young people signal in these days to the whole world that we don't have to accept the fact that some people are weakening democracy and poisoning the planet. Because of Davos, the Swiss activists become even more central to the international FFF movement. It is Thursday evening. Many long phone calls take place. Loukina, Jonas, Lena, Miri, Marie-Claire and all the others are well-read and think strategically. They report on the openings there might be in Swiss legislation that would allow big, sustainable changes to be made quickly, in the financial sector, for instance, which is shaped by Crédit Suisse and UBS and is responsible, through investment, for carbon emissions that dwarf the entire emissions of Switzerland as a country. How could laws be formulated to prevent this and reshape the financial sector so that the money goes to those who treat nature well? There could be a petition for a referendum demanding a ban on the financing of fossil industry. But how long would that process take? What counts as valuable in a society?

Meanwhile, in Germany – the coal commission is meeting

It is still Thursday evening. A Twitter notification comes up on the activists' phones which makes them all laugh for a long time. The Munich Fridays for Future group has published a letter written by the government, the education ministry to be precise, inviting the 16-year-olds to a conversation. FFF has established its power so quickly and unsettled the ministries of education to the point that they are compelled to offer discussions. The children can't be forced to go to school; how would that work? A few headteachers and cities want to introduce fines for the parents, but the public doesn't agree. A few schools insist that exams can be marked as failed.

Then I see FFF's answer to the Munich ministry: we would have liked to come, but we're away on Friday. We're travelling to the national strike in Berlin. And what a strike that is. The Berlin group has brought together the local groups which have been expanding across the whole country and delivered them all to the ministry of economic affairs in Berlin. More than ten thousand young activists confront Minister Altmaier who is leading the so-called coal commission (officially the Commission on Growth, Structural Change and Employment). Germany's coal power stations are some of the biggest carbon emitters in Europe. Who owns nature, and who owns the future?

"The house is on fire" – what would a prosperous society be? (On Kate Raworth)

In Davos, it is Friday lunchtime, time for the press conference. Swiss television is covering it live. Then Greta says the words that immediately become canonical: "I want you to panic. I want you to act as if the house was on fire. Because it is." The house is on fire, and we have to extinguish it. Later, the young activists will explain that people should obviously not be running around in panic but reacting sensibly. But sensibly in response to an emergency situation.

At Stockholm University, we ask ourselves: How should we react? What would it mean to extinguish the fire? In my institution, we use the ideas of Kate Raworth. The students are familiar with Kate, the economist, from a lecture she gave at the World Economic Forum in Davos last year, which I uploaded to their shared learning platform. She has been in contact with FFF and has inspired many young people. As usual, she talks at an incredible pace about her invention, the Doughnut Economy (Figure 4).

Raworth's *Doughnut Economy* (2018) and Maja Göpel's *The Great Mindshift* (2016) are two of the books on the reshaping of the political economy which inspire the Fridays for Future movement. They discuss them – in webinars, for example – and the young activists have a range of opinions on these approaches. Somehow, "our" way of managing the economy and teaching and thinking about economics at universities has terrible consequences, say both Göpel and Raworth. How can we change that?

Raworth explains her basic idea as follows, as I understand it watching the videos: The inside of the doughnut, is the 'safe and fair space' for a future society. That's where we want to get to. And the surface on the outside of the doughnut is the zone where we are ignoring and transgressing the limits of the planet. We want to get away from that. Away from global warming, from acidification, from the extinction of animal species, from the loss of biodiversity, from pollution. And in the hole in the middle of the doughnut are the dangers of neglecting people's basic needs: space to live, food, political rights, education, equality, and so on. We have to transform our societies so that we find our way into the doughnut. Doing justice to everyone's needs, and without transgressing the limits of the planet. How do we get there? That is the basic economic question: how do we organise the economy so that social and ecological sustainability are possible, meaning that societies flourish? The question is not how we create growth. We have to change the aim of society.

We know the phrase "planetary limits". One of its inventors was my colleague at Stockholm University, Johan Rockström, who headed the Stockholm Resilience Centre for years (Rockström et al. 2009).

This kind of research may always have emphasised that ecological sustainability is most important, and that economic sustainability must be shaped by it – and for that, social sustainability is required. But green growth and similar concepts were only rarely questioned; at most, they were up for debate.

Fridays for Future shake up this discourse within a very short time, with the help of Göpel and Raworth. They turn the basic question inside out.

Figure 4: Kate Raworth's Doughnut idea

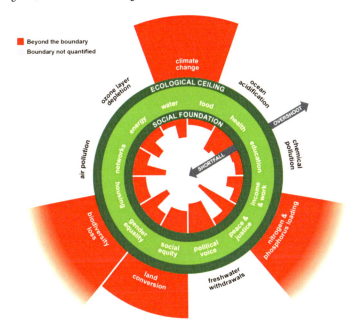

Image: Kate Raworth and Christian Guthier/The Lancet Planetary Health

Because the actual issue behind all this is the following: what is a dignified, prosperous life for all, a real global democracy, in which everyone has enough without exploiting others or exploiting nature? What would a society be like that offered true safety and security? And: which structures prevent us from making such a society a reality? In contrast with this, the question we've been asking so far is: how do we guarantee growth? Not that every kind of growth is wrong, says Raworth (2018; chapter 1), but prioritising this question is not just strange for activists. Most of them can see that we must first consider whether we have a shortage of water and whether the soil can provide enough food for everyone, and that we must treat each other respectfully and fairly and dismantle problematic power relations before we commit to other goals or to an economic dogma.

"But what would be the consequences of that for the economy? In particular: for the financial sector, for monetary policy?" Marie-Claire is one of the most active Swiss climate strikers. She is a student, four or five years older than Jonas and Loukina, Fanny, Lena, Miri, and Paula, who are still in high school. She is studying politics and environmental studies at the University of Zurich. And she met Greta two months earlier in Katowice at the COP24 meeting and went on strike with her. Not long afterwards, she visits us in Mynttorget in Sweden in the name of the "Glacier Initiative". She has just finished working on a book edited by Club of Rome president Graeme Maxton, called *Change!* (2018).

In the months that follow, we have long discussions about the question that is uppermost in her mind: what should we say when people from the UN or the Swiss banks ask: what is it that you want? We must start with the financial sector – that's a point we keep returning to. Or the logic behind it, meaning the logic of investments, but also that of "invented" money which could be replaced by "positive" money (Raworth 2018, chapter 5), and the logic of the current form of economy, late capitalism. At the moment, money and production flow to the place where investment is worth it, meaning the places where a profit can be made, regardless of whether nature is destroyed, either in the extraction of raw materials or in the consumption of these materials and the subsequent pollution. And for this concept of profitability, people's wellbeing is often irrelevant. As is the question of whether this supply and demand logic actually produces what is needed, what is necessary for us as a global population.

In these conversations, the Swiss activists keep coming back to the idea of an emissions budget, connecting it with the doughnut concept. Emissions budgets would define what a "safe and fair" space would be in relation to the climate.

As soon as this framework is established, it is clear how drastically the economy would have to change. If we are supposed to be making sure that within twelve years in richer countries (and within twenty years globally) there are no emissions worth mentioning, then we will not be able to avoid tightening standards and regulations by about ten percent every year from now on (see Anderson 2019).

Hijacking the university – "Rethinking Economics"

In the live broadcast of the WEF, hardly anyone talks about these ideas, which are so central for the young people. And nor do they talk about them at the university. Or do they? Sometimes, on these Thursday evenings in winter, after all the students have gone home, an international organisation meets in our university rooms. A colleague from our institute runs the Swedish branch, and I sometimes go to their meetings. It is called "Rethinking Economics" and its aim is to change how economics is studied at all universities in the world, to open it up and make it more scientific.

Those who do not study economics and don't have time to familiarise themselves with the syllabi of the most important universities might have the idea that the study of economics is structured in a similar way to that of politics, psychology, education, or sociology. Meaning that every student is offered a whole range of approaches and perspectives, in terms of theory and method. In political science, this could mean looking at liberalism, socialism, conservatism and other movements, feminism, ecologism, and posthumanism. But the study of economics is practically never like that, as I realise after a few meetings of "Rethinking Economics". Instead, a kind of monotonous uniformity is poured over everything, at almost every university in the world. This is known as neo-classicism. Of course, there are plenty of alternative theories (Kelly 2019; Felber's (2018) economy for the common good; post-growth approaches; Hickel 2020), but they are barely mentioned. In the textbooks of neoclassicism, which include most economic textbooks in existence, after the first three pages you find yourself in a world of mathematical representations of supply and demand curves and market mechanisms which are supposed to lead to prices. Planetary limits and basic human needs are barely mentioned. Why do universities allow entire courses of study to be structured so unscientifically, we ask ourselves. So that there is no discussion of different theories or different basic questions, or methodological exploration of these theories and questions?

And then comes the question that overshadows everything else: how are the students supposed to study economics in a sensible way – or history, education, architecture – if they don't know that the "house is on fire"? I walk along the corridors and think about how I can change the institution of the university, together with my colleagues: by introducing something like a "studium generale" for all students, in which they get to know the existential threat of the ecological crisis. But it is also necessary to explain the societal context which produces this crisis, the root causes.

The problem with the basic economic model – what is wealth?

Göpel and Raworth have similar ways of describing why mainstream economics and the current approach to the economy are problematic. The initial question is: what defines work – or being economically active? What is good "production"? The basic neoclassical model stipulates that there are a few ingredients, and the way in which they are mixed together determines what it means to be economically active, regardless of whether someone is making a pasta bake for children or producing an electric bike in a huge company. On the one hand, matter is required, nature, basic raw material; then capital, financial means; plus tools and knowledge, and finally human labour. Combined with economic knowhow, this produces a bake or a bike, meaning on the one hand a product or a service with a use ("goods/services" with "utility"), and on the other a kind of investment profit, something that is worth it. And if you want to do good work, you must ensure that all ingredients are good quality. Correspondingly, a business can fail because of any one of these components.

The two economists say that there is already a problem with this picture. But that becomes much more obvious when it is developed a bit further, as it is in the traditional textbooks. Already in the first pages, we are told: what is produced, the goods and services, can be sold at a price – they are often seen immediately as products for the market. And the market itself decides how high the price is, based on supply and demand. That makes it possible for a profit to be made when the product is sold, as well as allowing the product to be consumed or used. This step is anything but natural: many valuable things cannot be described in this way, for instance when we create something for each other without money being involved. Caring for our parents, raising children – in short, all the "care" work or core work which represents about thirty percent of GDP and is done often by women, leading to a form of systemic injustice in terms of gender (NEF 2010; Schmelzer/Vetter 2019).

And much of what happens when things are produced is not included in this basic model: the costs of the waste that is created; the exploitation of nature as a "resource". Sometimes these elements make an appearance as "externalities", but how is it possible to calculate the true value of nature?

At any rate, this is how the mainstream picture looks in all economics textbooks, according to Göpel (2016, chapter 3.1); what is known as micro- and macroeconomics. More precisely, all of economics is expressed by the neoclassical model as follows: there are two poles, households on the one hand and businesses on the other, and together they form a productive cycle. "Households" provide employees who produce things in the businesses; they receive a wage. And they consume, which brings an income to the businesses. A closed system. Voilà. That is economics. Then there are the factors in society which can intervene benevolently (or destructively): banks; the state. They provide money when it is scarce, and collect taxes to finance the infrastructure of the welfare state: educating people, keeping them healthy, providing capital for the banks, raising interest rates, avoiding unemployment and so on. And everything is aimed at keeping the wheel turning, making sure more and more goods are produced, so that incomes and the welfare state are possible in the first place. This is how societies understand prosperity and wealth: healthy, educated people should produce lots of goods and services, because this makes it possible for them to consume and spend their salaries and pay their taxes. The end of the story: the aim is to increase GDP, meaning the value of all goods produced, and this shapes all kinds of legislation, including climate legislation. If there are discussions between the powerful in Davos, these discussions remain within the framework of this basic model.

Criticising the basic model and outlining an alternative – what are needs?

At first glance, the model seems plausible, as Raworth and Göpel argue, but it turns out to be abysmal. Such an economy would collapse, for instance, if everyone only produced and used the things they needed for a dignified life, meaning durable products that could be used for years – the form of economy which we should now be introducing very quickly, according to UN reports. Some people call this "degrowth" (Hickel 2020); others call it a "post-growth model" (Schmelzer/Vetter 2019). Tax revenues would take a nosedive – they are tied to incomes, and these are tied to turnover and consumption. Jobs would disappear, and schools and hospitals could no longer be financed. Something is fundamentally wrong here.

How can we solve this problem? In their criticism of the model and in the alternatives they suggest, Raworth and Göpel once again have similar approaches. They say: what we need to change right now, both in our laws and in the minds and hearts of the population, is the aim of our economy. It should be about meeting the needs of all people, without going beyond the planet's limits (climate system, biodiversity, ...). But what are needs? Raworth defines them using the UN's 2030 agenda goals as the inner boundary of the doughnut: enough food, water, space to live, a political voice, equality, and so on. Göpel (2016; Chapter 3.2) uses a slightly different model drawn from Manfred Max-Neef: being able to live, be protected, be looked after, be understood, participate, enjoy oneself, be creative; as well as identity and freedom. But for both authors, these needs are defined as the primary focus of all economic activity. The model in the economics books is wrong: after all, the focus on growth need not have anything to do with the more fundamental goal that we should all have something to eat and a roof over our heads, meaningful work, and a politically equal say.

And the world would really be a different one if we focused on the needs of all people when we designed the political framework of legislation, nationally and globally. There is still enormous poverty worldwide; access to drinking water and sanitation is not a given; more than two billion people have no running water at home. The children across the world who join the Mynttorget group understand this. During these months, they document their strikes on Twitter.

So, according to Göpel and Raworth, the economy has to be sustainable, both socially and ecologically; "redistributive" and "regenerative" is what they call it, including such sectors as transport, the clothing industry and agriculture. Correspondingly, the taxation model must also be "redistributive" and "regenerative" (Raworth 2018, Chapter 5/6). A regenerative economy creates products that can be used for a long time, in such a way that they fit into a "circular economy"; when they are worn out, they can be reused to a large extent as raw material. And their production does not simply destroy nature: cutting down forests and burning them, digging coal mines and so on. On the contrary: forests must expand. So how can we reorganise the economy? Raworth and Göpel say: we must redefine the relationship between market, state, citizens, and "commons" (what is owned and produced jointly or cooperatively), including in law; the market alone will not regulate that.

On the peculiarities of the capitalist market economy

In her analysis, Maja Göpel also focuses on the fact that most countries are organised not only as market economies, but as market economies along capitalist lines (2016, chapter 3.3). That difference is established in political science and emphasising it does not immediately constitute a political value judgement. In Davos, the people Greta faces are some of the figures who most embody this capitalist approach to the economy. Again: around 60 percent of global emissions are produced by one hundred corporations. And those that manage and own the most capital emit disproportionately more than the poorer part of the population (Gore 2015). Political measures targeting their consumption of resources are therefore particularly effective (Anderson 2019). And the specific way in which our economic system is structured means that they become increasingly rich without having to do anything at all. Their wealth works for them. They can withdraw it as capital whenever a more lucrative investment appears. These few people and their decisions shape the living and working conditions of so many more people.

But we could also live, according to Göpel and Raworth, in such a way that everyone would have roughly the same control over what is produced and how – democratically. That is the fundamental difference between corporations and "ordinary" businesses: shareholders do not make investments or profits in the same way that independent dance studios or hair salons do; with the risk of losing everything. The terms may sound the same, but profit and investment are structured differently in a corporation. The business model is not even primarily about producing goods, but about the multiplication, the accumulation of capital, money (Göpel 2016, chapter 3.3).

In the form of capital, the notion of "property" is suddenly uncoupled from responsibility, care, and identification with that which is "owned". That goes just as much for the way in which banks are organised and managed as for the oil, gas, cement, steel, and coal industries; those corporations which are most responsible for global warming, and which absolutely want to hang onto the business model of increasing capital. (Wind and sun cannot be owned as easily as coal and oil, or sold as goods.) Apart from this, "fossil" businesses organised on capitalist lines only work when they grow exponentially and always draw more resources into their cycle (Göpel 2016). They are legally bound to this – they must primarily increase the property of their owners; which is structurally something different from making a profit through a successful business.

For that reason, it makes sense, according to one tradition in political science, to talk of several different classes in society. Historically, so much wealth has been concentrated in the hands of a few people; of very few people (Piketty 2018). The richest ten percent in Sweden and in Switzerland own more than eighty percent of all wealth; often these are white men in the middle and upper classes.

The temptation is very great to develop business models in such a way that nature and humanity are pushed further and further towards the limits of their resilience. And it is often women in the Global South who are working in insecure and underpaid jobs. And they often live near factories and power stations or in poorer areas which are most affected by air pollution. Every year, this kills seven million people (WHO 2014).

From this point of view, my students find that the enormously productive global economy is unjust in multiple ways. We talk for a long time about precisely the term justice. Injustices based on ethnicity, gender and class seem to be intertwined. It is the women from BIPOC communities who are often the most disadvantaged, and it is a few white men who are already rich who profit the most (Fraser 2022).

How could we organise all of that more democratically, we ask ourselves; more democratically in terms of people having a say at work, but also in terms of property relations – so that everyone would be on an equal footing and have roughly the same amount of power?

The "destructive" aspects disappear, according to Göpel and Raworth, when the economy is reorganised through social relations and property forms (on a post-capitalist basis) which focus on the needs of all people, for instance when communities have their own solar panels or wind farms (Felber 2018; Raworth 2018, Chapter 6). The logic of the relationship to nature is then at least not automatically connected with extracting or taking out natural and human resources – as must inevitably happen in the current economic system, quite apart from the "decoupling" of economic growth from the increase in other parameters (wearing out of materials; pollution; energy consumption etc.), which research results show is probably not possible (Hickel/Kallis 2019). Instead – and this is the constructive project – we should now develop everything in such a way as to strengthen our resources – rather than taking them out of the system. The question of how exactly that could look is one which is increasingly discussed by Scientists for Future, the group which is now beginning to form.

The economic causes of the climate crisis

I continue with the lecture at the university and ask the fundamental question: why is the climate crisis actually happening? What is the reason for it, and is the crisis connected with the way we structure our economy?

First of all, the climate crisis is obviously due to the fact that we are razing forests, keeping animals in intensive factory farms, and continuing to burn coal, gas and oil. Andreas Malm, my Swedish university colleague from Lund, says (2016): at the beginning of the industrial revolution, capitalism was established as a political economic system, and was then imposed and normalised – and machines were invented, with engines that take over the work of transport, heat and so on. And – and this is crucial – because the natural materials used for this literally come in the form of goods or can be transformed easily into goods. In contrast with wind, water and sunlight, lumps of coal and oil can be placed in containers and become the quintessential products. Mining them, distributing them and selling them can be organised perfectly along capitalist lines. For this, slaves are used, at first, and then mineworkers who toil away and have nothing but their labour (Hickel 2018). In this way, a product is created that can easily be transported and sold on the market in portions that can be decided at will, often leading to the exploitation of indigenous populations and the destruction of their ways of life.

This has been pointed out again and again by Jamie Margolin, the 17-year-old climate activist in the USA (Margolin 2020). She has also pointed out that this story has still not really be told. At this time, many activists from the indigenous population appear in the Fridays For Future chats, especially from North and South America, but also from Australia. Unimaginable riches were appropriated simply through the ownership of coal and oil extracting companies, through the investments of the big banks in this fossil infrastructure and the corresponding corporations. How was this legitimate? Why should nature belong to them, and to those who live in the Global North? Why should the work of the miners go to them? And above all: what are we to do with this history, which has damaged our shared "fabric of integrity", as I call it in my lectures? At its core, then, for Andreas Malm, responding to the climate crisis means rectifying this form of property relations and socio-economic relations, which means gaining power over investments and the flow of money, democratically, abolishing private ownership of shares, so that these resources can be focused on expanding renewable energy.

Criticism of Raworth and Göpel – what the doughnut is made of, and what ultimately holds the world together

We can change all of that immediately, say Raworth and Göpel, along with so many other researchers who are close to the young activists of Fridays for Future. States and communities (and even UN institutions) could adopt the doughnut model into legislation as their economic aim; the city of Amsterdam has done so (Boffey 2020). A universal basic income or "basic services" could be introduced, globally, because otherwise it would only increase the inequality between the Global North and South; this would mean that the care work that is often carried out by women would be appreciated, and everybody's dignity would be taken seriously. It could be partly tied to locality (through currencies, for instance), as suggested by Hornborg (2017), so that fewer resources go to transport and shipping. Emissions would have to be reduced by more than ten percent annually in Europe in all areas, with regulatory measures to drive society in a more sustainable direction in the realms of transport, food and so on (Anderson 2019; Hickel 2020). "Real money" could be introduced so that the attraction of making a profit and the financial sector would be subordinated to the doughnut goal (Raworth 2018): a more sustainable economy would gradually emerge. What slowly becomes clear in all these critical points: the economy needs a "mindshift", as Göpel calls it, when it comes to our relationship to each other and to nature, especially in terms of what is seen as valuable and worthwhile.

How are worth and value measured? That seems to be the central question. Göpel and Raworth answer: "worth" should be measured according to whether needs are met, rather than being based on market mechanisms of supply and demand, and also according to what they call "well-being" (Raworth 2018, chapter 1; Göpel 2016, chapter 3.1.5). This can be defined in different ways. Through the Human Development Index of the UN, or the "Happy Planet Index" of the New Economics Foundation, as suggested by Gough (2017; chapter 4): using a formula which includes the parameters of life expectancy, subjective life satisfaction, income equality and ecological footprint.

In discussions with my students, we ask ourselves the question: could there be a more fundamental measure for our future societies and economies than one based on people's needs and individual well-being? I try to capture the basic problem differently. The question of needs and of property forms, including whether or to what extent the fossil industry should be made into state property and then quickly scaled back, is an important one. But it does not solve all problems, because even states must be guided by a compass of some sort in their actions, and this compass is not simply defined by the property question. Vattenfall is a Swedish state company which was running the coal power stations south of Berlin. In summer 2016, some of us blocked those power stations with Ende Gelände, using civil disobedience, and prevented carbon emissions at least for a few hours. But Sweden did not then decide to keep the coal in the ground when it withdrew from ownership; instead, it sold to a Czech consortium (Reuters 2016).

Don't we need a better compass for our economies than that of GDP growth, or Göpel and Raworth's focus on needs and well-being? Well-being and needs do not completely capture what economic activity should really be about, I would argue – namely about building up and strengthening our resources, and creating a sustainable "being-towards-the world" (as Merleau-Ponty would say), a humane relation and exchange. What is missing here is the centre, the definition of the dough from which the doughnut is made; what holds the sustainable world together at its core – and what we explore in the course at Stockholm University every day in the theatre spaces, with our social interactions and our existential situatedness in a concrete environment. In research, the notion of being "connected" or "deconnected" turns up in this context, the experience of being in contact with oneself and with others, which can strengthen us (see the chapter on education for literature). In other theoretical traditions, this is called a non-alienated relationship with the world and with oneself, a real exchange and meeting; or "resonance", "democratic acceptance on equal footing", and so on (Rosa 2020).

A crucial insight here is: if we tense up, even slightly, we lose contact with what we really think and feel, but also our connection with others and the environment. Our view of the world becomes slightly diffuse, as if we were under a bell jar, no longer able to breathe properly or communicate. In extreme cases, we become ill, either mentally or physically or both; this has been analysed from a neuro-physiological point of view by Immordino-Yang (2015) in her empathy studies; physiologically, the Alexander technique can offer explanations. Philip Pullman (2001), in the *Dark Materials* trilogy, came up with the metaphor of the daemon which every person has: a living alter ego in the form of an animal, which jumps and plays around us as a kind of external soul, and can talk with us and advise us – from which the dark powers want to separate us, cut us off, by making us indifferent to it. The developmental psychologist Winnicott (2005) calls it the relationship to our "true self". Being cut off from it (often as a reaction of protection against different forms of oppression) can (but not always does) lead to what Adorno (1995) and others after him have called the "authoritarian character", which shapes the political rulers of so many countries which do not take the climate crisis seriously.

Then we become like the fictional miser, Scrooge, in Charles Dickens' *A Christmas Carol*, which our students are acting out during these days – and which reminds us very much of the powerful elite in Davos. "That was intense," they say later; "really working together and including everyone's ideas – we have strong personalities here..." But it is clearly fun to inhabit the character of a villain, and not to suppress the impulse to dominate others. At least as an experiment.

Beyond "well-being"

Again and again, we go back to the socio-psychological theories which explain how an authoritarian character develops, and how they can learn to interact with children in the playground in such a way as to give them security and freedom at the same time, and thus strengthen their characters and open them to the world. Playing with dominant high-status characters is a good way to do that, we realise, because this makes it clear to everyone what domination is, a violent form of behaviour (even if it is subtly hidden) – and what the alternatives would be, humane democracy (see for the following also the chapter about education in this book).

If we are really comfortable, we are able not to dominate; to meet on eye level, to connect, according to the research. In this sense, it is not even just about "well-being," but about a more fundamental phenomenon. It is a relational concept (being connected with...; or: really meeting), in contrast with "need" or "well-being", which are rather European, quite individualistic concepts. The research (Bowlby, Winnicott, Stern) also demonstrates something significant: this sense of being "in non-dominating, democratic exchange", without which so many things would be meaningless, cannot – contrary to

Raworth and Göpel's theories – be defined as a "need". In fact, our needs can actually prevent us from finding our way into this domain. This is the conclusion drawn by one of the most thoroughly researched theories in the humanities, attachment theory. The basic need for closeness and protection is so important for small children that they "sacrifice" real connection with themselves and others in order to receive them. If they see their parents or teachers look away in irritation when they express anger or joy, they will suppress, with time, the expression of these feelings, to satisfy their basic need for closeness and protection (Broberg et al 2006). Masks are created which become part of muscle memory, shape personalities, and prevent real democratic encounters on an equal footing. We subordinate ourselves or try to dominate.

However, we could perhaps say that there is something like a longing for affirmative connection which is even deeper than what we call a need. Yes, we can view needs and how they are met according to whether a real democratic exchange is made possible or prevented. This means that we can talk about whether we are "doing well" (see Fopp 2015), are "connected" to ourselves and others (and thereby to the idea of being humane), instead of "wellbeing", as a measure of economic activity which is not only individual but societal.

This "doing well" can come about above all if we have spaces in which a democratic exchange and real connection and meeting can happen. For these, I suggest in my lectures, we could use the term "humane spaces." What we need is not a dusty humanism modelled on the notion of reason developed by white male colonisers; but also not posthumanism, which doesn't even see humanity as an authentic value, but instead a form of "humane-ism". This means: it is impossible to be "in contact" if you are constantly hungry or fearful of not having enough to eat or a roof over your head, or of being unable to find work – or if you are in a state of permanent competition (in this sense, this viewpoint could redefine the "capabilities approach" to global transformation, see the work of Martha Nussbaum and Sen 2001). And this tendency to cut oneself off or to enter into a good exchange with others, the weakening or strengthening of resources, also takes place in university spaces, at workplaces and schools, in the way we listen to children and so on; in all areas of life.

Here, a criticism of Raworth and Göpel's economic theory comes into play which has consequences for how we understand university disciplines, not only economics. We cannot simply ignore the scientific and philosophical research which tells us what it means to strengthen resources; or become disconnected from each other and ourselves. And that applies at all levels, from social spaces to the spaces of the environment: starting from whether we are tense or alert as we move through, work in and interact with nature; up to the way we structure democratic institutions and finally to the way all areas are shaped politically, from food production to transport.

An overall picture emerges from the research: it is in a substantially democratic context that being "in contact" is possible. When we can count on everyone to respect each other and nature, we don't have to tense up anymore. And: we won't reach that point just through fine ethical and political declarations like "we are all equal," but by actually creating such spaces, in concrete terms, and in the process playing with and seeing through the things that make contact impossible: (violent) domination. If we cannot meet on an equal footing, it is difficult and in the long run impossible to remain in contact. This applies to the small-scale intersubjective relationships between people, but also on a large scale: entering into contact means that power relations based on gender, ethnicity, class, sexual orientation and so on are dissolved and replaced with democratic encounters at eye level, as part of the concrete structure of everyday life.

A new logic: freedom, and integrity

In my lectures, I use a metaphor to explain what is created or damaged in this way: the shared fabric of integrity. It is harmed when we destroy the soil and the forests, when we allow hunger and poverty; when we make children cramp up in school rooms and later establish dominating power relations in the world of work and discriminate against each other. This fabric of integrity which connects us with each other and with nature, or which expresses this connection, is strengthened when we look after each other, not just by recognising each other but also by sharing material resources. It has a historical dimension, too: the wrongs of colonialism still destroy our collective fabric of integrity now, and they have to be repaired; this would be a possible description of the situation in which we find ourselves, as vulnerable beings on this living planet.

What is "worthwhile" and "valuable" can – as it is in so many books for children and young people – be measured against whether it helps to weave this collective fabric of integrity, or whether it damages it. This measure is not freely chosen by us; it is rather a given, like the limits of the planet. Even if everyone in a society agreed to ignore it, it would still be there, nagging at us in the background and leading to suffering when it was neglected.

Theories at universities fail to take account of this when they rely on "deliberation", the joint negotiation of democratic goals. We cannot negotiate away the fact that domination can cause us to tense up and lose contact with ourselves and others. That is our inner planetary limit, we could say. But that doesn't mean that through this we simply become a part of nature, as claimed by posthumanist research and "new materialism" (Latour 2018), theories which are currently as influential at our universities as poststructuralism once was. We are "agents" in quite a different way from hills, lamps, or computers, even if it is important to emphasise how we are intertwined with nature. We have responsibility in a different sense. Alf Hornborg (2015) and Andreas Malm (2017) point out that there is a gulf between us, our social activities and decisions, and nature. And that it is dangerous to underplay this difference, as posthumanism/Latour (2018) do, and correspondingly play down our responsibility and our scope for action. Conversely, one could object that Malm and Hornborg exaggerate the gap between nature and culture and thus miss the phenomenon of "being connected" and democratic exchange as found in Gestalt theory (Merleau-Ponty 1974).

If we reshape our societies now in such a way that we reach this point, see through problematic power relations and abolish them, make affirming contact possible and establish a "humane energy system", we will find ourselves with something researchers call convivialism (Vetter 2021). Democratisation is then not limited to property relations but is focused on all the ways in which we organise the quality of our relationships with each other and to nature. It is not just about the formal "by and for the people", but about the substance; that we should not go against the outer framework of planetary limits (by cutting down the rainforests, etc), or against our own inner framework, because both determine our collective fabric of integrity.

In the university lecture room, we are once again watching the live stream from the WEF in Davos. Davos. The place of healing, the health resort. We see the incredible snowy landscape in the background of the recording. The air is special, the forests up here above a thousand metres smell more intense, the light and the colours are stronger. But this world of the "masters of the universe" in their limousines does not look healing. Not even when they are holding panel discussions about the health industry of the future.

Ever more urgently, the question comes up in our seminar rooms of how exactly "value", "freedom", and "integrity" are connected, and who should define them. If someone changes our notion of what integrity means, they will change the world, I think. Right now, an adult man can insist on flying to Davos in his private jet, even though he and everyone around him knows that he is harming the lives of others by doing so. The carbon emissions are enormous; a direct (small) contribution to global warming with all the disastrous lethal consequences. Like a three-year-old child in its tantrum phase, he throws himself on the ground, kicks, waves his fists on the ground and screams: but I want to fly, even if it harms people in Mozambique. Or I want to eat my steak. Or sit in a traffic jam in my SUV. I want to own this forest and chop it down. And and and. And this is seen as acceptable, because anything else would curtail his freedom. This is what the politically established concept of freedom looks like, which is inseparable from our concept of property rights and territorial integrity (von Redecker 2021). We imagine that we each own ourselves and can do what we like with ourselves and our own property; anyone who limits our autonomy is regarded as a threat to our individual integrity.

But this approach to integrity is not without alternatives. For instance, it abstracts away from every sensation (see the chapter on education). Many children sense that something is not right, that I cannot really own the forest. The UN programs emphasise again and again during these months how important it is to protect the forests, to switch to a plant-based diet, a shift in approach which would mean that we no longer mistreat nature and animals, but also other people (see e.g. the IPBES reports). This is the kind of integrity that must be preserved and strengthened; freedom can be seen as a result of such processes which create integrity; and this is what school and university education should be organised around.

But all that can come later. Right now, there is a crisis. Emergency laws must be formulated and introduced, to halt emissions and build a sustainable society within the next ten years. How is that possible? Across the world, we have to reinvent democracy.

At this time, we receive the alarming news that the glaciers in the Himalayas are melting (Carrington 2019). One and a half billion people are affected. The social conflicts that will result are scarcely imaginable.

Chapter 6: The Prelude to the Uprising
February – March 2019: The first international meeting, and the founding of Scientists for Future

The global uprising emerges – the young activists are connected across the world

In these days, new countries join the Fridays for Future chats one after another. They act as a climate movement for all countries – these young people who don't want to accept any relations of domination anymore. We can no longer accept this way of treating people and nature, they tweet, from Uganda, Uruguay, Argentina, and Japan. That is why the idea of the global strike makes so much sense to them. What is happening is not an issue in just a few countries, but a systemic problem for all populations. For a few children in the Global South, the connection between pollution, global warming and labour exploitation is visible every day, as they report in the chats. And they can talk to their peers in the Global North about this twenty-four hours a day, as well as about the suffering, the heat waves and the floods which they see. It is not about abstract "populations" but about their aunt who can hardly breathe because of the air pollution from the burning of fossil fuels.

All together, they prepare themselves in their own ways for the 15th of March. The map and the homepage of Fridays for Future, created by Jan and Jens a long time ago, turn out to have a whole unexpected potential. In one city after another, across the world, a new dot appears marking where people are striking, at which location exactly, and who the contact person is for locals. An English Twitter account called Dormouse helps to keep a record of all locations. The map is like a treasure map, and like the fictional adventurer Nils Holgersson, we can use it to fly across the country and then the world, visiting individual rebels. It will continue to play this role until a year later, when a new website is created by an international group of young activists around Simon from Mynttorget, with the help of Chris in Berlin.

Week by week, new countries join. Bangladesh, Pakistan, Kenya, and so on. To the delight of everyone in Mynttorget, it suddenly seems that a "global" protest action is really coming. What was initially nothing more than an idea in the minds of a few young people is gradually materialising. In many countries, we hear in Mynttorget, it is dangerous to go on strike; partly because it is forbidden by the regime, as in Moscow and China, partly because in many African countries the education system is not set up in such a way that strikes are easily possible. In many countries, such as Bangladesh, Friday is a holiday for cultural and religious reasons. Working groups emerge. Fridays for Future is on a good path internationally, I think myself, but it is still so unbelievably vulnerable. On the new communication channel, Discord, which is really a platform for gamers, it is possible for anyone to get involved and then suddenly act like they have a central position in FFF and suggest things that could mess everything up. That is also because many of the young people who are otherwise very active, the most active in their countries, don't have time on top of that to read what is being said on an international channel and take part in the discussion. And every week there is a terrifying new report on the ecological crisis. The Keeling curve, which shows CO_2 concentration in the air, rises higher and higher, easily passing 410 ppm. Tipping points are already being passed: the permafrost is thawing, and the Greenland ice seems to be melting irreversibly, at least close to the sea (King 2020).

A homecoming, and a farewell

There is widespread relief when Greta finally returns from Davos and appears behind her sign. Everyone gathers in Mynttorget. The magazine *Spiegel Online* is here. So is the TV channel ZDF. A few more critical media are here too, including *Svenska Dagbladet*, and the *Spiegel* is taking an investigative approach and attempting to follow up all the rumours that Greta is trying to make money through her strike. That's all nonsense, I say again and again; the movement is held together by Isabelle, Loukina and all the others. In the international chats, preparations for the 15th of March are in full swing, and more and more young people and their parents are hearing about the plan. Brice, Benjamin and Andreas, the French-Luxembourgian-Danish musketeers, are deep in discussion: Should we be presenting a unified list of demands? Now that we have the attention of the whole world for one day and can coordinate with each other, what should we ask for? The young people in Mynttorget are sceptical: let's focus on the collective uprising.

On this strike Friday in late winter, I go back to the university. The corridors are silent. The students here and at the other universities where I've taught still have too little idea of the ecological and climate crisis, existentially, I think to myself; and they do not learn the skills which are so bitterly needed in order to build up a global, sustainable, fair society. The heads of the universities ought to act. How to change that? The idea occurs to me of forming a centre for sustainability, a prototype which could be adopted by all schools and universities worldwide (see the chapter on education). I myself am more sad than anything else. My limited contract at Stockholm University is running out and I have soon to stop working with my students, who have become important to me. Maybe I can continue working at the next-door Institute for Children and Youth Studies.

Far away, in countless German-speaking lecture halls, academics in a whole range of subjects are having similar thoughts and are in the process of joining together. They will call themselves "Scientists for Future" – and they are composing a statement in support of the young activists. They have had enough of their institutions' passivity. Just in time for the global strike, the world is to hear about the most important research.

The conflict over the trip to the EU parliament in Strasbourg

The weeks of the global uprising have come. And at one focal point of this uprising are those who could change everything, the politicians in the parliaments.

We are standing at the edge of Mynttorget in a group of fifteen people: on the one hand, climate scientists from the Stockholm universities have gathered, and on the other, there are the environment and climate spokespeople of all the political parties in the Swedish parliament. After half a year of strikes on their doorstep, they seem to feel that they ought to take a stance.

In the centre stands a friendly, politically active climate scientist, a leader in his field: Douglas Nilsson, who had visited the strike already in the very first August week, and who is now speaking to a green liberal politician. "Yes, we welcome your research and also this strike," says the "centrist" politician. That's what all the politicians say who come by here, and then they take selfies. "But you're not even listening to what our research says; you have to make massive reductions in emissions, right now, every year. Where's the legislation? That's your job." "We would like to, but in politics it's necessary to make compromises; that's democracy." That is the main argument which is made again and again. It is also made by the green climate minister and deputy prime minister Isabella Lövin, who sometimes comes by Mynttorget. "I would like to, but..." The climate action plan presented by the left-wing and green coalition government soon afterwards is criticised by representatives of all universities (the climate advisory group set up by the government) as entirely inadequate; particularly because it does not even make clear how much emissions have to be reduced by, or which measures could contribute to this and how.

In Switzerland, the climate activists demand that the government present a climate action plan for "net zero 2030". Together with scientists and civil society, they begin to make such a plan themselves, for all sectors including transport or energy. But the social democrat president Sommaruga also only signals hesitant agreement: "We have to make compromises. Do you want to undermine democracy?" she asks. "We want to strengthen it," reply the young activists. "Across generations."

But then the letter arrives from the green and social democrat MEPs in the EU parliament, sent to all European national groups of FFF. "You are cordially invited to join the climate discussion in Parliament on the 13th of March. We will pay for your travel and a hotel – 20 000 EUR." How should we react, the young people ask. The whole strike is directed against these very MEPs. For many of the young people, the trip is attractive, especially for those who have never met each other in real life. But isn't this a kind of takeover move – a potentially damaging one – some of them ask. The whole movement has the force that it has because it comes as the voice of a generation. Not as the youth organisation of one or two parties. Wouldn't the press have a field day: Oh, look, this FFF thing is just a youth group of the greens or the left? Some of the young people try everything to persuade the other countries not to travel there; and particularly not to fly. The mood becomes increasingly tense. But this question about visiting the EU is only the tip of the iceberg. In February and March, there are more and more challenges. Anything else would be improbable. A new democratic form of cooperation is supposed to come about, globally.

But now, before the big strike day, such questions have to be decided. Can someone – or individual important groups – "represent" the whole movement? That idea is rejected. How can joint statements and texts be composed for newspapers and press releases? The most active young people now know each other better and better; they have been in a state of emergency for weeks, and they can argue and resolve their differences again. The easiest question is about the shared logos, the round green logos designed by Sophie in Germany. But already when it comes to the open letter to the Guardian, which is being worked on mainly by the Swiss activists with an international group, it becomes difficult. Who should sign it; can anyone speak for the whole of Fridays for Future? A heated discussion takes place on all the platforms and in all the chats.

The strength of FFF seems to me in these days to be the fact that the national groups have retained their specificities, including the different names such as "Klimastreik" and "Youth For Climate", and so on. A varied mishmash of "bottom-up" initiatives. And that these national groups are just one actor. There are also lone fighters who are active in the chats, and groups which include different countries. Keeping all the variation and still managing to forge unity; how is that possible? The best option is to talk to each other, all the time. Should FFF accept the EU invitation? Who can go? What is FFF?

A Zoom meeting is agreed to decide this. Even arranging the meeting presents democratic problems. How can everyone be informed; how many people can take part; are all countries represented roughly equally? Online, there can be a strange kind of group panic which would never happen at an analogue meeting. Who is paying for the invitation: the green and left-wing parties, or the whole EU? When it becomes clear that it is the whole EU, most people agree with the idea of the meeting. And soon, Isabelle and Tindra from Mynttorget, and Andreas from Falun are on their way to Strasbourg. They get stuck on the night train in Hamburg because of a storm, and only just make their connecting train.

The first international FFF meeting of sixty climate activists from across the EU is a success; so it seems to me. A protest action "against" – and not "with" – the MEPs is the focus. Generally, this remains the basic attitude: FFF does not cooperate but acts as an emergency brake, reminding people that real political change is needed. And so, the first important bonds are created which carry the European movement through the next years.

The sensation – the founding of #ScientistsForFuture

At exactly the same time as this journey, something happens which has probably never been seen before in the history of the universities or of academia. Thousands of academics across many disciplines, universities, cities, and countries agree to basic principles for grassroots cooperation in the future – and a central statement (see scientists4future.org; and Hagedorn et al. 2019). Scientists for Future is formed. "The concerns of the young people who are demonstrating are justified." They are not exaggerating. The media and politicians, who say something different, are not basing their statements on currently recognised global research but are distorting reality. It is serious. People everywhere have a right to hear this: without unprecedented drastic measures to make our societies sustainable within a few years, it is impossible "to limit global warming and halt the mass extinction of animal and plant species."

It is not just a few German climate scientists who have agreed on this document after weeks of hard work (also with the help of Jörg from the group in Mynttorget), but leading scientists and the directors of almost twenty of the most important research institutes across the world: from Philadelphia to Manchester, Potsdam, and Zurich. One of the most respected climate scientists, Joachim Schellnhuber, describes what is happening as follows: "The solidarity between scientists and the young generation in the struggle for a new society with a sustainable economy and way of life is as powerful as the big bang." The alliance between young activists and scientists is announced at a press conference at Tuesday lunchtime – three days before the global strike. The young activists Louisa and Jakob sit next to professors from S4F (Scientists for Future). Already at this point, more than ten thousand scientists from all disciplines have lent their support to the statement and to the children's global strike.

On this Tuesday evening, while the Swiss activist Marie-Claire is speaking on Swiss television about the difficult state of the planet, and while Isabelle, Tindra, and Andreas are travelling from Mynttorget to Strasbourg to meet their fellow strikers from across Europe and are stuck on their train in the storm, academics look back proudly at the day and think about the question of what a university's task really is. Are researchers allowed to get involved? Isn't that dangerous for the scientific status of their research? What is the role of students and academics?

A lot of very different points connect the team around Gregor Hagedorn at the Berlin Museum of Science when S4F is founded. And none of them can be taken for granted. What is actually happening in strategic terms? First, the idea of a statement is the focus. Later, there will also be a charter, with values and a description of the meaning and purpose of the grassroots movement in which all academics worldwide can participate from this moment on.

But now, at the beginning, the core of the S4F-movement is about presenting the best and most reliable research. This statement is published in April in one of the most important scientific journals in the world (Hagedorn et al. 2019). However, at the same time, it is also distributed across the whole scientific community. Not only climate scientists should be able to stand behind it, but everyone who is open to the results of fundamental research. This creates a sense of momentum, because now they can be joined by people who have been working for years in all different areas to research sustainable transformation (in the realms of energy, transport, agriculture, digital technology, finance, but also: politics, education, economics). Local groups are formed, mirroring those of FFF, and email lists are made with hundreds of participants who can be reached at any time regarding science communication, debates, but also events. As a grassroots movement, it does not have any formal hierarchies, although the coordination team which is soon formed and the advisory board with around a hundred members do both play important roles.

As a central point, the press conference on Tuesday also establishes a further principle. This is not about presenting research results from different disciplines side by side. It should mean that there is finally action on something the university seems to promise in its very name – the possibility of seeing and conceptualising all disciplines together, as part of a system. The assumption behind this: we can only respond to the ecological and climate crisis as a crisis, stop emissions in such a short time and delay global warming if we are able to see the larger context and connect the specific details (solar energy; plant-based diets and so on) with these overarching contexts.

From the beginning, the project of S4F is therefore a project of the university itself: the knowledge of the humanities on theories of global justice and intersectionality should be taken into account just as much as environmental and climate sciences. That is the reason why an economist, Maja Göpel, sits on the stage in front of the media, beside Volker Quaschning, a professor for regenerative energy systems, and Eckart von Hirschhausen, a doctor. It is about marking out the science-based framework in which politics must immediately find its way. How exactly this should look in concrete terms is still up to the democratic interplay of all the other actors involved. A vision emerges of a new, better, more sustainable interaction between science, politics, federal departments, the legal system, and the population.

In the ensuing months, my "double culture" comes in handy. On the one hand, I can establish a direct connection with the young people in Mynttorget. On the other, the German-speaking research institutions represented on the board are familiar to me, and cooperations emerge reflecting this (see the discussion of the EU climate law in the chapter on corona). What I hardly succeeded in doing six months ago in Sweden with the S4F project – or only to a very modest degree – now develops at an entirely different scale, to our delight in Mynttorget.

At the same time, I do feel a certain amount of scepticism. Aren't there also researchers who are not guided by FFF's compass? Who prefer to talk about 900 Gt rather than 420 Gt as the emissions budget in 2018; who do not want to take note of any criteria of justice when calculating budgets – and all because they appeal to the idea of what is "politically and realistically" feasible, rather than what is truly feasible and what the Paris Agreement demands on democratic grounds. Didn't the young people begin their strike because of precisely that attitude? And this criticism is not only catalyst for the new grassroots movement, but also that for the arguments over its direction (on this, see the chapter on Smile For Future).

How can we guarantee at all, I ask myself, that we, as S4F, are not going to ignore the young people in the name of some kind of "realism"? The scenarios have to be possible, but who can say what "realistic" politics is? We are in a crisis. And Scientists for Future, which soon has more than 25 000 participants, only exists as a huge academic movement because those five regular strikers have been returning to Mynttorget every week for months, through ice and rain. We owe it to them to make sure a "principle of caution" guides us, and also to create intergenerational justice (on the concept of intergenerational justice: Bidadanure 2021 and Wolff 2022). It is about reducing emissions now, with university science as the starting point. And because politicians are not doing that at all, the children are on strike.

And us researchers? There are weekly climate strikes by researchers in a few cities and in front of universities, and from Mynttorget we share their tweets with the world: in the long term, we need more than just statements.

The newspapers are now writing almost every day about the new environmental movement. My gaze wanders to the "global map" on the Fridays for Future website. More than one hundred countries have announced that they will be taking part in the strikes. Including China, Russia, Cuba, and Svalbard. A few are in contact with researchers in the Antarctic. Then all five continents would be on board. The whole world, united, for a fundamental change in the way we live together.

It has to be possible to stop the absurd global warming, to prevent the death of thousands of animal species and to ensure that there won't be struggles for food and water among millions of people already within the lifetimes of those who are preparing so feverishly for Friday. The idea of a global political movement, which has been on my mind for years, comes up again. Isn't this the first step, what's happening now this Friday? Global solidarity between more than 150 countries, supporting FFF's basic demand: keep to the Paris Agreement. Maja Göpel, the academic from Scientists for Future, implied at the press conference on Tuesday that a new political movement was needed. And we have been working here at the university for years to outline a new perspective on the economy, health, education, and the energy system. FFF is the young people's movement. Can we as Scientists for Future – together with groups of Psychologists for Future, Artists, Workers, Teachers for Future and so on – come together as People for Future? So that everyone can join us?

Chapter 7: The Uprising
March – April 2019: The first global strike,
and the London occupation

The 15th of March – the day of the global strike

The day starts undramatically in New Zealand: "Wellington is on the march" says the brief message on Twitter at eleven in the evening, when it is still the 14th of March in Europe. Smiling faces of primary school children. Then: Australia, easily more than 100 000 people are out protesting. The films from Sidney and Melbourne are impressive. Soon afterwards: Kathmandu, Nepal: hundreds. That must be the highest-altitude strike. It comes closer. The earth turns.

And in Stockholm, a hesitant rain is falling. How will the day be? Have the ones who travelled to Strasbourg even returned safe and sound? While we receive the first photos from South Africa at around 9 in the morning, a few activists from Stockholm are on the biggest newspaper livestream, "Expressen," giving an interview on breakfast television. The evening before, they arrived back at midnight, happy after their EU demonstration. Now, on the morning of the global strike, they look out from the TV screen, just like their fellow activists from Mynttorget, who have been invited to the biggest private TV station, TV4.

"You're really worried?" asks the TV presenter. "Yes. If politics doesn't change, our life will become a catastrophe." "And you've been on strike every week?" asks the co-presenter in admiration. They are suddenly – as they are again later on-stage and backstage – the ones who are looked at by thousands of their peers not just in recognition but in excitement: heroes who are reported on in the media and whom children talk about in the cafes where they go to warm up with hot chocolate and cinnamon buns on the day of the strike.

Week after week, they sat on the ground and hundreds, thousands of people just walked past them.

But in Mynttorget it is still quiet. A few more TV teams and journalists are there than usual, and the first school classes also appear amazingly early. Slowly the square fills up, it will be midday soon – and everything is going wrong. On the posters, the young people had written: we will meet in Mynttorget from 12. But a few of them had the idea a week ago that the high schools of the city could march in a star formation to gather on the main square in Stockholm, the Sergelstorg. Then they were to march onward to Mynttorget, which is close by.

Shortly before 11:30 some of them set off, curious to see what is happening in Sergelstorg. Are the school classes actually leaving their classrooms? No one has really organised this part, or even thought about it. The Drottninggatan runs directly from Mynttorget to Sergelstorg, through the parliament buildings, past the government building known as the "Rosenbad", and then past the education ministry and up to Åhlens, the biggest shopping centre in the city. Microphones and cameras jostle each other in front of Greta, Ell, and the others.

The Drottninggatan is half empty. The only obstruction is the press – until they all walk round the corner and stand in the middle of Sergelstorg. There are not eight school classes there, but ten thousand young people, chanting and yelling. All of them are just going forward in a kind of trance. And soon they are standing up on the Gallerie in the centre of the square. Chaos reigns, just as it does on other big squares across the world in Manila, Milan, Paris, Kampala, Berlin, Bern, Sidney, Buenos Aires, and Tokyo where Loukina, Mitzi, Sommer, Bianca, Nicki, Erik, Saoi, and Dylan are gathering with their friends.

Days ago, years ago, they thought that they were the only ones who were concerned, sitting alone in their classrooms or at home; haunted by nightmares of a burning planet. Now, they realise suddenly that they are not alone, and haven't been. That there are many more who care.

They stand still. Everything is blocked. For a second, it is as if time stands still and the sound diminishes.

"What should we do now?" "We should just go back and take the rest of them with us", someone suggests. "We can't go back. The police won't allow it." The only other possible route is a long detour through the Kungsträdgården, the most popular and biggest central park in the city, and then over the bridge in front of the palace, round the back way to Mynttorget. That means passing the junction which we blocked five months ago with Extinction Rebellion.

Many have their "There is no planet B" signs with them. And they will yell back when the young people half an hour later shout from the stage, "Keep it... in the ground. Keep it in the ground" – referring to coal, oil, and gas. That is what they care about, this whole generation. They know that ultimately coal, gas and oil must stay in the ground. In Germany, but also in China, Venezuela, and Saudi Arabia. And that politically no path has been established to guarantee this in global cooperation, even if UN General Secretary Guterres tweets that night that he is going to organise an extra meeting in New York in September because of the global school strike, to guarantee that the Paris Agreement is followed.

Hours earlier, when the day began, the young people sat on the bench in front of the wall they all lent against in September, when it was so absurdly warm in Mynttorget. Swedish TV had finally arrived. "What do you propose?" the reporter wanted to know. "How should things continue after the strike? What should happen next?" "Everything has to change." "If we take the Paris Agreement seriously, everything has to change." What has been legitimised during entire lifetimes, the exploitation of nature and of other people, all of that has to stop, I think, it has to be seen as irrational and violent; the soil has to be ploughed sustainably and animals cannot be eaten. People cannot get rich from this violence. It is as if the whole of humanity has to wake up from a dream.

"There are a lot of people here." Greta looks at the masses of people in front of her. It is now after twelve and the march has finally arrived in Mynttorget. There are more than 15 000 children and young people standing there and listening to the beginning of her speech; in exactly the place where otherwise on Fridays about fifty people stand quietly talking to each other. Next to her on stage: the whole Mynttorget group, the "original crew", as they are now called. Many parents describe weeks later what the day meant to their children. "First I would like to thank you. Thank you for coming." Greta turns to the Stockholm young people who have dared to strike. They are supposed to be at school. Along with them, on this day, 1.6 million children and young people in 125 countries on this planet have left their schools. "Thank you to those of you who are standing on this stage." In her yellow raincoat, Greta looks at her companions, who have been by her side for seven months. "And thank you to everyone who has helped to organise this strike." Then a short pause for breath. "We children did not cause the problem of the climate crisis, it was just there. [...] We do not accept that! We don't let this happen! Therefore, we are on strike. And we will continue!"

The global Facebook chat is completely overcrowded. Everyone wants to know how it's going; new numbers are coming from countries we've hardly heard from before. Chaos breaks out, the Excel tables have to be started again twice. In the middle of the night, everyone in the main chat agrees that there is reliable enough data for us to use the figure of 1.4 million when talking to the press; later, this is corrected to 1.6 million. Paris and Madrid – around 50 000 people. The police confirm it. Montreal and Milan – around 100 000 children and young people... Jonas is there with the Swiss numbers, Luca with the German ones, Janine and Benjamin are trying to keep track. When we pass the threshold of a million, at around ten in the evening, the news spreads like wildfire; also to the strikers in all their suburbs in Stockholm. And everywhere in the world, children and young people are going to bed late, full of experiences which have changed their lives. They have power. If they seize it.

Looking at the map with the thousand cities on all continents, I feel for a second, how a world could be beyond all structures of domination, a world where everyone, every child, is treated as equal and free. And then, at the same time, it is clear: all of this makes no sense if we older ones do not react, tenfold, and reflect back to all of the young people what we have seen in their hearts and heard from their lungs during this day – and act.

"Everything must change" – a new foundation for the UN charter

"Everything must change." On the morning of the 15th of March, two articles are published in The Guardian. One of them is by George Monbiot, the climate columnist. The other is by Greta and the seven or eight well-known European faces of the climate strike. Both agree that there is a basic problem with our society. Rather than caring for and sharing nature and its resources, we regard them as the property of competing individuals and states, as property which is also defined as an object, or as Monbiot expresses it: as something we can use however we want. We can burn it. We can waste it. We don't have to care for it.

But the gap report by the UN (UNEP Production Gap Report 2019) shows that within this political approach and setting, so many infrastructure projects (coal power stations, oil towers, refineries, gas pipelines etc.) have been set up and planned on a contractually binding basis that in the next ten years these projects will already produce double the greenhouse gases allowed by the Paris Agreement and the IPCC-SR-15 report. How can we break away from this path to a nightmarish world that is three or for degrees warmer – a path which has been legitimised by Swedish, German, and Swiss governments? All of us have to gather around a table, I think on the evening of the strike day, in order to stop the fossil projects together and replace them with renewable projects. But this depends on a new way of thinking, a new way of seeing nations and property, of defining resources and nature, for all of us. In his short article, Monbiot refers to the ideas of John Locke, who still shapes our modern economic system today (see also von Redecker 2021): that nature can be private property, according to Locke through and for those who first claim it and "mix" it with their own work, by cultivating the soil, as Monbiot says, or by having slaves cultivate it. So it becomes our own; it becomes a thing under our control. More or less every one of the 1.6 million young people who are walking through the streets of Kathmandu, Melbourne, Stockholm or Rio will find this idea absurd, I think; at least the ones I talk to. It is a philosophical construct.

The intuition which many of the activists probably have is this: nature is something we find ourselves in; we can prune its trees to improve their fruit, and we can harvest the fruit, but nature is not a product. The modern concept of property is hardly appropriate when we think about forests, fields, and mountains, let alone the inside of the earth. If there was any message on this day, I think to myself, it's this: we are a population who depend on each other on a single living planet, and we want to look after each other in such a way that we can live together, as human beings.

Chapter 7: The Uprising 153

The strike is really becoming global. Suddenly young people from all countries are standing together in the streets at the same time, close to their parliaments: in Europe, America, Australia, and also in countries in the Global South, which was exploited for centuries by colonial powers, dominated by western elites (see Margolin 2020, Hickel 2018), which extracted their knowledge, their work, their mineral resources, and, you could say, their humanity. It is possible and necessary that the populations of this world can come together and see all the riches of nature as something shared, that is the idea that now comes to the fore (similar: Dixson-Declève et al. 2022); and not as abstract property that can be used. There would have to be a global contract, I think to myself, setting out how we can leave oil and coal in the ground. Then there could be a new article, article "zero", in the UN charter, I go on imagining the utopia. Of course, you'd think that the general assembly and the security council would never allow a change to the foundation of the UN. But the basic premises have changed as the Amazon has burned. Even for governments, the climate has become one of the most important security issues (Hardt 2020).

After the unimaginable horrors of the Second World War, when the world picked itself up and set about creating a shared foundation, the UN charter was created, meaning the basic understanding, shared by all humans, of how states should behave towards each other, as a basis for what is legal and what is not. What was important and sensible at the time now reads differently. A crucial dimension is missing. Article 1 states that the global community must be organised in such a way that there is peace for everyone. Article 2 then explains that every state is responsible for its own territory and that everyone must respect this territorial integrity. Clever, when you think about the wars of aggression and about colonialism. This purely "negative" concept of our shared life on this planet, defined by the idea that we cannot interfere in other people's affairs, reaches its limits when states can cut down forests on their own territory and burn coal, disturbing the whole planetary earth-system; and when we consider our shared history. The next step is to put this idea into practice (see the chapter on a new global order): to sit down together for a conference and make a binding agreement as "one people", defining which stores of coal, oil and gas have to remain in the ground where; how we can work together in a fair way to reduce emissions by more than ten percent per year; and how we can provide enough resources to everyone so that they can live a good life – this would be a possible program under the heading, "We, the people… for future."

We shouldn't be allowed to take the toxic substances out of the ground; all of us are convinced of that. For that, we need to have some sense of a world community; and precisely that is embodied by the global unity of the climate strikers.

The occupation of London

In the week before Easter, everything comes to a head again. After almost a year of preparations, the "rebellion week" arrives, the uprising for which Extinction Rebellion was founded last summer. Now it is not just about a demonstration or a strike. It is literally about an uprising, with all the consequences. The centre of London is to be occupied, disrupting society in England until the government responds to three demands: telling the truth about the extent of the climate crisis, setting the goal of zero emissions by 2025, and doing so democratically through "citizens' assemblies". This form of uprising by a large section of the population for future generations, prepared for years, has probably never been seen before in a western democracy. Part of the population is prepared to go to prison to insist on a different politics.

In Stockholm, we know which protest actions are being planned in the centre of London, and we study the map to see where they will be taking place. Four critical locations in London are to be blocked, and above all permanently held, for days or weeks, so that the population and then the government will have to react: Oxford Circus and Marble Arch as the two centres of the shopping district, Waterloo Bridge as the central axis, and above all Parliament Square, right in front of the centre of power. For that to work, thousands of ordinary people must be prepared to block these places peacefully with their bodies and be arrested for it. No one knows how the police and government will react. How will the rebels be able to hold the open space in front of parliament, against hundreds of police officers?

We have also prepared ourselves for our blockade in Stockholm, which is taking place at the same time. There is to be a smaller, twofold action, straight away on Monday, at the same time as the beginning of the London occupation. First, there will be a "die in", with everyone lying down as if dead in the middle of the parliament passage, and then the parliament bridge will be occupied and held for at least a day, even if there are arrests.

At the same time, Greta makes a speech at the EU parliament at the last session of the climate committee before the elections. She calls on everyone to take part in the coming elections, but also points out in doing so that none of the parties has an approach which really takes the Paris Agreement seriously. She has to bend over the microphone to speak. After a few sentences, she begins to list everything our civilisation is destroying, from the forests to the animal species which being exterminated by humans. In her voice, there is sadness, and a huge sense of grief comes over us who hear it. The regular strikers in Stockholm begin to write a message, sending love and support.

Many of them and even of the older activists have a sense of despair in these weeks. Some also call it eco-anxiety: not just a feeling of fear for all that will happen in the future; not just nausea because the older generation doesn't care, breaks the attachment and leaves them alone; not just a feeling of being powerless, but all of this combined into a diffuse anxiety, as if a black hole were eating up every joy and the trust in humanity. Tori Tsui, a young climate justice activist from Hong Kong, points out how important it is in such situations to find a community that can offer support (Tsui 2023). And I think again: it shouldn't be the young people who have to fight. It should be the generation that is now standing up and leading the fight in London.

We watch eagerly, following Twitter and Facebook to see the Londoners taking over the squares. Everything unfolds surprisingly peacefully (Taylor et al. 2019). At Marble Arch and Parliament Square, there are people dancing. Many of them have brought plants with them and turned the depressing tarmac designed for cars into a colourful living room. Musicians soon arrive, especially in the evenings, including classical performers, and begin to play. Oxford Circus blossoms into the centre of the speeches: a pink boat is driven along and screwed down in the middle of the street, the "boat of truth." The most difficult part is defending the four entry roads onto Parliament Square against the police, who keep trying to clear them.

While we set up our blockade on the bridge, we hear about the first arrests in London. There will be many, hundreds, day after day. London's centre is becoming a battleground without violence. Parliament Square is surrounded again and again by endless chains of police. But it still doesn't fall, not on Monday evening, not on Tuesday, not on Wednesday. The lawyer Farhana Yamin, who was one of those who worked on the Paris Agreement, glues herself to a building so that she cannot be removed (Mathiesen 2019). For a day, in Stockholm, we hold the bridge which connects parliament with the government building. Most of the young people who are striking for the climate want to take part in the blockade, even though they know that they can be arrested. The adults from the Mynttorget group stand on the side facing the parliament, with Mynttorget in front of them, and next to them the Mynttorget young people with about a hundred rebels behind them.

In front of us are the police, and for a long time it is unclear how they will react and whether they will arrest people. We are standing in an area of the city which is regarded as requiring special protection, meaning that the police have extra powers. We keep the bridge despite all the uncertainty and fear, and after an hour in which we let a few passers-by through, the young activists block it completely for the rest of the day.

And that is how a key moment comes about: the spokesperson for the green party on climate policy is literally standing in front of us, separated from us only by a huge banner. The parties of the red-green government, dependent on the neoliberal parties, promised before the elections some months ago that they would implement the Paris Agreement. But now, they talk about "net zero 2045" as the goal – when we all know that the CO_2 budget for 1.5 degrees will be gone around 2030. And on top of this, the government's own climate council with around ten universities represented says that the established policies are not at all sufficient to reach this very inappropriate target of 2045 (Klimatpolitiska rådet 2019). Everything is wrong here; it goes against the fundamentals of democratic institutions and processes. If we know that 2045 is too late, we have to change the agreements even if these were made years ago by most of the political parties. And again, the most vulnerable in Sweden and globally, the children, are affected the most.

Should we just observe all of this, we ask the green spokesperson. And we say: it is legitimate to use methods of nonviolence to get the society on the road to a sustainable future for all. Many more should join. During the next four years, the emissions must already decrease drastically. We are getting nearer and nearer to more tipping points. But emissions are not going down (Urisman Otto in Thunberg 2022). The Swedish banks and pension funds are still financing the fossil industry. But even after the discussion with the climate policy spokespeople during the blockade, it is unclear to all of us how Sweden is in any way thinking of reducing its emissions by ten or twelve percent per year; how the ten tons of CO_2 per person could decrease to two. Flying, driving "fossil" cars, eating huge amounts of meat, cutting down the forests, investing in oil shares – barely any limits are placed on any of this by political means; nor is there any debate how this can be done with justice and care for the ones with the least resources. The rich who produce the most emissions are even supposed to be receiving a huge amount of tax relief in the next four years. Why should people in China or India stand up for more climate politics when they see the inaction in Europe?

At the same time, it becomes so palpable that those who are intervening in the power relations with their bodies or making speeches about biodiversity at the EU parliament are vulnerable. There is no point imagining how we could shake off that vulnerability. That is not possible, and it is not a path we can take. That is precisely the path taken by those who are bringing the world to the brink of destruction: they build tanks, barriers, walls, weapons, SUVs and try in vain to hide their "condition humaine". We are endlessly vulnerable and reliant on each other, on the care of others for our vulnerable bodies when we are small or very old, on help, on encouraging words. That is what Kierkegaard described as "absurd": that we humans live our lives and even love other people and animals, wanting to build relationships, even though we know that at the next moment a branch can fall on our heads. We have to learn to deal with that. The society, the economy which we must build together in the next ten years will have to be founded on that, on the realisation that we need each other.

Some days later, in Rome, Greta focuses on this in her speech: for decades, some powerful and rich people have accepted the fact that the whole environment is being destroyed and that thousands of people are dying – in floods like the ones in Mozambique a month ago, in droughts like those in Syria, in storms and wars for oil. In Paris, Extinction Rebellion occupies the "Défense," the area where the headquarters of the big banks are located, as well as oil companies such as Totale.

In London, meanwhile, the police are acting ever more decisively against the protesters. Emma Thompson, the Oscar-winning actor, stands on the pink boat in the middle of Oxford Circus and encourages the rebels to stay, surrounded by a circle of police. In addition to this, the whole square is now surrounded by a further circle of police in yellow. Kate Raworth, the economics professor, tweets: hey, they're forming my doughnut, the shape of her vision of a fair and sustainable society. On English TV, Rupert Read, a philosophy professor at East Anglia University, is describing how serious the situation is. A week ago, George Monbiot was on television calling for the abolition of this system which stuffs more and more into itself because it is aiming for exponential growth – how is nature supposed to bear that? And in the evening, David Attenborough's documentary film is shown, which begins with him visiting us in Mynttorget, and which shows clearly how dramatically the situation has already deteriorated.

We all breathe out slightly. It is possible, after all, to spread this information in the media. But in contrast with this, the rest of Europe says nothing at all about the dramatic events in London. The Swedish papers don't report on them. The German papers barely do so. Where are they now, ZDF, *Le Monde*, all the familiar journalists who have excitedly stopped off at Mynttorget? Why are they silent now, when a real conflict is happening and it's not just a fuss over celebrities and children?

Again and again, the occupation threatens to collapse, especially in Parliament Square; on Thursday evening it is especially dramatic. But then suddenly a line of cyclists appears with the XR symbol on their backs. And when everyone thinks that the square is being completely cleared, a group of drummers marches into view, hundreds of people, reinforcements. And the police have to start again with their arrests. The protesters do not give up the square, for the whole of that first week. They hold Oxford Circus too, without the boat now. More than a thousand people are arrested (Swenson 2019).

Here, too, on the bridge we're blocking, the situation comes to a head. The police announce that they have the power to arrest us immediately. We take this in and say that we need to discuss it, to the astonishment of the police. And now the process starts which has been such an outstanding feature of FFF and XR in the last months. In the individual small democratic groups which we have formed, heated discussions take place and a consensus is sought and voted on, using the thermometer system: hands in front of you means "I don't know," hands up means "yes," and hands down means "no." A delegate is sent to the general meeting. None of this can be observed by the police, because we have turned the banners into a wall. We delay the decision. But then the band of rebels from Mynttorget decides to go all the way with the blockade. They lie down in the street as a human carpet, and the police back off. The young people are the ones who behave most bravely, with the greatest solidarity, as a group.

In the UK and other countries, polls show that by the end of the week, people have changed their minds in enormous numbers, and now see the climate crisis as the main problem, and protest as legitimate (Kountouris/Williams 2023). The UK government declares a climate emergency.

The results of the Australian elections: against all predictions, the centre-right parties win. They are the ones who want to continue with coal production and expand it, and this makes it possible to build the Adani coal mine, one of the biggest fossil infrastructure projects worldwide which will affect the global climate in the next decades. The parliamentary discussion about this project already provoked the first international strike back in November. At the same time, temperatures of 49 degrees are recorded in Australia.

Part Two: The Adults Respond

Contents

Chapter 1: The Second Global Strike and the Preparations for the Week for Future
April – August 2019: The task of civil society

Chapter 2: Smile For Future in Lausanne and Scientists for Future
August – September 2019: The task of science

Chapter 3: The Week For Future
September 2019: The coordinated uprising of eight million people across the world

Chapter 4: COP25 in Madrid
October – December 2019: How can we end our fossil society in a fair way?

Chapter 5: Corona, #BlackLivesMatter and the Climate Justice Movement
January – September 2020: The crisis and the intersectional, sustainable, global democratic project

Chapter 6: Many Fights, One Heart – #UprootTheSystem
October 2020 – October 2021: Forests, agriculture, banks, courts, and the political manifesto – all the way to a new theory of democracy

Chapter 7: The Idea of Social Movements and the Journey to Glasgow – What Is the Right Way to Live?
November 2020 – December 2021: On a theory of democratic grassroots movements that can change the world

Chapter 8: The War, Fuel, and the Global Social Contract
January – August 2022: Towards a new world order to lead us out of the crises

Chapter 9: Education in Times of Crisis – Learning from Young People on the Way to "Centres of Sustainability"
January – August 2022: How to change schools and universities – (eco)philosophy for a sustainable democracy

Chapter 10: Global democracy, the Elections, and the Future
September 2022 – June 2023: On forming a global, democratic, sustainable postgrowth society

Chapter 1: The Second Global Strike and the Preparations for the Week for Future
April – August 2019: The task of civil society

The first big wave of rebellion is over. While a certain exhaustion spreads among the activists, the movement is now really spilling over to the wider public and changing political events. Influenced by the global strike and the actions by Extinction Rebellion, by Greta's speeches to the EU and to the British parliament, but also by the new BBC documentary film with David Attenborough on the climate crisis, governments and individual cities begin to declare a climate emergency. City after city follows suit in Switzerland and Germany (starting with Constance), and then globally, under pressure from the young people's climate strike. A large part of the movement's energy is currently flowing into this project; new "climate emergency" Facebook chats are formed. Statements on the emergency are also formulated with concrete proposals for change for schools and universities (see https://www.cedamia.org).

In Mynttorget, the realisation is taking hold: something fundamental has changed. The question is often no longer whether there is a climate crisis, but how to respond to it. The global group of activists has succeeded in making a first crucial step, a historic breakthrough. Right? The EU elections are coming up in May, a measure of whether European populations are really taking the project seriously. The young people cannot vote. The hole in our democracies can hardly show itself more clearly than it does in these weeks: it is their future which is at stake, and yet they have no formal say. And the task which the Mynttorget group took on last September has not been completed. The newspapers report every day that the Paris Agreement is scarcely being upheld by any country. We read in the UN reports that the global community is heading for a world which is three degrees warmer (Emissions Gap Report 2019).

The basic question is still: how can the world be united in the struggle for a safe future for everyone? How can we establish political rules worldwide which will force societies to transform every sector, in such a way as to establish greater justice and more democracy, and to dismantle the structures of domination? The first global strike was a historic step in the right direction, many people think. But what comes now?

One thing is clear: the adults are needed. They can no longer stand on the margins and applaud; and broader groups in civil society must take action, not only those who have already been committed to protecting the environment for the last forty years. But those are the ones who are now being talked about. In May, those who are globally responsible for the big environmental associations and NGOs begin to interfere in the situation. The question becomes more urgent on the day of the second global strike and a risk emerges, as well as an opportunity: the idea of a much bigger global "movement of all movements" comes up for the first time, one which can be joined by all people everywhere, of all ages. But this also means a threat (which will never disappear) that these NGOs will undermine the networks of young activists and make their democratic processes impossible.

It is early morning on the 24th of May, and birds can be heard twittering in the park. The whole group has gathered in a circle in one of the subterranean theatre spaces which are otherwise used by world-famous performers. The atmosphere is tense. These rooms are located under Stockholm's central park, the royal Kungsträdgården. Up on the huge stage, the second global strike is about to take place, because the EU elections are in two days.

"We have to stick together," say some voices. Everyone nods. What has happened? The previous day, the older "superstars" of the established climate movements, many of them linked to NGOs, published an article in *The Guardian* which has caused an uproar. This is not because of the basic idea: it is true that the young people are now asking the whole of civil society, including trade unions, to join them in September as part of a huge strike day. Older people cannot simply watch and applaud but have to take responsibility. But the NGOs didn't consult the young people properly before naming the 20th September as the date of a monumental global general strike. As would become clear, this was not agreed with "the" movement (that is difficult; there are still the discord platforms in parallel with WhatsApp groups and so on), but at best with a few individuals close to the NGOs. Most of the strikers had already agreed months ago on the 27th of September as the next strike day, together with the organisers of Earth Strike, who are all young volunteers, not paid NGO employees. And many countries had already communicated it that way: we're going on strike on the 27th of September, us young people, and we call on the unions to help us. Many adults had taken the day off or informed their workplaces.

And now they are suddenly there: the much bigger NGO heavyweights, announcing their own climate strike campaigns, even with a website of their own, which many of the young activists see as a provocation. It is their strike, after all, and they don't want to lose control over it, especially not over internal democratic processes. Many of them writing in the global FFF communication channels feel caught out when the message is quickly spread – on the night before the global strike, of all things. A kind of collective nervous breakdown sets in; and during the strike day, everyone is so exhausted that the events surrounding the EU elections pass many of them by. The unity of FFF is more important; all misunderstandings must be cleared up. The adults have to take their cue from the young people, and not the other way round.

But in these months, there is a sense of despair over whether the democratic autonomy of the movement can be preserved in the face of the individual, often male NGO workers in the Global North, who are trying ever more obviously to put together a controlling "elite group" within the youth movement. What happens in the next weeks and months represents a historic opportunity, I often think at that time, but it also brings the danger that everything the young people have built up during this year will fall apart.

In general, it is enough to change something small in a group, a small shift: adults only have to take control slightly more, and then many young people withdraw who used to be active. That's how humans are. Especially young people. That is the subtlety of democratic relations: that we have to pay attention to making sure everyone feels motivated to take part; and that they are also confident in doing so – and that everyone is heard; that there aren't individuals who express their opinions so loudly that that others retreat and think, "Okay, I guess that person knows what they're doing; what do I know about it?" This is the threat in the coming months, and a conflict takes place in which the young people ask the NGOs again and again to explain themselves – and I do the same. They can't actually make individual young people into spokespeople, or form groups whose composition they decide, or influence strategy so as to create more and more demonstrations instead of strikes and more campaigns than disruption.

How to find a balance in the cooperation between young activists and adults, between grassroots movements and paid workers? At the university,

I begin a research project and compare the ways in which the interplay between NGOs and movements including children and young people has been organised historically. Do we need new global guidelines? How might they look?

But from this day on, the question no longer gives me any peace: what would a globally organised uprising look like, including people of all ages? It wouldn't be a mishmash with unclear power relations, that much is clear. FFF should belong to the young people; they have to be in charge. In the dispute over the date of the next strike (the 20th or the 27th of September), I try to mediate between the two sides of the conflict and to give the young people the last word.

In the process, I come across an idea mentioned by some of them: connecting the two days with each other. A few of us, including strike activists, NGO workers and the Earth Strike organisers, write a paper: there should be a Week for Future, a gigantic global protest which governments will not be able to ignore. We can keep both dates, and soon we can all concentrate on the biggest of all projects so far. And still, we cannot possibly suspect that three months later eight million people will go out and protest. In Bern, where Loukina is, there will be 100 000. Despite all this: in the end, it can't just be about ensuring that adults join the strikes, which are becoming more and more like ordinary demonstrations – they also have to organise themselves into a political force which will make change happen worldwide.

And so, the work begins on planning the joint September strike, together with the established organisations and grassroots movements like XR on the one hand, and the young people on the other. In Germany, for example, the Together for Future platform is created, gathering the whole of civil society. Every week, on Wednesday evenings, everyone comes together in a global Zoom meeting to discuss the situation.

Will people dare to leave their workplaces? In Stockholm, the young people set out and go to all the important unions. Others approach the universities and the students. But what should be communicated as the content of this protest and strike week? What is it now about, one year after the start of the strike? It has to be about a dignified life for everyone, including future generations.

This is our chance, I think to myself, and send a paper to everyone: let us use this week globally to plant the idea in everyone's minds of a shared global movement with a shared goal, and make it clear that this is a global goal and a political one in the broadest sense. After all, some are already going on strike in thousands of squares in 160 countries, meaning almost every country; we have to make our unity clearer and establish a very small number of demands, especially now that the young people are themselves saying that they want us adults to be involved. We are a single world population on a very sensitive planet, and we have to care for everyone together. The first Friday of the Week for Future could be dedicated to this idea; the Saturday could then be focussed on the idea of a carbon budget and action plans, a global budget (of about 350 Gt CO_2) and a national one, so that it becomes clear to all people how incredibly quickly we have to stop emissions if we want to keep the temperature at a halfway humane level. The third day could then be dedicated to the aspect of "equity" or social justice, which would mean distributing these budgets fairly, so that the richer part world supports the Global South and no longer exploits it. And fourthly, big "citizens' assemblies" based on grassroots democracy should give everyone the opportunity to think about how society would have to be transformed, which rules are needed in which sectors (shifting food to non-animal products, transport away from cars and planes, the financial sector to more democratic structures and away from the financing of the fossil industry, energy towards renewables, and so on). Every day would have to communicate knowledge, give space for discussions, and highlight concrete images of a shared global transformation. All this ought to mean that existing power structures in which certain population groups are disadvantaged (intersectionally: gender, class, ethnicity, sexual orientation...) are replaced with non-dominant relationships. Democracy must be strengthened.

This idea for the demands relating to individual days remains a dream, however. Instead, a joint, simple paper is put together, sketching out the basic idea of the week and emphasising the two Fridays: the first as a school strike and the second as a general strike; the unions have to be organised, and halfway through, the UN special climate summit will take place with Greta in New York. We send out the press release, and the world begins to prepare what then brings millions of people out into the streets in September.

We cannot continue to strike and march for an undefined length of time, I think to myself, without establishing a global movement which is also there in the periods between global events. Otherwise, many people will burn out. With these thoughts, the summer holidays begin in Mynttorget. But some are preparing for a longer journey...

> In the EU elections, after the global strike, the German greens gain massively in support. Among young people under 30, they become the biggest party, with well over 30 percent of the vote. The European Parliament does not see the shift to the right which many were predicting. In Sweden, there are barely any political changes.

Chapter 2: Smile For Future in Lausanne and Scientists for Future
August – September 2019: The task of science

The meeting begins

So the summer begins, and with it, a further group of adults becomes even more active: the scientists. At the same time, the Swedish activists find themselves in Switzerland.

A gentle breeze blows from Lake Geneva across the fields of stubble and up to the modern glass building of the University of Lausanne. A few young people are standing in front of the main auditorium, still sleepy, looking down at the water. It is the beginning of August, exactly a year since the first strike, and behind the glass walls of the university building, more than four hundred young activists are scurrying through the corridors: the participants in the first big international FFF meeting. They have come from all European countries, including Russia, Ukraine, Afghanistan, and Israel, as well as from Stockholm. Loukina, who has just finished her final days at high school here in Lausanne, came up with the idea together with Isabelle and all the others during the Strasbourg trip in March. She runs the organisational team and finds funding for the 400 young people from more than 28 countries to cover their travel and their board and lodging. With a small team, she puts together a program, finds a core group to take over media work, and leads the press conferences. So many questions came up back in March, which are still burning issues for all of them in their individual countries: how should decisions be organised? When and how often should strikes take place? And: should there be global demands, European demands?

Temperatures are almost unbearable, and not only in western Switzerland. In country after country, the highest ever temperatures are seen since records began (Eumetsat 2019). A large proportion of glacier water will have melted away, and billions will have to flee because they live in places which are too hot and dry, researchers are now saying (Xu et al. 2020), already in the middle of the century, when Loukina and Isabelle are as old as I am now – if politicians don't reorganise the entire economy together with the people who are sitting next to us in the tram in Lausanne and listening in bemusement to the shouts of "climate justice."

At the university, Loukina sits on the stage. On the benches in front of her, they are all sitting, the heroes who are around her age, who for such a long time have only been able to discuss with each other via chats: Ianthe, Erik and Balder from Holland, who visited us in Mynttorget; Saoi from Ireland, Dylan from Scotland; Anna from England; David from Turin; Jakob from Kiel, who was one of the first people we noticed from the German-speaking region; Jonas, Fanny, Lena and the group from Zurich, and many of the young people from Mynttorget in Stockholm, surrounded by four hundred other activists who organise the strikes in their own countries.

A few hours earlier, the young people were greeted by the president of the university and a research team. Now the atmosphere is expectant. In the various rooms in the labyrinthine building, there will be parallel workshops and seminars on questions such as social justice, organising the communication channels on Facebook, Discord and Telegram, and the basic demands being brought to politicians. A "Declaration of Lausanne" is to be composed: which words, which principles, which organisational guidelines and which demands should – at least as a proposal from these hundreds of young people – become accepted and shape the European FFF movement from now on?

In the midst of the hustle and bustle, Isabelle, Greta and the other Swedish activists sit on the floor, deep in discussion. How strange, I think to myself, they are sitting here in a university which has placed its main building at their disposal – for the sake of their fundamental idea – for a week, surrounded by hundreds of their peers, who are discussing the details of Fridays For Future as a complex movement and planning the global rebellion in September. Some of the regular strikers are among them, who already joined Greta in the first week, exactly a year ago; could they have imagined such a development at the time?

I stand outside in front of the entrance and see, reflected in the glass façade, the lake and the French alps rising behind it like a picture on a postcard. Behind the glass, they are working hard, the young people who wrote to the professors, spoke to the university on the phone and organised the program with all its workshops.

In July 2019, temperatures are recorded in European countries which were barely thinkable before. In Germany, they climb higher than 42 degrees (before that, the highest ever temperature was 40.5). There are similar records in France (more than 40 degrees in Paris) and Belgium. The month of July is the warmest since records began.

The European network and Mont Pèlerin

My gaze shifts over to the Swiss side of Lake Geneva. Only a few kilometres away from the campus, really a stone's throw away, there was another gathering of politically interested people seventy years ago, who met and looked down at the lake with the will to change society, not only here, but across the world. They, too, discussed their concept of humanity and of economics, and they founded the notorious Mont Pèlerin Society. Many of the main participants were later awarded the Nobel Prize for economics, including Friedrich von Hayek, Milton Friedman and so on. They are the inventers and developers of the neoliberal model of society which has shaped life and economics in Europe and the world more than any other ideological movement in the last fifty years. State regulations should be abolished, most things should be privatised, accessible to a global market, exposed to competition between all people, with minimal social safety nets, if any; competition should encompass all areas of life and all areas of nature, too (Mirowski 2015). Nature is defined and used as a giant supplier of free raw materials and – implicitly – as a rubbish dump. Politicians since the 80s, spearheaded by Reagan and Thatcher, will listen to them, employ them as consultants, award them prizes and in doing so strengthen the oil, gas and coal industries, the car industry, agribusinesses like Monsanto and Bayer with their mix of monocultures and pesticides, and above all the whole finance industry; the astonishing development of Exxon, Shell, BP, J.P. Morgan, UBS, CS and so on would not be conceivable without Mont Pèlerin, the ideology and the legal framework of the corresponding politics. However, the greatest "achievement" of this elite network, this Mont Pèlerin group, was establishing the idea in the minds of the majority of human beings that there is no alternative to this economic and social model. It suits us, as humans, they say. It can be applied in all areas, even in schools and universities. It defines who we are, what is seen as valuable and what is not. And it has led to immeasurable wealth for a small number of people – as well as the systematic destruction of much of what these young people see as the most important things.

My gaze returns to the main auditorium. Greta and Isabelle are just standing up and joining the other Swedish activists to discuss the situation. Loukina comes over holding some food which she has been carrying around all day because she never has time to eat it. I think: this is the alternative model. Here at Lausanne University, with all the young activists, it is at least possible to get a faint idea of it. Instead of a collection of men without a scientific grounding, without media, with their own economic interests, in strict secrecy, here a predominantly female group of young people are hurrying around. They have invited everyone from the media, including the tabloids, and in principle, the press has free access to all events. They are well-read and are in constant contact with scientists. And they are just as determined to change our lives. They have neither founded an international association with statutes, nor do they have finances, nor a political program. But now they are searching for structures that could make them stronger.

Quickly, I go back through the corridors of the building and sit down at the back of the plenary meeting, beside the scientists from Switzerland and Austria. We want to form a team for Scientists For Future, a group offering scientific assistance. All the young activists' questions are collected and passed on to the existing network of 25 000 German-speaking scientists, who came together in the run-up to the first global strike. Several hundred of them reply to emails within a few minutes, including some of the most renowned climate researchers in the world.

But then the week begins, and with it the problems and challenges among both young people and scientists. Debates arise, existential ones, which basically relate to strategy. At heart, the question is whether society can rid itself of the toxic legacy of Mont Pèlerin. And beyond that, the biggest question of all: how do we really organise democracy? How do we make sure everyone is involved and everyone is heard? And that everyone's basic needs are met?

The fundamental conflict

On Tuesday evening, the debate comes to a head. The young people's strategy group meets. Outside, in front of the doors, some of the most hard-working Swiss activists are sitting; they have been active already since the movement's beginnings in December, and now they are enjoying the evening sun on the floor on the upper storey of the university. To many people's surprise, the Swiss organisers have planned the conference in such a way that there are already statements and demands on every imaginable political topic, including energy policy, agricultural policy, transport and so on, and now everyone is being asked to respond to those statements and demands.

The door opens and we are greeted by furious faces. One faction is sceptical: we do not regard it as a sensible idea to go into these detailed demands, they say. We want to organise strikes, and not behave as if we were a youth party attempting to get involved with governments on an equal footing, with policy demands, and trying to play the politicians' game; we want real change. This conflict is a running theme throughout the history of FFF: on the one hand, those who make concrete demands and discuss policy suggestions (such as carbon pricing) with governing politicians; and on the other, those who emphasise that this steers the whole dynamic of the movement in the wrong direction. Two needs emerge: the need to know how a zero emissions society in 2030 could look in concrete terms; and the need for a dynamic of real change.

On Wednesday morning, the situation becomes even more heated. There is to be a vote in the plenary hall, with the hundreds of young people taking part in front of the press. For some of them, the situation is absurd. They don't want to vote because they are not for or against the specific demands, such as the demand for a zero emissions society in 2030 with nutrition being mainly plant-based rather than animal-based, with a price on carbon emissions, without subsidies for fossil infrastructure, and so on. They agree with all those ideas, but they are against any such demands being made in the name of FFF. For them, most of the demands are right, but making demands is wrong. How should they behave in such a situation? They simply withdraw, marking the fact that something here is wrong, and force the plenary meeting to break early for lunch.

To outsiders, the delicate situation might seem strange. Rarely have so many people who agree with each other been found in the same room, right down to the details. The beginning of the 2030s should be the date we reach net zero emissions, not 2045 as demanded by the Swedish government, and not 2050 as the Swiss government has said. That means that emissions must be reduced by more than twelve percent per year in all sectors, starting now (Anderson et al. 2020). There is barely any conflict in terms of content or substance. The fundamental principles are shared by most of those present: we must listen to climate science; the transformation has to come about on a socially just basis, locally and globally; and the rebellion has to take place without violence.

Even after a turbulent year, the biggest youth climate movement on the planet is as united in terms of its ideas and values as a movement can be. Precisely that gives a youth movement unbelievable strength, which is clear to us scientists in the university rooms. The only thing that forces discussions to happen is this question of strategic direction. Some of the activists believe that agreeing on a catalogue of demands would mean conforming to the established political discourse and losing the whole potential of the strike movement, which is after all directed against and towards those who rule; they are the ones who have to change. And only because it is possible to address themselves to those in power is it possible at all for millions to join them with so much momentum and force, distinguishing FFF from every other environmental movement, NGO and youth party. However, the difficulty in these hours in the Lausanne auditorium is that this alternative does not seem to be clear to most of them in the hall, because they are under such pressure from adults to come up with solutions, as 15-year-olds. The adults simply aren't up to it, I think to myself, sitting at the back of the hall with the scientists; they can't take the task off the children's hands: coming up with a well thought-out, global-local plan covering all sectors as well as the overall economic approach. They would rather ask the children to come to them with a catalogue of demands and then point out small mistakes or make bad compromises.

And so, the young people end up with an impossible task. They have to use civil disobedience to break through the status quo in the first place, show that the governments are negligent pyromaniacs, pull the emergency brake – and at the same time work out what should take the place of a fossil society. The tension in the air on this Tuesday and Wednesday in the first week of August in the hall of Lausanne University comes from the failure of my generation, I think to myself, not just from a conflict among the 15-year-olds. This is only a symptom of society's silence, which is not their fault. We adults finally have to act.

The situation is coming to a head. How should they go on? Some are disappointed, having worked for months on these demands. Two o'clock comes, all 400 participants arrive at the plenary hall, but only the strategy group of about 30 are allowed in. The two groups are becoming ever more entrenched in their positions. Some "facilitators" are brought in from outside. And suddenly, as in a film, they sit there, gathered on the stage at the front: all of those who have created the youth movement in a tour de force in the last eight months. Really, they should be running across the fields, I think to myself, all of them together, but they've been stopped from doing that by us, by the adults who have no political plan. And so, they are standing there, on the stage, conjuring up the essence of democracy, power distribution, political strategy and a shared future. A democratisation of structures would certainly be a good idea. They can agree on that.

In fact, two problems are being negotiated at once: on the surface, this is about the catalogue of demands with specific political solutions; at a deeper level, it is about democratic decision structures, or much more simply and more fundamentally: about the hundreds of thousands of young people who have the feeling that they can contribute something relevant and be seen as such by everyone else. How can everyone be included without the compass being lost?

The conference agrees to present the detailed demands on energy, agricultural, economic and transport policy as possible paths that could be taken, rather than as ultimate demands. Both sides can live with that. On the one hand, these detailed demands can be found written down in the "Declaration of Lausanne", the first core FFF document, and for a long time the only one. On the other hand, the movement can finally focus again on its central point, the strike, putting a stop to business as usual; the rebellion. On Friday, they all stand side by side at the strike in the centre of Lausanne. But I have the sense that an uneasiness is making itself known behind the conflict, perhaps without the activists realising it. It is not possible to cover up the scar caused by the fact that these children have to rebel against the whole of their own society, including their parents' generation.

The curse and blessing of the scientists – the facts

Another conflict is being played out behind the scenes, among the Scientists For Future. The young people, particularly the Swiss activists, had asked us in the middle of the hectic final preparations for the conference to re-familiarise ourselves with the fundamental problems and describe them to the strategy group in the form of a short presentation: whether we have to transform our society within ten or fifteen years (around 2030) or around thirty years (2050), by stopping emissions, makes a difference.

What are the arguments? I do some research and then write a group email in which I ask leading scientists for arguments, particularly those whom I know have differing views of the question. I receive a corresponding variety of answers, but those answers are still within a clear range: the richer countries have in 2019 about twelve years to reach "almost" zero emissions, if they don't want to pass the limit of 1.5 degrees global warming, assuming you follow one of the central calculations in the IPPC-SR-1.5 report. The criteria become increasingly clear: the basis is formed by the budget calculations made by Kevin Anderson (2020) and by Rahmstorf and Schellnhuber (2019), among others, which practically all scientists agree on, and which coincide with the IPCC SR 15 (i.e. the special report on the 1.5 degree goal from October 2018).

Again, the idea is that there is a budget, a certain amount of carbon dioxide we can emit if the earth is not to become more than 1.5 degrees warmer, because there is a direct connection between CO_2 emissions and global warming. These calculations are of course not completely exact, but at the beginning of 2018 the budget is around 420 Gt CO_2. This is what the UN says according to the first scenario in the IPCC report (figure 5). And currently, we are emitting between around 45 and 50 Gt each year, at least. The climate prognoses of the last forty years have been incredibly accurate. If we continue like this, we will use up all the oil, coal, and gas that we can ever emit in 8 years if we don't want to pass 1.5 degrees of warming. Irrevocably. For that reason, the whole world has to reach zero emissions soon, in 2040 at the latest, according to Schellnhuber at the Potsdam Institute for Climate Impact Research (2017): so emissions here in Europe have to be reduced by more than twelve percent immediately.

On Wednesday morning, my presentation keeps being delayed; the temperatures in the university rooms are unbearable; it is well above 30 degrees and the strategy group once again splits into smaller groups. My attitude is this: I try to present the scientific situation and debates in the most accessible way, but not simplistically, and then I try to get out of the way. I have nothing to do with the young people's decision on the FFF movement's possible direction. This is the line I've followed since the first day of the strike last August, and I want to remain true to it. We adults cannot decide for them, we cannot even "advise" them, as some scientists call it, and as more and more of them do, especially in Germany, which worries me. We can only go into their questions and describe the state of research – taking account of the perspectives which are most important to them, in the most wide-ranging and transparent way possible. They have to find their own compass.

The scientific dispute in the replies of all the professors regarding the question of whether zero emissions should be reached in 2030/35 or in 2050 for European countries develops around the following issues: what percentage of "risk" can we accept; or more precisely, should we follow the IPCC 1.5 SR 67 percent scenario, or the 50 percent scenario? "Risk" is a misleading term, however, since this is really about the margin of error in the calculation of the correlation between warming and emissions (see MacDougall et al. 2017).

The second question is even more controversial: what exactly is social justice, or as the Paris Agreement says, "equity"? How should we calculate that scientifically? For reasons of fairness, European countries, especially those in western Europe (known as the II Group), because they are a rich continent which is historically already responsible for a disproportionate share of the emissions, cannot have the same amount of the emissions "pie" as poorer countries (added to this is the legacy of colonialism). The Swiss activists place particular importance on the fact that we should also apply different standards within Europe, so that poorer eastern European countries are not disadvantaged. A famous scientist writes in the email replies that reach us that weighing this up is not a question for the natural sciences but an ethical question which cannot be answered scientifically. Other researchers object and point out that we can also pursue questions of justice and fairness. That has to change, I think to myself in the heat of Lausanne; the fact that natural scientists do not see ethics and political philosophy as being part of their universities at all.

A third question also leads to disagreement among the scientists: should the young people take account of something like political feasibility, as some people demand? The argument is: when the reduction of emissions by 12 to 15 percent per year is mentioned, or when the year 2030 or 2035 is mentioned as a goal for zero emissions, many people will object that this is not "realistic". The production of new diesel and petrol cars would have to be stopped immediately, the replacement of oil heating with gas heating would have to be banned, meat production would immediately have to be reduced by ten percent every year, a large proportion of cement would have to be replaced with wood as a building material, and so on. For that reason, we have to fall back on 2040, some of them write. Here, there are also different opinions in the replies.

I go through the speeches by Greta and other activists again and see that they have answered these three questions clearly from the start: we cannot play with the future of humanity, so we can't take very big risks. We have to take account for equity properly, and take on more responsibility as western European countries in the global transformation of energy systems and the economy. And we cannot say beforehand that something is too demanding politically, but have to stick with what is really needed in order to avoid exposing entire countries to unbearable heat and forcing hundreds of millions of people to flee, and to prevent the catastrophic floods that are already taking place. In addition, European countries ought to be paying substantial "fair shares", massive amounts for the socially, economically, and ecologically sustainable green transformation of poorer countries, far beyond what has been planned so far (see http://civilsocietyreview.org). In other words: the Swiss, who on average emit 14 tonnes of CO_2 per year according to the Swiss Federal Office of Statistics, would have to reduce this within a few years to a level of around 1.5 tonnes, which would be globally fair.

But how on earth, the young people ask, do governments come up with the idea of focusing all their policies, from agriculture to transport, on a "net zero emissions goal 2050", meaning a 20-year delay, which means a much bigger emissions budget which might lead to 3 degrees of global warming? Well, the scientists answer: they are accepting enormous "risks"; they are more or less ignoring the justice and "equity" aspect of the Paris Agreement; but above all, they are completely ignoring Scenario 1 in the UN report IPCC SR 15.

Instead, they are coming up with numbers which can only be reached by calculating huge amounts of so-called negative emissions, meaning future processes through which a technology that doesn't yet exist will capture carbon dioxide from the air. That is why they all talk about "net zero": the sum of the excess greenhouse gases and the negative emissions should be zero. (There are also "natural" sorts of negative emissions such as reforestation, which no one objects to – on the contrary – if it is achieved sustainably and doesn't transform old forests into "bioenergy" and "biomass"). It is as if governments were to build a waste-water treatment plant to produce drinking water and say: we're building the plant in such a way that thirty percent of the toxic particles can pass through it, but we will capture them again afterwards from the drinking water. With a brilliant technology. That doesn't exist yet. As school pupil Leonie of FFF says after a discussion with the German Minister for Economic Affairs Altmaier: governments are not simply failing drastically to meet their own climate goals year after year. These climate goals are themselves deliberately skewed; and that is the much bigger problem. The movement has been drawing attention to this crucial point since Greta sat down with her famous sign a year ago, with her A4 factsheets weighed down by a stone, and handed them out to passers-by.

This argumentation seems reasonable, but I do not want to anticipate what the working group will say or manipulate them, and so I send them the whole discussion and explain it at the conference, including possible objections. The most important thing for me is that these numbers should be visible for the whole population in the first place, so that everyone can see how tiny the remaining emissions budget is. All coal power stations and oil pumps in Germany, China, America, and Norway (where at this time the biggest oil field is being opened, designed to run for 40 years) would soon have to suspend operations. Even the most conservative of the researchers ultimately writes in our email correspondence: "The implication of all my remarks here is that we have to pull the handbrake immediately and introduce drastic measures."

I move away from the team of scientists, take a plastic chair, sit down in the middle of the stubble field and let the sun shine onto my face. How bizarre, I think to myself, that there are people who have voluntarily immersed themselves in these facts and found out all these connections: the greenhouse effect and the albedo effect, the connection between carbon dioxide emissions and global warming, and so on. Someone could calculate the albedo effect – that is, the ability to reflect light – of the tip of my nose. How bizarre, I think to myself again. How can someone express the experience of the sun on the tip of my nose in terms of physics? The world is full of stories, after all, so much richer and more magic than the natural sciences imply. Like so many of the young people here, I miss my world of books, my students, who run around the theatre spaces in Stockholm as Emil or Ronja. Imagination. This year with the climate movement is also a year which threatens to reduce us all to beings with a purely scientific worldview, I think. Why do the rays of the sun follow a law; why can't they fly freely through the air however they like?

> That same day, the nearby UN offices in Geneva state in a central "IPCC land" report (IPCC 2019): overall, the world population must shift very quickly to a vegetarian or vegan diet because of the ecological and climate crisis and the loss of biodiversity.

"Tipping points" and "feedback loops" – what is the state of the world?

The next day, all these thoughts can once again be seen in quite a different light. Many older activists and scientists have come to Lausanne to give seminars on the strategy of FFF and XR, on the structure of our societies and activism in general. The room is huge, so that the young people are almost lost in it.

The activists giving the workshops remind us once again how social change takes place and how it doesn't, according to some traditions of sociological research. Not much is changed by marches, clicks on petitions, flyers and so on. What has worked, historically, has been massive disruptions of the status quo, directed straight at governments. That was shown by the civil rights movements of Martin Luther King and Rosa Parks; the suffragettes of the women's movement, Gandhi's rebellion, the workers' movements, and so on. Real change happens when the main squares in capital cities are occupied for several days with blockades, or when children keep refusing to go to school and the ministries start cracking down, so that the population turns against them.

Some of the activists and scientists begin again to list all the examples of ecological collapse which have become familiar to us during our work in the last months. From the unbelievably rapid melting of the Arctic ice to the catastrophic droughts. According to the picture they paint, there is a massive risk that within twenty or thirty years there will be so many crop failures and food production will be disrupted so badly across the world that there will be enormous social conflicts, famines and possibly wars over the most important resources. They describe a simplified chain of causation: coal power plants are being built or kept running in order to maintain our society of fossil, car, heating and cement industries; governments are not really adopting any measures; the Arctic ice is melting; the jet stream is growing weaker; we are seeing more extreme droughts and weather events; that knocks out food production in some parts of the world; drinking water becomes scarce; and so the social conflicts over distribution begin very quickly.

One of the most important points is that the whole discussion about emissions budgets needs to be called into question, which means all the work that we have been presenting so carefully during the last few days. We are, according to their claim, already well past the 1.5 degree "goal". That is why radical measures are needed. Once again, I bring in the scientific community via the email lists and ask them to check this claim. The unanimous result from these hundreds of climate scientists: there is the problem of "committed emissions" and air pollution ("aerosols"), which ensure that global warming that is already in the system is not perceptible. Estimates diverge on whether this could mean 0.2 or 0.4 or even 0.8 degrees – disastrous numbers. Because with the warming that has already been measured (1.1 degrees), we would then already have passed the goal of the Paris Agreement – even if all of the fossil infrastructure, all coal power plants, and all oil refineries stopped running immediately, and all cars, flights, steel and cement factories came to a standstill.

But the even more confusing point which some of the older activists now emphasise: most IPCC scenarios (which all governments rely on) do not take account of tipping points, which could come into play soon or which already have done so. For example, the permafrost in Northern Russia is melting drastically; and the methane being released exacerbates this dynamic. On this, too, all the scientists who reply to us agree. Perhaps these processes are even already going on without us realising it. And as soon as these effects are calculated, talking about still being able to emit so many tonnes of CO_2 becomes problematic: there is already much too much in the air, and there is a real danger of effects reinforcing themselves. This brings with it the horror scenario of two to three billion people having to flee in little more than 50 years (Xu et al. 2020). In my inbox, I receive a few explanations by scientists including Stefan Rahmstorf, pointing out that this information about "committed emissions" and aerosols (air pollution) is correct, and so are the calculations of global warming, but that these processes could in turn bring about complementary processes that might have slightly mitigating effects (Rahmstorf 2019).

So when governments talk about 2050 as a goal, they are ignoring all these aspects, including "committed emissions", tipping points and feedback loops, and they are reckoning on huge negative emissions with a technology that doesn't exist, and they are betting on a principle of incaution, and therefore accepting enormous risks.

At the same time, even at the current level of global warming of 1.1 degrees, millions of people in Bangladesh and China are losing their homes because of floods. And yet there is consensus in the scientific community that the Keeling curve shows a concentration of carbon dioxide in the air of 415 ppm. 350 ppm are regarded as somewhat stable for the current climate. So, we ought to be investing massively right now in protecting forests, in addition to stopping emissions (Röstlund 2022). That is why the idea of "offsetting" is so inadequate, according to many of our Stockholm scientists. In any case, we have to invest huge sums in "rewilding", in protecting and expanding forests. That cannot be outsourced colonially as compensation for our lifestyles.

One of the young people from Mynttorget lays her head on the desk in front of her and sighs. Most of them will still be going to school for years in Stockholm and maybe meeting on Fridays to strike in Mynttorget. These scenarios worry them: what are they supposed to think about them? What arguments are there? It is possible that huge famines and social conflicts will come about abruptly in a few decades? David Wallace-Wells gathered research on this in his book *The Uninhabitable Earth* (2019), and new results are coming in all the time. I sit on the bench in the auditorium and think about these scenarios. Actually, when talking to the young activists, I hardly mention the most depressing variants of the "hothouse" effect (see Lynas 2020). I also feel that scientific caution is required.

But it is just as important to point out that governments and parliaments are basically deceiving populations worldwide by not taking all these risks seriously and by playing tricks with the numbers. I flick through the speeches and papers by Greta and the other activists. They have said most of this from day one, the part about the aerosols, negative emissions, feedback loops and tipping points: that all of these are missing from governments' calculations. For me, it has taken a year before I've really understood it, along with the real dimensions of the dangers. Governments know that they are already very unlikely to be able to keep the agreements they've made, including the Paris Agreement. Something fundamental is wrong.

The curse and blessing of science – the conflict over the basic principles of Scientists For Future

If a few thousand children are punished for taking part in ever longer school strikes, say some of the older activists from the blackboard at the front of the room, political constellations will begin to change. Then there will be a confrontation which could lead to real change, as in the case of the civil rights movement in America.

And suddenly, a bad feeling comes over me again. The scientists have the idea of offering the young people the possibility of asking us more questions during the congress – and forwarding those questions to Scientists For Future. From this, a list of questions is created, and a document which is soon around eighty pages long, a little book on the most urgent specialist questions which the FFF activists have on the ecological crisis. And problems are connected with that which ultimately relate to all the big questions of our time: what counts as scientific; how does science relate to politics, and how should the adults behave towards the children and young people?

The questions brought up by the young people are sent first of all to the email lists. More and more scientists send their answers back. And those are – naturally, how could it be otherwise – shaped by their individual personalities. I hesitate. Some things are unproblematic, simply facts of physics, such as the mechanism behind the greenhouse effect itself, or behind tipping points. But there are barely any questions with no political slant, meaning that they not only have political consequences but also a political undertone. Can a "circular economy" work? What do we have to do to stop the climate crisis? And so on.

The problem is not that a political component is suddenly coming into play, but that those who are answering seem to ignore this and blithely send along their personal opinions: maybe the population needs to be reduced – to name one example of an argument which most of us in Lausanne regard as wrong or even racist; after all, the richest ten percent of the world's population produce more than 50 percent of all emissions (Oxfam 2015). They have to change their behaviour.

But what are the criteria we can follow? "This isn't going to work", I tell the others, who are working day and night on the document. I feel like a spoilsport. "It's neither scientific nor sensible. The young people here are just being flooded with ideological assumptions, without that being pointed out." They agree. We have to develop a process which reflects this problem.

Questions arise from that: what are the principles which should shape us, the people who call ourselves Scientists For Future? Scientific rigour, of course: calling results into question; working methodically; and so on. But it doesn't make any sense to send the young people's questions to some scientists who are not familiar with discussions and facts on topics such as social justice and the climate crisis. What distinguishes us as Scientists For Future in relation to our colleagues? Long discussions take place, in Lausanne, but above all behind the scenes in all the scientific networks. These conversations are spurred on by another fundamental development in S4F: in March, first 15 000 and then 25 000 scientists supported the young people and announced: what you are saying is the truth (Hagedorn et al. 2019). It's true: "The house is on fire." But S4F do not leave it at that. This stance remains the basis for everything else, but more and more the specialists feel the need to formulate in their own fields what would be needed in terms of societal transformation: describing the zero emissions world in 2030/5, which they have been researching for a long time, an incredibly multifaceted and powerful project. So, some of them write a paper about carbon pricing. Others "help" young activists to write catalogues of demands which include all sorts of things: from renovating buildings to building wind turbines, from transforming the transport sector to sustainable agriculture.

It is Thursday evening, and these discussions are making me panic more and more, given that we've promised the young people that we will come back to them with answers the next day. The individual suggestions all seem sensible. But the dynamic behind them seems suspect to me, and dangerous for the whole movement. On the one hand, there is our relationship to the young people. It is their movement, and we can't put words in their mouths. I insist on that. We can give them information, and point out debates, but we are not consultants for catalogues of demands. Consulting work is for clients or governments, not children. We ourselves have to stand up for what we believe in; we cannot foist it onto the young people.

But what seems much more important: more and more, from the jumble of policy suggestions, a false picture is emerging, as if we only had to tighten hundreds of small screws in order to reach a new society which would really be sustainable. That is precisely the ideology of the green liberal centrist parties which have been calling the shots in the Swedish parliament for five years without anything actually happening. It was their policies that the activists were protesting against when they first went on strike, because they precisely do not ensure that emissions go down, or question the economic approach which destroys biodiversity. Actually, almost none of the scientists seem to want to put forward such an idea explicitly – that is, the idea that we only need hundreds of policy proposals to deal with the climate crisis. It is simply the effect of the hyper-specialisation of the academic world, and the division of measures between different sectors.

The systematic view is missing, meaning a perspective on how all this hangs together and which parts of society are affected. The carbon tax as proposed by many would affect those on benefits most severely, in a socially unjust manner, one specialist calculates for us. None of the young people gathered in Lausanne want that. And so on. We need a systematic approach. We cannot discuss political measures individually without viewing the economic system as a whole, for example: what if we needed a basic income or basic services, globally too, tied to local currencies; or something quite different, different measures which would reshape many different sectors, such as quickly bringing the fossil industry under state control, so that it could then be cut back, or a democratisation of the economy: shifting the focus to the "care economy", looking after children and the sick, feeding and bringing up children; finding our way out of the processes in which we use up nature and wear out products – processes on which everything relies; and also a way out of unjust working conditions. Should we keep all of that and instead fund electric cars? Something is wrong here, and a danger is lurking for us specialists.

Charta von Scientists for Future

(English version below)

Selbstverständnis

Scientists for Future (S4F, auch Scientists4Future) ist ein überinstitutioneller, überparteilicher und interdisziplinärer Zusammenschluss von Wissenschaftler*innen, die sich für eine nachhaltige Zukunft engagieren.

Scientists for Future reagiert auf die historisch beispiellose Klima-, Biodiversitäts- und Nachhaltigkeitskrise, welche die Menschheit vor globale Herausforderungen stellt. Die notwendigen Wandlungsprozesse erfordern entschlossenes und unverzügliches Handeln auf der politischen, wirtschaftlichen und technischen, sozialen und kulturellen, wissenschaftlichen sowie der privaten Ebene. Denn die Zeit drängt. Als Wissenschaftler*innen sehen wir uns deshalb in der Pflicht, öffentlich und proaktiv die Stimme zu erheben.

The feeling also creeps over me – that might be the worst thing about the situation – that these scientists are not imagining at all how these legislative proposals will actually be heard, or considering the fact that they themselves might also have a negative influence on that. Why should governments which have held on for decades to laws which above all help the established fossil industries and banks suddenly accept these catalogues of demands from young people or from adults and say: "Oh yes, thank you so much, that makes complete sense, we will now force through these detailed laws which will change our industries fundamentally"? This problem comes up in Lausanne in more and more dramatic ways. The scientists are limping along a year behind FFF's strategy discussions. We're probably not going to get around system change and a rebellion, many of the young people are thinking; and the scientists are doing them a disservice by passing them catalogues of demands and thus reinforcing the status quo.

The basic principles for science and politics

What we should do instead, I think, is mark out a framework for crisis legislation. I set out to sketch a solution for the question of principles. After going back and forth for days, four such principles seem to be justified. Firstly: the cautious interpretation of the emissions data outlined above (IPCC-SR-1.5, Scenario 1 with 420 Gt in 2018) is appropriate, and as a result of climate research, it can act as the basis for all argumentations; we cannot retreat from the measures applied by the young people, and we have to uphold a "principle of caution", including intergenerational justice. Secondly: we can only take account of existing solutions; future technologies which have not yet been invented and which lead to negative emissions, cannot play a role, and nor can "solar geoengineering". Thirdly: the aspect of social justice and "equity", which is also mentioned in the Paris Agreement, should be the starting point of all arguments; including when it comes to calculating reductions in emissions and reshaping societies. And this goes just as much for the aspect of historical justice with regard to greenhouse gas emissions so far, as it does for the just distribution of wealth in a way that emphasises the responsibility of richer countries – who have also often built up their wealth through (colonial) exploitation of poorer countries; both between countries and within societies. Fourthly: we cannot lose sight of the systemic perspective, meaning a sustainable society globally and seen as a whole. And this systemic side, in turn, must satisfy the principles which are central for the young people: social justice and ecological sustainability, or in the words of my lectures: we don't accept power relations based on domination any longer (in terms of gender, class, ethnicity...), but instead use the transformation for a humane democratisation of our societies (see appendix).

However, as the young people emphasise again and again: in the short term, all it is about is awakening some kind of appropriate level of awareness of the crisis. And this is obstructed by the thousands of individual suggestions from scientists implying that we could actually leave everything as it is. Behind this problem lies a much bigger one: how can we reshape universities quickly in such a way that this systemically and socially relevant way of thinking is possible, one which combines our specialist knowledge from individual disciplines and sheds new light on it? In their courses, too, all students from all subjects must receive a crash course on this most crucial existential knowledge. (Hundreds of FFF activists across the world are specialising in this question, and are working together during these months in a specially created chat concerned with "Climate Emergency Declarations" in schools and universities, aiming to incorporate these declarations into the curriculum (cedamia.org).) I have to get all of this established among my colleagues at the universities, I think to myself, as soon as I travel back from my homeland to Stockholm.

A strike and a farewell

And so, the summer comes to an end. The Mynttorget group stands on the Lausanne station square. Soon, the strike will take place to mark the end of the week; after all, it is Friday, and on Friday they strike, regardless of where they are. If only it was not around thirty degrees. The previous evening, they stood on the terrace of the Lausanne Cathedral, the Stockholm group, and looked across at the French Alps. We talked about the year which had passed and what would come. Some of them would soon be studying, perhaps philosophy or human geography, biology? What would they become? In the 70s and 80s, there were four subtle Swedish comedians who were philosophical and incredibly funny; they made films, were politically active and wrote songs. That wouldn't be such a bad life; they can agree on that. But we are also aware that the climate crisis looms like a dark omen over all such plans.

With these thoughts, the final demonstration begins in the centre of Lausanne. In the middle of the strike procession in the flock of French-speaking people, the Mynttorget activists hold up the banner with the Swedish slogan, "Skolstrejk för klimatet", and yell: "Enough empty words, they do nothing to help the climate," which rhymes better in Swedish. They give me a nudge. "Hey, you have to join in." That's true, I'm part of the Swedish crew, even here in Switzerland, where I can vote. In three months, the Swiss parliamentary elections will take place, and the media are talking about a possible Greta effect which could shift the balance of power after decades.

For a long time, they go on chanting like this in Swedish through the streets of Lausanne. Then Greta has to go back to the station, to the Hambacher Forest and continue the next day to Plymouth. From there, she will sail to New York. We stroll back through the empty lanes. A year has passed since the first strike mornings, when everyone enjoyed the quiet and lent on the wall in Mynttorget; when journalists popped up from every different country – and often seemed to us like caricatures of their national stereotypes. The Dane with a Hawaiian shirt, gesticulating incomprehensibly with his arms; the orderly German who had planned and prepared the interview for weeks beforehand; the French journalists who just came walking over to try their luck...

Then the Twitter notifications about the Friday strikes start arriving, as they do every week in these months: hundreds of groups of young people can be seen in Afghanistan, Japan, several groups in Bangladesh, Kenya, Sierra Leone, the Philippines; there is no end to it. For a long time, I look at the picture of the young activists walking through the streets of Kabul, behind a huge Fridays For Future banner. The European activists here in Lausanne are part of a much stronger, global movement. And that movement is rising now.

Chapter 3: The Week For Future
September 2019: The coordinated uprising
of eight million people across the world

"The house is on fire" – returning to Mynttorget

The forest is burning. Two weeks have passed since the meeting in Lausanne. It is the end of August 2019. The Amazon rainforest is on fire while Greta sails through the currents of the gulf stream. Countless fires are raging and destroying the lungs of the world.

The tundra in Siberia is burning (Cormier 2019). The Arctic. Is burning. Hurricane Dorian destroys the Bahamas completely and turns around shortly before reaching Florida. Miami is spared once again. The rainforest is burning because governments are letting it burn, and because Europeans eat meat (Mackintosh 2019). For that, the forest has to be cut down. It happens quickly. Worldwide CO_2 emissions continue to rise, and a strike year is coming to an end.

On the train back to Stockholm, I sketch out a text for the third global strike on the 20th of September, the beginning of the Week For Future, which is quickly approaching. "We have to change. We are the ones, the adults, who are being told by the children that we are destroying their future, daily, by carrying on as usual. We must do as they are doing and pull ourselves together. We must finally take responsibility."

Preparations have already begun, with designs for placards and flyers. Some of the young people are talking to the unions. The 27th of September, the closing day of the week, should be a workers strike in which all adults will take part, and a new intergenerational cooperation emerges.

The Week For Future, from the 20th to the 28th of September

On Friday morning, they are sitting there again, as they did exactly one year ago, the ones who were the first to join the strike.

Sometimes you feel that a preparation is almost complete, a festival, a theatrical performance, something is about to come, and you have forgotten what is at the heart of it all. Something is missing.

It is the 27th of September, a beautiful autumn day in Kungsträdgården, blue sky, the leaves of the trees already turning, cool but not cold. The first Scientists for Future and researchers from "Researchers' Desk" are standing around trying to attach their question-answer cardboard signs to the streetlamps. I look at the huge stage and start to wonder. What can it be. Something is missing. I look at the schedule of the sixty-minute the young people have put together and are presenting. First comes sustainability researcher Line Gordon, the director of the Stockholm Resilience Centre. Then the Sami singer and activist Sara-Elvira from the indigenous people of Sapmi, the area taken away from the Sami by the Swedish state. They have felt climate change much more than we do here in the capital. Their vegetation is changing very quickly, very drastically, and with it their income opportunities, their culture, their way of living together. There is something magical about the singing. Finally, seven representatives of different religions, then another band, the young people with their speech and at the end Robyn, the world-famous musician.

I walk slowly past the stage, cross the bridge which we blocked back then with Extinction Rebellion, a year ago, and reach Mynttorget. The hours fly past. Final preparations. Yellow vests for those of us who will be protecting the march. Getting water. And always this feeling that something is missing.

The Week For Future is coming to an end today, and it is already a huge success, in terms of numbers. In Germany, 1.4 million people took to the streets, already on the previous Friday. My scepticism towards the NGOs has not dissipated, but the basic idea of a shared "movement of movements" has borne fruit and can be expanded.

The young people have painted new posters, bright yellow, which hang everywhere in the city. I have written to the heads of all the Swedish Universities and published the open letter in the magazine ETC. In it, I called on the academic world to join the strike and change their approach to education, take the climate crisis seriously, change their didactic approach and the content of their courses, in all subjects, from architecture to economy (see the chapter on education). Most of the university heads at least replied personally, with a similar argument to the unions, saying that they supported the strike day but did not want to call for a strike. This part of the world of adults does not yet understand how it could and should act, I think to myself angrily. There is not enough information about the crisis and not enough awareness of what kind of action would be possible. The people running the unions and universities ought to see their responsibility. This is about job security, about solidarity with those who are most affected worldwide, and about waking up the research institutions.

And finally, it starts, one little group after another gathers from all different parts of Stockholm's population. Big and small, young and old, with all their colourful signs. The march begins. When we turn the corner at the head of the procession and walk from the Mälaren lake past Mynttorget on the way to the Kungsträdgården, there are around 50 or 60 000 people following. It is the biggest political event in Sweden since the Iraq war demonstrations, one of the biggest in post-war history. More and more people come and walk over the Slussen bridge into the old town, a seemingly never-ending column of people.

We stand behind the stage and look at each other. "What have we done?" one of them asks again and again. The strange thing about the situation is that such a small group organised everything in the last weeks, a team which has become extremely close-knit, which has brought about the three global strikes. Without a budget. The costs run to around 7000 euros, for the march and the program on stage for an audience of 60 000. The young people have run a crowd funding campaign, and the biggest NGOs take on costs of about 1500 euros each, which are vanishingly small costs within their annual budgets.

They have spent countless afternoons and evenings in the Greenpeace branch, including the Monday before the large-scale strike. They watch Greta walking on stage in the UN office in New York and beginning her "How dare you" speech. And although thousands of kilometres lie between them, it is clear that she is one of them, appealing to the passive governments. A profound sense of shock and grief makes itself felt. What a good thing they have each other, I think, as they comfort each other until late in the evening and at the same time work on press releases.

The hours in Stockholm with thousands of people are magical. The speeches, the singing, the dancing, and the music. The atmosphere is cheerful, peaceful, but also full of expectation. The world can change. But still, something is missing. A few hours later, Greta will lead the march of 500 000 people in Montreal. Half a million in a city of 4 million. The populations in thousands of cities are giving their governments a signal which could hardly be clearer. We will not accept your hesitation any longer. We are ready for a transformation of our societies. We are striking, leaving our workplaces, and joining together. Altogether, it will be 7 to 8 million people who go out to protest on the two Fridays, in more than 160 countries. For the first time in 30 locations in Russia.

I am again responsible for passing on news from the world to the young activists, as at the first global strike in March, so that they can then announce it from the stage. But now there are so many more places. In Bangladesh alone, countless locations appear on my Twitter feed. Once again, the strike begins in Wellington, in New Zealand, when it is still the middle of the night in Europe, and comes closer, via India and Nigeria to Stockholm. In Bern, Loukina stands onstage in front of almost 100 000 people and calls on every generation to get started and unite in 2020.

And I begin to sense what was missing. Suddenly two of the young speakers in Stockholm rush past me. Isabelle says: "I'm nervous." Then they go out, leave the shadow of the little roof over the stage, and stand at the very front, facing the people of the city.

Their speech begins. And I feel an enormous sense of relief. This could easily have become a beautiful day, a magical day, a Woodstock day with a 1968 atmosphere. But here they stand, daring to address the politicians directly with all the anger that they continue to feel. You must change your policies, you're doing the wrong thing, you're deceiving people. We will not leave. Their speech becomes ever more vehement, and ever clearer. "This is not a demonstration. This is a strike. A rebellion. We will make sure that you reshape our societies. We do not accept that you are holding onto the wrong rules." In this anger and in this courage lies also the love for the strange creatures who stumble around on this planet, I think to myself.

In the days afterwards, the governments and rulers of the world, including those in Sweden, respond with silence. It is as if nothing had happened, as if the capital had never transformed itself – on an ordinary Friday – into one big protest march. Mynttorget feels cold in the ensuing weeks. That is not just because of the arrival of autumn and winter, but because of the parliament building, which stares down at the square, motionless, and because of politics, which remains the same.

But in this week in September 2019, a global climate movement has formed from all parts of society, young and old, workers and students, in thousands of cities in the world, and found a shared goal. Millions of children are becoming aware of the meaning of the ecological and climate crisis.

How can we not only mobilise but bring about a change in the rules? How can politics change if those who rely on "campaigns" and demonstrations set the tone – and thus push those into the background who have had the courage to organise real disruption on a large scale: a real school strike including solidarity strikes at workplaces? They do not see themselves as a part of a well-behaved "civil society"; but as a rebellion against the rules which jeopardise their future. Demonstrations can be ignored. Professional campaigns can be ignored. Striking children and workers in solidarity cannot, not in the same way. Three tasks are emerging: how do we maintain the continuity of the movement, even between these largescale events, in such a way that the children, those most affected, and the Global South can take on a leading role? How do we agree on demands, a political framework behind which the population can steadily unite? And: how do we find the means, with the help of civil disobedience and strikes, to push this new politics through beyond demonstrations and campaigns?

At the national elections in Switzerland on the 20th of October, the Green Party makes huge gains: no party in Swiss history has ever achieved such an increase in seats in one election. The Green Liberal party also increases its support by several percentage points. Meanwhile, the Social Democrats lose out, as does the right-wing Swiss People's Party (SVP). The media talk about a green landslide and a Greta election. The conservative majority in parliament is replaced by a green-centrist majority. Meanwhile, in California and Australia, the forests are burning as they never have before. Venice is flooded.

Chapter 4: COP25 in Madrid
October – December 2019: How can we end our fossil society in a fair way?

Changing direction

It is a cool Friday in October. As usual, we are on strike in Mynttorget, when Andrea joins us. She often helps the young people and belongs to the core group among the intergenerational strike organisers. "Look at these videos from Chile," she says; that is the country where many of her relatives live. Large parts of the population refuse to accept the enormous social disparities in the country any longer, and thousands are protesting in the streets of Santiago. They want to see political change, a democratisation of their society.

At the same time, after strike days with Indigenous people in North America and Canada, Greta and other activists are heading for that same city, Santiago – for the next COP meeting of all countries. The unrest in Chile is increasing every day. The government is intervening with increasing brutality, and goes on to cancel the COP summit. Madrid steps in. And the activists begin their journey back to Europe.

Many young activists from Europe travel to Spain. They want to develop the movement, because the official delegations are doing almost nothing: but without international cooperation, oil, coal, and gas will not stay in the ground, and the forests will not remain intact.

The global group comes together

December has arrived. Outside in the streets of Madrid, the air is mild. In the huge halls of COP25 (Conference of the Parties), the governments of every country are supposed to be finding political answers to the climate crisis with the help of their specialised delegations. It is exactly a year since Greta's speech in Katowice, a year since her meeting with Jonas, Marie-Claire, and Luisa; exactly ten years since the violent police measures against the protests at the COP15 summit in Copenhagen. But now the situation is different. It is no longer adult activists who are facing the police, but hundreds of children, young people and adults facing the politicians who are responsible. Something new is in the air.

"Where is the place for the strike?" some of the young people whisper to each other, among them a few from the Swedish group. They look around the anonymous trade fair hall in Madrid and then join around fifty other strike activists around their age, who are sitting on the floor in the foyer of the UN climate summit.

The movement has really become global, I think to myself as I sit between a pot plant and a security barrier. In front of me, for half an hour, not only the Europeans from the Smile meeting in the summer have been sitting silently, but also Hilda and Leah from Uganda, Fernando from Mexico and Canada, Xiye from New York, and Arshak from Russia, who will be imprisoned two weeks later for striking in Moscow – the young activists literally come from all continents and all corners of the planet. These are the faces which have become familiar to us from the Twitter feeds. Many of them have built up strike movements in their countries, including Hilda Nakabuye from Uganda, who makes an important and influential speech in Madrid (Nakabuye 2019). She has long been a supporter of FFF and one of the strongest voices of the African movement for the climate.

And now it is Friday, strike day. The delegates from all different countries have been walking past the activists for the last thirty minutes, often without looking at them, especially those who come from the countries playing a central role in the negotiations: Brazil in particular, with the Bolsonaro regime, Prime Minister Morrison's Australia, and the USA, still run by Trump. These three are constantly slowing things down at every juncture.

The conference focuses above all on one aspect, "Article 6" of the Agreement: an unclear exchange of emissions rights is supposed to be defined more clearly, so that richer countries would be able to buy their way out of their own emissions reductions by giving money to poorer countries. Unfortunately, the mechanism itself is not up for debate. What is being negotiated is only the question of whether both countries can then count this as a reduction – which would make it impossible to reach the goals of the Paris Agreement, because suddenly there would be far too many emissions rights in circulation. As such – according to many representatives of big NGOs such as Greenpeace and Germanwatch – the summit cannot succeed. Only the worst-case scenario of "double counting" can be prevented.

In this sense, the global climate movement which has gathered here has to keep its own goals in mind. If there are people, younger and older, who want to take part in a global political "one people, one planet" project, then they'll be here, I think to myself. We just have to find them in these huge halls.

Then there is a sudden commotion. Greta, who arrived this morning at nine with the night train from Lisbon, has seen the tweets and is planning to join the strike in the foyer. "Hey!" shouts the UN police officer and pushes a journalist back who has hit him over the head with his tripod in the melee. Immediately, hundreds of journalists arrive. The young activists slip under the barrier, join their peers and only avoid the horde of cameras with some difficulty. Somehow, some of them end up in a room which is more reminiscent of a presidential palace, with a thick carpet and expensive pictures on the walls. They are at the edge of the trade fair grounds, having passed through halls 1 to 10, past the individual countries' pavilions and the big plenary hall in which the governments are holding their negotiations.

I am curious to know everything that has happened in the last months in America. But at the same time, I am worried about the group dynamic among the young activists, who are treated very differently by various NGO workers. These conferences, along with the attitude of NGOs from the Global North, always deepen the gulf between privileged activists and those who often build structures in the background without any recognition. And the mental health of all young people has to be the main concern, I say again and again to all the adults who are going about their business in the halls in Madrid.

Instead, the conversation soon focuses on the question: what is the plan for Madrid and the global climate conference? And once again, the problem becomes evident which was already obvious at the COP conference in Katowice the year before. Theoretically, the delegations could, in consultation with their governments, develop overarching solutions here (where and when else could such solutions come about?): a global master plan to stop the ecological, social and climate crisis. They could really change the rules, establish laws to reduce emissions by more than half across the world during this decade, as Hans Joachim Schellnhuber and other researchers have long been demanding, following the IPCC. Instead, it seems that these emissions are going to rise. Norway has just started operating the new, gigantic oil platform Johan Sverdrup, which will extract oil for decades. And Saudi Arabia is taking its oil business to the stock market. Aramcom immediately becomes the biggest company in the world, bigger than Apple, Amazon, and Microsoft.

The delegations in Madrid could agree to call an immediate halt to the building of fossil infrastructure, including the financing of such infrastructure by Swiss banks, for instance. And they could agree to build a global renewable energy system with help from richer nations (Jacobson et al. 2019; Teske et al. 2019), which would take power from the hands of a few companies and transfer it to the population. And maybe, I think to myself, maybe the populations of all countries, "the people", would themselves support such global political change – if anybody asked them.

Then our phones ring in the Madrid conference centre. Isabelle and the other Stockholm activists are on the other end. Because on this same Friday, the fourth global strike is taking place in Stockholm, and now the other activists want to have Greta live on stage with them by phone – this time not in the elegant Kungsträdgården Park in the centre of the city, but in the suburb of Rinkeby. Climate justice is the theme, and the group has contacted people in the suburbs, which are so segregated in Stockholm; the rich white areas are separate from the poorer BIPOC areas. If we don't work together globally on a just basis, the sustainable transformation of society will not succeed, I think to myself, and look across at the hundred young people from around the world.

They, too, the young activists of Fridays for Future in the COP halls, are forging plans for the coming days and months. The world has to see that young people are not going to play along anymore, and that they will not let the government delegations have the power to ignore their future. But how should they make that happen? A big event is planned for Wednesday, immediately after Greta's speech. Daily meetings are called. Official COP rooms are booked through the UN system. The activists soon find their way around in the hubbub of the ten trade-fair halls, they know where the free coffee is, where they can charge their phones, and where they can get the latest information about the negotiations. They update each other at top speed via WhatsApp, and they hardly eat anything.

On Wednesday, Greta makes her speech in the big plenary hall, which is filled with hundreds of delegates and ministers. She has barely put down the microphone when the big event takes place. For the first time in the history of the COP, 100 young people storm the podium in the plenary hall and occupy it, before the very eyes of the delegations. I stand at the edge of the hall as they all suddenly stand up and dare to dive past the police, over the barriers, and onto the stage. A few of them begin very hesitantly to sing their climate justice song, and gradually more of them join the familiar tune. Through this, they make it clear that they will no longer accept what those in the hall are doing in their negotiations. They are literally taking a stand against these regimes which are threatening their future. And shortly afterwards, the UN police expel them from the congress centre, along with the rest of us – all those categorised as representatives of civil society.

The conference was supposed to be about mitigating climate change and the ecological crisis. Its decisions were meant to reduce emissions drastically and help poorer states, not only with the transition to renewable energies, but also with what is known as "loss and damage", the destruction which is already now being caused by droughts and storms, bringing misery to thousands of people. None of that has happened. In the end, the leader of the Brazilian delegation even threatens to reject the feeble document and has to listen to a tirade from Tuvalu's delegation before he agrees at the last second. All the crunch issues are postponed to the following year's meeting in Glasgow.

Suddenly, we sense how huge the task is, and how omnipotent the fossil industry seems to be. And still, perhaps for the first time ever, the activists are united, carried and led by the young people and the indigenous people of this world. That unity comes from hours of discussions, joint actions, and an atmosphere of agreement which becomes ever more prevalent and links everyone together, even when there are still plenty of undecided strategic and organisational questions. Already in the lead-up to the summit in Copenhagen in 2009, there were signs of greater unity; now, decades of fights between climate justice movements are put aside. This is not because they have found the lowest common denominator that they agree on, but because they have the greatest possible common ground.

The FFF activists stand by their analysis, together with the scientists of the IPCC: we do not need some abstract goals for "net" zero in 2045 or 2050, but real measures which guarantee immediate massive reductions worldwide, every year. And real global justice, concrete support for people. And the realisation becomes ever clearer: for that, we need "the people," large parts of the world population.

The band of rebels doubles in size – the fossil fuel-treaty idea and the three pillars of political change

But then so much changes for us. A new perspective opens up. If these governments don't react to the crisis, what are our other possibilities? How can we translate our vision into a political reality? I continue to be convinced that for this we need a global political movement; not a movement of NGOs and not a party political movement, but a political movement in the broadest sense of the word. One which enough people from all populations can join, including in the interim periods when a global strike is not taking place; beyond marches and petitions. A movement which lives global democracy.

I look around the halls. And in the middle of these conversations, a plan starts to form on the horizon. These are individual people who don't know each other well, from different continents, with different backgrounds: researchers, activists, lawyers and so on, who have significant social (and limited financial) resources. And they are all committed to the same idea. On the Sunday between the two conference weeks, a rainy, cold day, we sit in a dark theatre in the centre of Madrid, outside of the COP site. They are sitting in front of me: the grown-up global band of rebels. Some of them are dressed smartly. But they are a band of rebels nonetheless. They all come from different countries, just like the young people in the COP halls. Australians and Canadians, people from Nigeria, Sweden, and Ecuador. For this group, led by Tzeporah Berman from Canada, a core idea is that of a "treaty", a global social contract organised from "below" on a fair international basis, which individuals, organisations, cities and countries can join, since global COP cooperation "from above" is barely bringing any results (www.fossilfueltreaty.org).

The UNEP Production Gap Report (2019) has just been published, and is shaping the conversations taking place in the breaks at the COP. It says that with the fossil infrastructure being built and currently planned up to 2030 (oil drilling towers, coal mines etc.), it is impossible to uphold the Paris Agreement. The world has already planned much more fossil fuel into its energy system. And because the absolute numbers are the important part here, this is not reversible: if these fossil fuels are extracted and burned, the temperature will rise by two, three, or four degrees.

But this means that the crucial strategic course is set: reducing emissions is not enough. That is key. And building new fossil-free infrastructure is not enough either. These two processes may be pillars of the change that needs to occur – in fact, they are the two central pillars. But another pillar must be added: preventing the building of infrastructure which has already been planned and contractually agreed, and immediately reducing existing fossil infrastructure – in a fair way. And for precisely this purpose, the group has developed an idea during the last year. What if global society agreed on a contract which would closely resemble the one which led to nuclear disarmament, the "Nuclear Non-Proliferation Treaty": a "Fossil Fuel Non-Proliferation Treaty" (Newell & Simms 2019)? Nations, cities, and governments commit themselves to leaving coal, gas, and oil in the ground, first by placing a moratorium on the building of new fossil infrastructure globally, and secondly by reducing existing fossil infrastructure by at least 7 percent per year. And this should thirdly – and this has already been outlined by this group (Hällström 2021) – take place on a socially just basis, in terms of the situation within a country and between different countries; not only for workers in the fossil sectors, but for everyone who is affected by the transition.

What is needed now is pressure from the streets, so that cities and countries, at first just a few but then more and more, create and ratify this treaty, which defines all of this globally. But do we really need another treaty? We already have the Paris Agreement. And why should we once again rely on the governments of nation states, which have prioritised their short-term economic interests so many times? Already in 2017, I made a website to present (after describing the history of the COP15 in Copenhagen) the seemingly utopian idea of a conference to determine once and for all which stores of coal, oil and gas could be extracted in which locations, and which must stay in the ground. But at the time, I didn't know what kind of process could be used to make such a calculation and guarantee that it would be followed worldwide – apart from the fact that a global grassroots movement could make it happen.

The crux: we can ensure that no more coal, oil or gas is extracted than the Paris Agreement stipulates, if we radically protect forests at the same time and reshape agriculture along regenerative lines. This quantity can be calculated. This is the only way for us to have a small chance of handing over a liveable planet to future generations; that is the concept. Of course, there are big challenges on the way to such a new agreement across society. It has to be designed so that many people see it as fair and reasonable. For that, we need a list of all fossil projects worldwide which this group has to take on (www.fossilfuelregistry.org). But the gain would be enormous: we could all assume that the poison stays in the ground and that global warming is kept in check. At least the idea is on the table. And so that the idea is also in people's heads that extracting oil, gas and coal and maintaining coal power stations and oil fields is similar to pressing the red button of a nuclear weapon. They are toxic. Building them needs to be forbidden worldwide, and replaced by renewable energy, as well as by a different global political culture. This could complete FFF's focus on emissions budgets as the central idea of a global movement.

Components and processes of the fossil society - a system theory

On this morning in the theatre in Madrid, we work together to try to understand the core of "fossil society", to understand what it really consists of. What is it exactly, this system which began decades ago when the focus on fossil fuels was connected with the capitalist organisation of the economy? And what could replace it?

Who influences which processes so that they flourish or wither; who profits; which forms of domination and power are enabled? And where can we intervene and steer these processes in a democratic direction from which everyone will benefit (cf. Stilwell 2019)? Or on the level of values and ideas: how can we ensure that people no longer accept this fossil society, and the energy and finance industry behind it? That politicians who accept applications to seek new sources of fossil energy and exploit them are revealed to be irresponsible?

With the help of system theory (on system theory as an idea, see Göpel 2022), we trace the backbone of our fossil society, and distinguish components, on the one hand, from the processes which affect them. The components are the search for new fossil fuels, meaning the strange, dead, organic matter deep in the ground; then the mining or extraction of this matter; its transportation; its purification in refineries; its export and import, often from harbours; transport and storage; creating products (petrol, plastic...); the distribution of these products; their sale and finally their consumption. All of this leads to the production of greenhouse gases and to profit. Between these individual components, there are intensifying mechanisms and mechanisms holding things back: the more fuel is extracted, the bigger the transports; the more consumption, the greater the profit and the CO_2 emissions.

All these components are affected by processes: political processes such as approval for prospecting and exploitation and for the building of the infrastructure itself (the Nord Stream pipeline, for instance); legal processes; media processes (such as normalisation through the Murdoch press, or advertising for fossil products such as cars and flights); and economic processes such as private and public financing of these processes and investment and speculation in them; and finally, in turn, processes based on politics and values, which regulate these economic processes. And behind all of that, the overall view of what counts as valuable.

This means that the question of why banks and even public institutions such as universities, pension funds and central banks still invest their money in the fossil industry, several billions per year, is connected with the logic through which profit is made on the financial market (Vogl 2021).

At the time of our meeting, the Harvard Professor Naomi Oreskes publishes an article which continues the work of her book *Merchants of Doubt* (2012), on the sheer incomprehensibility of the lobbying work that maintains this fossil society with nothing but lies, financed by corporations such as Shell and

Exxon. What the biggest fossil corporations, supported by banks and politicians – in countries like Switzerland, Germany, and Sweden – are doing to the living situation of the children who are storming the stage in Madrid, will one day be seen as a crime against humanity; that is what comes out of our discussions.

Time passes much too quickly. The people gathered in the theatre agree above all to focus on a global "Non-Proliferation Treaty": the global population has to be aware of what the UNEP Production Gap Report says. We must leave fossil fuels in the ground, worldwide.

We sketch out which processes can be stopped very quickly, and which other ones can be stopped in the longer term: banning the building and financing of new power stations, but also the search for new fossil energy sources and infrastructure, can be decided tomorrow at a global level. It will take a bit longer before we have an economy up and running which is no longer based on dominating other people and nature (Raworth 2018; Hickel 2020; Göpel 2022). That seems to be the only really sustainable way to live, especially for the ten billion humans on this planet. The idea of social relations also existing without domination must therefore be anchored in people's minds and hearts in a transformed education system, I say to myself.

The people sitting in the room have close connections with the global grassroots movements that have already been fighting for decades for social and climate justice and for "keeping it in the ground", especially in the Global South; they are connected with hundreds of similar movements, big and small – with their history of colonial exploitation and with today's processes of renewal. They share networks with the leading university researchers, not only in climate research, but also in environmental systems, transformation studies, the new systemic approaches to economics, and so on: a band of rebels with potential.

The justice aspect is central to this. What is the use of young people or other movements striking for a ten percent reduction in emissions, say the members of this treaty group from Nigeria, when there is no energy at all in the region which could be reduced, but instead a need for basic infrastructure? I think back to Greta's speeches. Already from the very beginning, she has emphasised this: that the whole strike idea and the transformation of societies only works if we don't just take account of reductions in greenhouse gases, but also work "holistically" on building dignified living conditions for everyone, everywhere. That is the common project in my eyes; unconditionally guaranteeing everyone the resources we need to live.

At the same time as the Madrid conference, Mark Jacobson and his team at Stanford University publish a detailed paper, similar to the approach of Teske et al. (2019), showing how a global, renewable, fossil-free energy system could be created in the next twenty years: country by country for 150 countries, with information about the type of renewable energy and the energy network that would be needed for the distribution of electricity.

We can't just ban the fossil energy system, we say to ourselves, without building an alternative globally. If we combine these two basic ideas, we already get a long way: a global contract to ban and dismantle fossil infrastructure; paired with a global process of building a renewable energy system, financed on the basis of solidarity.

Only the third pillar is still missing: stopping emissions by establishing national, regional and individual budgets, which is what Fridays For Future have brought into the foreground, more clearly than any other movement, together with the idea of "fair shares", meaning that richer countries (with their historical accumulation of emissions and of exploitation) have to contribute financially to the transition in the poorer ones, to repairing "loss and damage", and even to debt relief.

If we combine all three pillars (global contract to stop and dismantle fossil infrastructure; alternative renewable energy system and "basic services;" pushing through emissions budgets at a political level) and do so in a fair way by transforming the most important sectors (agriculture, transport, building and so on, supported by changes in the financial sector and the structure of the economy; see appendix), we would at least have a vague idea of how we can face the biggest challenge humanity has ever had to deal with. If it takes place in a socially just way. But that can only work, and this is something we all agree on in the Madrid theatre on this Sunday in December, if we can rely on something I call the non-domination principle, or: global, humane democratisation. How can a sustainable society emerge if we still allow discrimination or structures of domination, whether in relation to gender, ethnicity, or class (Fopp 2020/2)?

Changing democracy

It is now Wednesday, and the two-week meeting is drawing to a close. Wearing her stripy green cardigan, Greta stands at the front on the stage in the huge, crowded plenary hall and makes a speech. While the first part was about warning people against letting themselves be lulled into a false sense of security by net zero 2050 promises from the EU and other states, as well as corporations, she now reminds us that it is also dangerous to assume that democracy consists only of going to vote every four years. This image of ourselves as citizens is something we have to change, because change has to come now. We don't have four more years. Democracy is also something else. It is also public opinion and activism. That means us, every day. We can be loud and clear, through civil disobedience too, about demanding rules and laws which protect us and ensure that we are cared for.

From a systems theory perspective, I think as I stand in the audience, it is about finding the crucial lever, the "leverage point" for change (Göpel 2022). For me, this lever is about pointing out the logic of domination (meaning not just that we are burning nature down, but also that power is concentrated in a few hands) which is behind this fossil model of society. So it is about making the societal paradigm itself into a topic of conversation, and discussing why it is there. One aspect of this is to question the apparent separation of politics from the economy in late capitalism (see Fraser 2022). If democracy is only seen as a form, as the institutional organisation of political decisions, but is artificially separated from perspectives on the economic system (property structures, distribution of wealth, power and so on), everything becomes distorted. Violent, unsustainable relationships which are actually illegitimate when measured against the definition of democracy can be presented as the "will of the majority" and preserved on that basis. As soon as this problem becomes visible and comprehensible, not only does the legitimacy of fossil practices fall away, or so I hope – it can also become clear what is pushed out of the way in those circumstances, "substantial democracy": humane relations beyond domination, and care for one another and nature, which has always defined the foundations of the "economy", without really being valued (for an alternative, see the appendix).

The basic model for making grassroots democratic decisions

In the halls, we climate justice activists from all countries and generations begin to talk about what might happen to the global climate movement and Fridays for Future next year. The winter/spring semester will be starting soon, and this time I'll be working at the institute at Stockholm University which specialises in education, youth studies and intergenerational living. And so, I think to myself in the middle of the COP halls in Madrid, this daily struggle for non-dominant relationships obviously applies to our climate justice movements themselves. They, too, can become more democratic.

Far away, in foggy Zurich, Loukina is walking to her lectures at around this time, shortly before Christmas. She has begun studying environmental sciences; she could not free up the time and so she is missing from the big Swiss crowd in Madrid. The Swiss activists are also working on a democratisation project at the moment. The Swiss population is to be involved in a democratic process through citizens' assemblies or councils.

In Madrid, we begin to discuss different organisational models. Which is the best one? A democratic model that many of us adults in Madrid particularly value is based on small groups. For each of these groups ("affinity groups"), which work on a shared problem, there is a rule that first everyone has to have the same information about the issue, a similar level of knowledge. Everyone must then be able to express their worries, their fears and expectations. Only then does the brainstorming begin to decide possible solutions. These are then assessed with a temperature check; when people can see each other, it's still the best solution for them to put their hands in the air to show the extent to which they see a solution as good, bad or indifferent. A "facilitator" leads all these processes without intervening with their own opinion, and tries to guide them towards a consensus. A good rule is that someone only has a veto if they see a solution as being entirely inappropriate. A spokesperson is chosen who then represents the group's solution at a higher-level group which is formed from the spokespeople of all the basis groups; this person has the mandate to seek a new solution with the other spokespeople if there is no consensus; this solution is then reported back and discussed in the original group. This process can be repeated at any number of levels, so that thousands of people can come to democratic decisions. A few additional mechanisms have to be introduced, but that is the basic grassroots model. And this would be a way for the local and national climate justice groups to come to decisions. A collective intelligence comes into play and is allowed to develop. This is the counter-model to the fossil society and its logic, I think, in which power, resources and influence are concentrated in only a few hands.

There are 500 000 of us, young people and older ones, when we walk through the streets of Madrid during the COP meeting on Friday evening; one of the biggest climate strikes of all time. FFF Madrid, with the help of Alejandro, who has long been in contact with Isabelle, Sophia, Ell, and the other Swedish activist, have achieved something huge in organisational terms. Children can be seen everywhere with cardboard placards: here, once again, there is no planet B. The almost unimaginable sea of people is led by the global strike activists who are so familiar to me by now. They are carrying a gigantic banner ("Climate Justice") and they dance their way along for three hours through the narrow streets. They are followed by half a million people. They say to themselves: enough is enough. We want change. We are rising up.

Chapter 5: Corona, #BlackLivesMatter and the Climate Justice Movement
January – September 2020: The crisis and the intersectional, sustainable, global democratic project

Mynttorget is empty – the strike has been stopped

And so, the year begins in which we all learn what it means to treat a crisis as a crisis. And ourselves as incredibly vulnerable creatures, as part of a shared "fabric of integrity" which connects us with nature, a fabric which is so easy to damage and so difficult to patch up again.

The Mynttorget group has gathered once again. It is mid-March, and winter is gradually coming to an end. Or rather: it never really started. It was five degrees warmer than normal during the winter months; for the first time ever since records began in Stockholm in around 1750 (SU 2020).

Now the young people are putting the finishing touches to the plan for the biggest strike action of all time in Sweden, the global strike on the 24th of April. One hundred thousand people might come together; that is what they are hoping for. On the third floor of the trade union building named ABF, in the centre of Stockholm, the whole group is standing in front of some huge sheets of paper. The activists are sorting out police contacts, security arrangements, stage management and finances.

For the first time, the route will go right along the seafront, under the windows of the richest people in Stockholm, and up to the wild city park, "Gärdet". The team are well-practised, and the discussion focuses on the final details. Where exactly is the power socket in the middle of the park? Should they go public by announcing the musicians and call on civil society, "the people", to strike, finally forcing the politicians to act? This year must bring change. A few activists who they are in contact with have joined together in a movement combining FFF, XR and other civil disobedience movements and calling itself "by 2020 we rise up", because now the year is coming in which global politics must change and emissions must fall.

And suddenly everything changes. A virus spreads across the globe, along with so much suffering and pain, in the form of the corona pandemic (on the connection between the spread of viruses and human domination of nature, see Malm 2020). The young activists immediately call off all their strike days and shift to the internet, with #ClimateStrikeOnline. A new kind of activism begins: they update each other twenty-four hours a day in Zoom meetings, arrange further training in countless webinars, and wait impatiently for further developments. A few of them organise help for the indigenous young people who are most affected by Covid19 in the Manaus area, in the Brazilian rainforest, who have long been active in the FFF chats. Donations are collected and painful experiences are shared.

And suddenly it becomes clearer what it could mean to treat a crisis as a crisis. I think back to the day in August 2018, when I met the Mynttorget group for the very first time, a year and a half ago. "What do you want to say to us?" "Treat the crisis as a crisis." At this point, in early summer 2020, I sense what that could mean in concrete terms. Back then, I didn't really understand. In a crisis, it is about taking extraordinary action, not undemocratically, on the contrary, democracy must be reinforced. New laws are now being passed in the face of the pandemic, at least in the states which are not ruled by authoritarian presidents. It is about protecting every life. And yet some lives seem to be worth less than others. Those of the roughly seven million victims who die every year due to air pollution, meaning due to the burning of fossil fuels (and forests), and which belong in particular to BIPOC populations far away from the Zurich Lake or the Stockholm seafront (WHO 2014).

At the time, I think to myself: the same thing needs to happen in response to the climate crisis. And many people say the same in our Zoom meetings, including older activists in the group focusing on the idea of a global contract to keep fossil fuels in the ground. We cannot wait for years for new elections in which environmentally conscious parties might manage to get a few percent more, meaning that they can then negotiate over several years for laws which will only have an effect years later. That would mean before anything happened, the ten years would pass in which societies across the world have to "change drastically", according to the UN. We must treat the situation as a crisis and guarantee sustainability, dissolve contracts and establish a new global, democratic, fair societal contract so that emissions fall immediately worldwide.

The corona crisis highlights two aspects. Firstly, the fact that such a drastic interruption of the global economy does not even reduce emissions by the amount that would now be needed every year. Such a reduction would in any case have to be organised in quite a different way, democratically and sustainably. And: the economy is structured in a such way that it falls apart when people only buy what they actually need. It becomes easier to notice the badly paid core work of our economy (often carried out by women): the "care economy". Plus, when energy is saved and less is consumed, tax revenue disappears, and with it the welfare state. Within this system, we cannot so much as start to tackle the climate crisis. But what should this systemic change look like? What does it mean to see ourselves as part of a fabric of integrity, woven together with each other and with nature?

BlackLivesMatter, racism, and justice

The transformation has to be fair. The young activists agree on that. But what is the right way to think about justice, and how can it be implemented in everyday structures and in laws?

One aspect becomes obvious in these days in the most painful way. On the 25th of May, when George Floyd is murdered by a white police officer in Minneapolis, choked to death, protests also take place in Stockholm: thousands of people take to the streets with #BlackLivesMatter and fight for justice and for equality between all people on a living planet. Enough is enough, as many people in the climate justice movements also say: police violence against Black people must be stopped, and so must discrimination and structural racism in societies worldwide. The history of the last four hundred years has to be on the table; the authoritarian thinking and murderous actions of white supremacists, who have more and more momentum due to the Trump regime, must be replaced with real democracy. In hundreds of cities across the world, thousands of people join demonstrations for weeks on end. In Stockholm, too, some of the young people from Mynttorget take part in the demonstrations, wearing masks. But how should we go about this shared struggle for a fairer world?

Both for the young people and for the older activists and scientists, this draws attention much more urgently to the fact that the ecological crisis and the way out of it are tied to another process of change, leading away from racist structures and oppression. The fabric of integrity which is so valuable, from which we are made and which we want to continue "weaving" by building a sustainable global world, has a historical dimension which has led to its destruction in many places; the dimension of colonialism and racism. How can it be repaired? Or at least approached and addressed? Who leads this process and how can it look?

What is intersectionality? A new perspective

At the institute for youth studies and education at Stockholm University, my students are discussing theories of democracy (see chapter on education), using the tools of drama education. What would it mean to have democracy for everyone? Not discriminating, some of them reply, not based on gender, ethnicity, class, or sexual orientation. A few say: making diversity possible. Others say: including everyone, not just being "inclusive" or "diverse" by allowing people in, but ensuring that everyone has the resources they need. Meeting everyone on an equal footing. Distributing power. A debate arises over the question of how these structures and ways of oppressing people or acting unjustly relate to each other, "intersect", and how we can deal with our problematic shared history.

Almost all students within education, social sciences and humanities in Sweden take at least one four-week program on "intersectionality", no matter what they study, often designed together with the institutes for gender studies (see Fopp, 22.9.2021). This is very helpful to understand the relation between different dimensions of domination (gender, ethnicity, sexual orientation, class etc.; see Collins 2019). It is above all Black and indigenous working-class women who often suffer the most from the consequences of climate change and the fossil industry in the hands of the Global North, as research shows (Sengupta 2020; WHO 2014).

Sometimes, the students ask if there is an internal logic for the way in which racism, heteronormativity, patriarchy, and the social class system are connected through the economic system. They wonder whether one of these intersectional dimensions is more central than the others (see the chapter on the "many fights", and Fraser 2022)? Or are they based on one and the same system error? For some people in the group, this discussion is more of a thought experiment. For others, it is their life, and that makes them all the more actively involved in these seminars. They know what it means not to be able to rent a flat just because their names sound foreign, or because the market is tailored to the upper middle class. Or what it's like to study at an institution as a Black person when all the management positions are occupied by white people.

During these debates, I wish that these courses had two additional parts. As important it is to understand the influence of one dimension of oppression (such as ethnicity) on another (such as gender and class; as Judith Butler, Stuart Hall and Angela Davis have shown), the discussions can easily become academic – and the basic phenomenon can disappear: the experience of being dominated, pushed down, neglected, hurt. Teaching about it, the "intra"-sectional core of oppression, would mean teaching in a transformative and regenerative way (see the chapter about education). It is more difficult and challenging than learning in an abstract way about intersections between dimensions of discrimination. And the same goes for the other aspect which is lacking: learning to treat one another in a humane way. Why should it be enough to be able to read texts about discrimination without learning the practice of creating humane, sustainable spaces (Fopp 2016; and see the chapter on education)? Spaces in which people affirm each other, and share resources and care? This is not only about theories in books, I say to myself, but about changing lives. It is one step to articulate injustice and another, equally important, to create together circumstances in which everyone feels affirmed as a person, beyond specific actions, attitudes, and structures. Or in terms of moral and political philosophy: the idea of (intersectional) justice needs – as a political compass – a complementary twin idea, that of being humane, and making everyone visible as people with dignity, even those who we see as perpetrating injustice. That is the foundational idea behind the most important concept of all, universal dignity, or the intact space and fabric of integrity as I try to reframe it; otherwise the fight for justice threatens to result in its opposite, one could argue with Nussbaum ("on love": 1992).

How much justice is there within the climate justice movements?

The BlackLivesMatter demonstrations also bring about a reaction within the climate movements themselves. Isn't there also sexism, racism, colonial thinking, and oppression among activists? How can structures be built which don't rely on domination? What does "allyship" mean, and what does it mean to say that the "most affected people and areas" (MAPA) should take the lead?

This question already comes up in January in Davos. Greta is once again on her way to the World Economic Forum, but this time she is joined by Luisa from

Germany, Loukina from Switzerland, Isabelle from Mynttorget, and Vanessa from Uganda. Vanessa had started her climate strike in Kampala almost at the same time, literally in the same weeks of winter 2018/9, as Loukina left her school in Lausanne and Isabelle went for the first time to Mynttorget. In her texts and public speeches, she often links the climate crisis to dimensions of poverty, of children's and women's rights, but also to questions of education and a globally unjust system which relies on fossil fuels (Nakate 2021). Together, they are taking a stand against the passivity of the economic and political power elite and trying to draw attention to the fact that it is not enough to talk about "climate neutrality" by 2050. Emissions really must be reduced immediately; and that is something we have to organise together worldwide, through measures such as ending the financing of fossil infrastructure. That is what they write in a joint article in the *Guardian* (Thunberg et al. 2020).

From far away in Stockholm, I try to stay up to date with the local climate activists who are striking in Davos and preparing for their citizens' assembly. As a counterweight to the elitist concentration of power, Loukina and other young activists want to make real democracy happen, and involve the inhabitants of the village outside the conference centre in the big questions of the future. How can we reach a sustainable society together in ten years? In the meantime, in France and the United Kingdom, government-organised assemblies are taking place: a representative group of people are chosen at random and asked to suggest, with the help of scientists, changes to the law in the light of the climate crisis (Chrisafis 2020).

The young climate activists Loukina and Isabelle switch between the rooms – the glamorous world of the WEF and the citizens' assembly – and end up in a crucial picture. The international press agency sends a photo around the world which has been cropped to remove Vanessa, who was sitting next to Greta, Luisa, Loukina, and Isabelle. With her dark skin, Vanessa Nakate does not fit into the picture of the white climate movement. This expression of racism becomes visible to everyone when Vanessa refuses to accept being made invisible and shares the unedited picture on Twitter. She describes this experience, the reaction to it and the societal forces involved as well as her way into climate activism in her book *A Bigger Picture* (2021). And she talks about it when she visits Mynttorget some months later.

Many of the Swedish activists travel north during these weeks to their friends and fellow strikers who are exposed, as members of the indigenous population, to similar forms of structural discrimination and oppression.

In the weeks that follow, it becomes increasingly clear that Black activists – and BIPOC activists in general – are not given enough space in the movement, even though many activists are aware that the climate crisis affects people in the Global South much more severely and that racist structures, including within the economic system, have partly caused the whole problem (Hickel 2020). "No climate justice without racial justice", the Mynttorget activists also write. Many of us try to change, sometimes feeling ashamed and guilty, reacting to the underlying structures, every day. And so, young and older activists from the climate movements work together across the generations to read texts from the workshop book *Me and White Supremacy*, by Layla Saad (2020), and the ideas of Mary Annaïse Heglar and Leah Thomas (communicated via Twitter) on an intersectional climate movement. They read classics by bell hooks (2000) and Angela Davis (1983), and try to understand their own privileges and their own contribution to structural racism more clearly and – as a life-long task – to fight against it.

Sustainability and democracy – a systemic approach

Even during the debates at the university, it seems more and more clear: sustainability is not just linked to ecology or economics, but also to social relations and society. And democracy is not only about formal processes of collective decision-making but about the same "substance" of treating everyone on eye level as free and equal.

Sustainability seems not only to be – as the Brundtland report proposes already in 1987 – about "development that meets the needs of the present without compromising the ability of future generations to meet their own needs" (Brundtland 1987). It should be redefined as a way of relating to each other and the world, as a "being towards the world" (Merleau-Ponty 1974). In this sense, "sustainable" and "democratic" share the core "substance" of actively affirming each other, treating no one as more "valuable"; providing together the resources which enable a dignified life. It is about so much more than psychological categories such as "empathy" or "compassion" (even if the use of them is so desperately needed): it is about a stance or attitude and worldview of being humane; this is what I argue for in my lectures.

For society to become socially sustainable, just and democratic, don't we need much deeper changes? Once again: the destroyed fabric of integrity must be mended at so many levels: historical reparations, if they are even possible; structural changes to power relations; ethical processes; individual statements. Where is the space in which all this can be discussed? Who can lead these processes? How can we make sure they are rooted in institutions?

Then Tonny Nowshin gets in touch and helps us with this search. We had met in Madrid at the COP conference and had begun similar discussions at that time. Tonny comes from Dhaka, in Bangladesh. She grew up there, came to the West as a young adult and now works for a small climate NGO in Berlin. She is closer in age to the FFF activists, and could be seen as one of them, but at the same time she is able to interact with them as an experienced person. She, too, has had bad experiences within the climate movement because of her dark skin. A larger NGO excluded her from a picture of the protests taking place against the brand-new coal power station "Datteln4", which was approved by the German government and built in cooperation with the Finnish state. How can we change this injustice within the movements and find our way to a different, more supportive culture?

And so, on a Thursday evening in the virtual Zoom room, some of the most active young people from FFF worldwide meet with Tonny. There are so many questions: who should they approach when organising local protests? How can they set up the FFF network in general so that BIPOC activists are really part of it and have leadership roles?

During the conversation, the following thoughts come up (see also Tonny's own article in the *taz*: Nowshin 2020). Tonny tells the FFF activists that they should start early, forming a working group to focus on this specific topic. That they could approach BIPOC people and actively listen to what they say. And then one thing will lead to another: they just need to start and publicly form a group, make the first steps, and consciously approach people and pass leadership roles on to them. The activists talk about privilege, feelings of shame and guilt, and above all about the prejudices which many of them have, whether they want to or not, images which they have to change actively. They ask Tonny how she reacts when someone asks her about her skin colour or where she comes from, or when she hears someone else being asked this.

She tells them that inner values count, and explains that she does talk about how hurtful it is, and also talks about the climate crisis. She doesn't want people to have a distorted view of her as only being an expert on racism. She wants to talk about the climate crisis and about the activism she is involved in, so that people will see who she is. In this meeting, some of the concrete work of change begins, as well as a kind of joint search for a way of living together beyond structural domination. Striking on Fridays does not work in Bangladesh, for example, because there is no school on Fridays. So, everyone has to show consideration when planning global protest actions. The young people express the fear that in a few weeks everything will go back to the status quo and oppression will again become invisible. Have the BlackLivesMatter demonstrations really led to a social "tipping point"? Structural oppression might be named as such. People who simply want to carry on as normal cannot do so quite as easily. As with the MeToo movement, values are beginning to shift, across society. Could that also happen in relation to the climate crisis and its effects?

Tonny talks about her fear and grief for Bangladesh, the unimaginable floods which destroy the border region with India, especially during these summer months; and the new coal power stations which are to be built with the help of German firms. Bangladesh is literally under threat of being swept into the sea and is at the same time afflicted by terrible heatwaves. During these weeks, an article appears in the respected journal PNAS (Xu et al. 2020) which seems almost absurd. It says that by 2070, up to three billion people could be living in regions which are uninhabitable, and the area around Bangladesh is one of them. Tonny is asked one more question: What should we do about the dilemma that the clothes we are buying are made by people in Bangladesh for low pay and damage the environment, but at the same time guarantee jobs? Does that make sense? Should we be boycotting clothes from H&M?

And that brings us back to the question of justice and what is right. Tonny replies that something is fundamentally wrong here. Something about the system itself; something we cannot avoid. It is about something deeper than the question of whether to buy or not to buy. A racist element is built in: millions of women work in Bangladesh and make a few white men unbelievably rich, in Sweden, America, and so on. Tonny says that we can't correct that only by buying or not buying the clothes. We also have to work even on a whole different project together to point out these structures and replace them, so that everyone is looked after equally and power is shared. After all, there are one hundred corporations, often owned and run by white men in the Global North, which are responsible for most of all emissions and are therefore at the core of the ecological crisis. This shows that all the dimensions of domination (gender, ethnicity, class) and the climate crisis are connected with each other (see Fraser 2022; Bellamy Foster 2010).

That is why it is important for me to be working at the universities on theories, for instance on the concept of a "shared fabric of integrity," which could throw new light on the concrete processes through which we can weave this fabric together and stop destroying it. Ethics, politics, sustainability, movement organisation and history must be viewed in conjunction with each other, in very concrete processes. For me, these conversations lead to the realisation: if we want to develop a program that could be called "People For Future", we have to take on the challenge and make it clear that reshaping our fossil society is also a movement away from these systemic mistakes. But how can we formulate this "democratisation principle" more precisely?

What is the core of the new politics?
The young activists and scientists write their manifestos

Suddenly all the activists are writing statements about how the world should look – partly because the opportunities for real protest actions are limited. We are still in the time of corona. Use the chance, many people are saying: if billions are being invested in getting out of the corona crisis, they should be invested in "green" projects. These strategy papers by NGOs have titles like "Build Back Better" and "Green Recovery." But why should some fine words from a few NGOs suddenly get politicians like Merkel, Macron and Bolsonaro to change their way of thinking and acting? I ask myself this at the weekly meetings with the movements and the "global communication directors" of these organisations. It should be about building a real movement which believes that we can change the political framework. But how? How should the manifesto of this movement look? We need a platform to gather everyone, I think; and we need to focus on two or three core points of a new framework which everyone can get behind.

During these months, two different groups come up with ideas for how this text could look, working in parallel. On the one hand, some young activists across the world. And on the other, a group of older activists and scientists. The results are presented by the young people in their open letter to the EU; they call it #FaceTheClimateEmergency (climateemergencyeu.org). Within a few days, it has been signed by more than one hundred thousand people, including the most well-known climate scientists. And Scientists For Future produces several texts, including an official statement to the EU, which is developing a new climate law at this time.

Quickly, it becomes clear that both approaches share the same core. Both groups aim for a middle way between overly specific (party) politics and abstract demands. Both present two or three principles which ought to guide policy, as well as two or three ethical values for how policies should be realised. What is this core which defines them both and underlies a movement that could be called "we, the People For Future"?

There should be fair emissions budgets, globally, nationally, and locally, starting from the 350 Gt which is roughly what we have left (in 2021); they should also be fair in historical terms. And all governments must immediately present plans for how these budgets can be realised in all sectors – a new framework must emerge and shape all political decisions. Secondly, as of now, fossil infrastructure should no longer be financed and developed, but should be reduced and replaced by a sustainable energy system, financed through "fair shares", in accordance with the Emissions Gap Report by the UN. And thirdly, this should take place fairly. The Global South should move into the centre of this global reshaping of society, in quite a different way. Because fourthly, structures of domination must be dismantled in all these steps, when it comes to gender, class, and ethnicity. This is roughly where the different papers overlap. If this framework is established, there must be further discussion in democratic arenas of what that means in concrete terms: for instance, how it could be combined with debt cancellation, a global basic income (an overview on the political theories of basic income: Bidadanure 2019), and so on.

Most importantly, what we need immediately is a realisation that this is a crisis, the young activists write, and we need systemic change. Greta talks about that on her summer program on Swedish radio, with a million listeners, and about a democracy in which we look after each other globally (Thunberg 2020/3). And once again, I think: it is one thing to make injustice visible and dismantle it. The other complementary project is actively guaranteeing a dignified life for everyone. It is not enough to make sure that people are not exploited. Again, the argument is that we need more than justice. The idea of (intersectional) justice must be complemented by that of caring for everyone's dignity and being humane: we all have to have enough to eat and a roof over our heads which will not be swept away by floods; we can provide each other with enough resources as global universal "basic services" without people having to pay for them individually. That is why I talk about this fabric of integrity and keep coming back to its historical dimension and to how fascinating and valuable it is. (The appendix of this book develops these joint principles further into a shared program.)

The S4F statement on the new EU climate law

But these thoughts also have to be expressed in concrete policies. As Scientists for Future, we have also been working on a statement that aims to change European policy directly, for the next ten years. This year, the EU is planning to introduce a new climate law which will shape all of policy in this part of the world (Jakubowska 2021). Judith Hardt, one of my research colleagues from Berlin, gets in touch to say that we as research institutions and academics have a humanist task, and with that also a responsibility within global society. If we see that policy does not match the Paris Agreement or our research, we ought to react. And after all, the two of us belong to the advisory board of Scientists For Future, which now consists of about a hundred professors, among them the heads of university institutions and gigantic research centres. We also have contacts among the climate specialists in the EU parliament. It is the EU commission which will propose the new law in March, but it is the parliament which then has to change and approve it in a six-month process. Here, a point of leverage emerges.

And on top of that, a specific event is coming up: some of the young activists are travelling to Brussels to meet their fellow activists from across Europe in front of the parliament. As Scientists For Future, we could publish a statement at the same time; that is our idea. And while the virus spreads round the world with terrible consequences, we get the hundred scientists on the S4F advisory board on side. A working group is formed. We all read the EU commission's suggested law and come up with concrete changes. The political framework must match the science. A challenging working process emerges under extreme time pressure. We all edit a google doc at the same time; we frequently have to lock it for hours in order to summarise all the comments; then we continue working; we circulate the final version; change it; circulate it again – and find our way to a common position.

Of course, the specialists, world-leading climate researchers, have the biggest say, but this is still about a systemic way of thinking which connects all academic disciplines, climate science with ethics, economics with political philosophy. The core points are: the goals have to be much stricter; a 50 or 55 percent reduction in greenhouse gas emissions by 2030 is not enough; especially not since these numbers only actually mean 35 percent, because the comparison is with the past rather than with the current level. And Europe needs to make a more drastic transformation than the one in the proposed law, for reasons of global justice.

Thus, we put the aspect of justice in the foreground. And for us, it is crucial that the text of the law includes the idea of a scientific committee which will constantly assist in EU policy development. Gone is the time when politicians could avoid facing up to the facts of the ecological and climate crisis – or at least, it should be gone. The statement is published and the activists tweet it into the world. Meanwhile, Kevin Anderson and Isaak Stoddard of Uppsala University calculate (Anderson et al. 2020) that with the official goals the EU would emit about double the amount of greenhouse gases that would be compatible with the Paris Agreement – if it even stuck to these goals. We have to make that clear to the public, I think to myself. This would lead to a world that is two or three degrees warmer, with all the disastrous consequences.

All these thoughts lead to an idea. We could become the People For Future: we scientists could join together with parents, artists, workers, economists and so on; each of these groups have set up their own networks by now in various countries, with thousands of active members. So that a gigantic global, continuous popular movement emerges. One which is guided by the two or three shared basic principles which the young activists and the scientists have come up with in two years, and which we will defend as long as it takes for the political framework to change.

Debt cancellation and reparations for the Global South would be a possible beginning of the work on weaving the shared historical "fabric of integrity." But then comes a change which is just as important: how can we reshape our economic approach so that the Global North no longer makes a profit at the expense of the Global South? More and more ideas about this come out of the movements (Hickel 2018 and 2020). Through the demonstrations and the court case against the police officer who killed George Floyd, a discussion also develops over these months about the role of the state, the exercise of power, and violence. A discussion about abolition emerges (Loick/Thompson 2022), which also becomes topical due to stricter laws affecting activists, which are being introduced in many countries during these weeks, including Switzerland and England, so that climate protests can almost be treated as terrorism. Arrests are becoming possible based only on suspicion, without a warrant.

What is becoming palpable during these months, in relation to the reaction to corona virus and to racism: how important it would be to put the dignity of everyone in the centre of political and economic action, to create structures which are not treating some lives as less valuable, and to make the shared fabric of integrity visible, as our common task.

Back at Mynttorget

At the end of August 2020, the time has come. Exactly two years have passed since the beginning of the strike. The activists gather once again in Mynttorget, between parliament and the palace – with safety measures, including social distancing and masks. They lean on the wall, as usual, and look across at the parliament. And in that same August week, all the climate movements join together, led by Extinction Rebellion, to block the centre of Stockholm, first the bridges around the parliament, and then the central shopping streets. This time there are not 100 of us, as there were two years ago, but hundreds.

And finally: on the 25th of September, the first global strike takes place since the outbreak of the pandemic. The young activists have agreed in their chats that they will change the communicative structures of FFF in such a way that people from the Global South and BIPOC communities can take over leadership, under the name MAPA ("Most Affected People and Areas"). The central theme of the strike: the climate crisis and global social justice. Tens of thousands of young activists across the world return to the streets of this wondrous planet.

Chapter 6: Many Fights, One Heart – #UprootTheSystem

October 2020 – October 2021: Forests, agriculture, banks, courts, and the political manifesto – all the way to a new theory of democracy

How is everything connected: forests, finance, and fuel?

There is a gentle breeze. A small group of younger as well as older activists are walking through the forest north of Stockholm. Here and there, a pinecone falls. Perhaps the trees want to play with them, I think to myself; also out of gratitude that someone cares about whether they survive, and about whether their relatives survive, the last real forests in the north of the country. The pinecones are stuffed into the bags which the group have brought with them. They are to play an important role at the next global strike in front of the Swedish parliament, the first really big international strike since the start of the pandemic.

Fridays activists across the world are swarming out in every direction to slow down climate change and mass extinction. They divide themselves into countless groups dedicated to all different projects. They are faced with many different political struggles: over the forests, over agriculture, soil, financial flows, court cases and climate goals; I wonder whether there are too many. And how exactly are they connected? How can we win them without getting bogged down? Maybe it is not enough just to try to stop emissions and keep fossil fuels in the ground. Maybe we have to rethink what defines democracy.

We all sit down on a fallen tree trunk and look around. To be precise, this is not a forest, just a small wood, close to the edge of the city. And the people sitting here know what a forest is. Many of them have specialised in forestry, because this plays almost as big a role in the climate crisis as the fossil fuels which have to stay in the ground, especially in Sweden. The emissions created by non-sustainable forestry and the burning of "biomass" have been so significant in the last 30 years that one could say that Sweden has not actually reduced its overall emissions at all; a lot of them do not even appear in the statistics (Urisman Otto in Thunberg 2022). How can we then demand that the population of China, India or Nigeria reduce their emissions drastically? CO_2 is released from the ground, but also from the burning of products such as paper and cardboard cups. Some say that it is to some extent absorbed by the trees as they grow back, but that takes decades. Time which these children and young people don't have; not if we soon reach tipping points.

In conversations with scientists from the agricultural universities, it soon becomes clearer to me: it is barely possible to understand anything about Swedish history, the emergence of banking, the power structure of the country or today's politics, including EU politics, if you don't know who owns the forests and what is being done with them (see Röstlund 2022). It isn't trolls or other fantastic creatures that shape the forests, and it's not their stories that accompany us on our outings, but very specific property relations, technologies, concepts of nature, and economic approaches. And I think: maybe our understanding of these connections could be something like a blueprint, a key to understanding much more about society and the necessary changes ahead of us?

But in these pandemic months, the activists are not the only ones going out into the open air. The global band of rebels also sets off in different directions, from Brazil to Finland. Careful work begins, sometimes also "pushed" by NGOs. They form new working groups, find each other online, form friendships – and they work together with specialist scientists. Many of these are connected with the Scientists For Future movement. In this way, behind the visible Friday strikes, a hidden structure emerges, unseen by the public, involving many different overlapping actions initiated by the global FFF network.

Some work together with their peers on the topic of forests and agriculture. Others get into stopping the fossil industry; and so on. They don't just collect knowledge about the deplorable situation, but also about what would be needed instead, in politics and on the ground: how regenerative agriculture and forestry could enable the woods and fields to flourish, as well as the people working in them, and what a sustainable financial system would look like.

Meanwhile, I try to keep an overview and excitedly follow the academic research into what might be called "regenerative life energy". This is the energy of living, blossoming, fruitful socio-ecological systems – or whatever we want to call the complex socio-ecological context. Anyway, I think to myself in these months in my seminar rooms: at university, we teach too little about sustainable processes of exchange and synergy, including when it comes to humans' interactions with each other and with nature. These themes are missing from our teaching, despite all the ground-breaking research which has long become established in social psychology and physiology (see Fopp 2016; Raffoul 2023). And as so often happens, this work makes me aware that I have a limited European view of the topic and of the movement, which has become so diverse and broad, and which also takes account of research into forms of knowledge which have been passed down and dispersed for years (Solnit/Young Lutunatabua 2023).

The young activists are becoming familiar with struggles over property relations, with the conflicts of interests behind EU politics, but also with the worlds of lobbying and trade unions (on the problems of the fossil lobby, see Götze/Joeres 2020). For every topic, a website is created, coordinated globally, and a campaign is posted on https://www.fridaysforfuture.org.

The work continues rapidly, because decisions are approaching, in these very months, often at the European Parliament. What should be defined as "sustainable" energy – for instance, should that also include nuclear energy and gas, as planned in a worrying German-French "pact" (Rankin 2022)? Which climate goals is the continent setting for itself ("FitFor55"; Jakubowska 2021)? And in particular: the biggest budget allocation of the EU will be soon be defined for the next seven years in what is known as the CAP, the "Common Agricultural Policy". This decides how money is distributed for agriculture, which farmers it goes to, how the soil is cultivated, what should be planted. This not only decides the level of (methane) emissions, but also whether species diversity will suffer even more (on this problem in general: Chemnitz 2019; Westhoek et al. 2014).

What should we eat in the future? Animal proteins, which would mean accepting suffering and emissions, or plant proteins? How can we guarantee food security for everyone? How can we protect thousands of smaller farms and ensure a good environment for workers? Values are at stake, not only financial but existential; and also the whole Paris Agreement, the question of whether Europe is making a fair contribution, in global terms, to stopping emissions.

But when they return from their protest actions in Brussels in front of the EU Commission or the Parliament, often despairing over the decisions that have been made and even more strongly connected with each other, I am worried. Isn't there a danger that the movement will be splintered between too many projects? Isn't all of this NGO work? What has happened to the school strike, to the mass movement? Above all, however, a central question keeps coming up: isn't there a shared basis, a Gordian knot, a political and economic chess move which would mean that we could suddenly organise all areas sustainably? That we could win the many different fights by seeing them as one big fight?

As a motto for the imminent global strike in autumn 2021, the global youth movement has chosen #UprootTheSystem, after a long international process of discussion. For that reason, the Stockholmers are now combing the woods and collecting pinecones which they want to use to "write" the hashtag on the ground in front of Parliament. But what is "the system"? I email Scientists For Future. We have to meet in the coming weeks. We can't just watch the young people getting entangled in struggle after struggle.

But in order to see something that connects everything, we have to consider the details of the individual groups, so that we can analyse the different sectors and find a general pattern that transcends them. One of the other goals of my research is to reach a better understanding of democratic means – and of democracy itself. Because even if a deepening picture is emerging more and more over these years when it comes to political economy (the problematic interplay of capitalist markets with other forms of domination), and alternative political visions are being developed (by Hickel, Göpel, Schmelzer, Raworth; see the theory of the three pillars and two principles in the Appendix), there is still one big question: how is it possible that in so many areas the basic necessities of human life are being damaged or even destroyed, apparently on a legal and politically legitimate basis? What would a successful democracy be?

Deforestation and the life of the forests

Many of the Swedish activists join the working group for the protection of the forests. Again and again, they travel to the area of Jokkmokk and visit the Sami activists who are the same age as them. The biggest problem: these forests are being cut down to a disproportionate extent, even the oldest, using damaging methods, and then often end up in factories manufacturing cardboard cups which are immediately burned (Röstlund 2022), instead of the forest being protected.

Above all, though, Sweden allows a technique which is banned in other countries and which could hardly be more brutal: clearcutting. Across a large expanse of land, everything is suddenly razed. All trees will then always be the same age; often they are planted artificially as monocultures of invasive species (Westberg 2021). A tree plantation emerges where the forest used to be; life is cut short. Through this, the soil also releases extra emissions. There are obvious similarities between the images of Sweden and of Brazil, where the rainforest is being razed faster than ever before (Spring/Kelly 2022). And this has political consequences far beyond Scandinavia: the Swedish and Finnish governments are doing everything they can to legitimise these methods both within the EU and indirectly worldwide (Lind/Sanddahl 2023).

The young people familiarise themselves with the material and ask self-critical questions: is this about thousands of jobs which would otherwise be lost, or is it about economic collapses which would endanger the welfare state? But that does not seem to be the case. Renewable energy, for example, also needs be developed further. Workers are in demand; regenerative, careful forestry is certainly possible; and all of this quite apart from the cost to the whole of society from the climate and biodiversity crisis.

But above all, we all think to ourselves, sitting on the tree trunk: what a contrast with what a forest can be, a home for many in northern Scandinavia, particularly the indigenous population, which has been treated so badly; and of course one of the most fascinating places, as the quintessence of life itself. That is what we keep hearing in talks by leading scientists, who explain how trees communicate with each other and help each other, and describe the interplay between older and younger trees. Yes, the tree trunk we are sitting on is rotten and disintegrating – but it still may be the most important part of the "ecosystem" in which we find ourselves, and not simply biomass to be burned. Very particular plants and animals nest in it and enable biodiversity and carbon storage (Wohlleben 2017).

What do we need instead, we ask: what kind of politics, what approach? We need to see the whole picture, across different times and different cycles, the scientists say, when the older activists who now call themselves People For Future sit down with them (on thinking in cycles, see von Redecker 2021). Forests and the soil are complex structures, and they are alive. Like us, the people who grew up in them or close to them – like the other species of ape. And it is this living forest that the young people want to protect.

We all try to see the power structures and patterns which connect this struggle for the forest with the struggle for other areas of life, for example with agriculture, with the fossil industry and with the financial system. The journalist Lisa Röstlund (2022) has spent years collecting the knowledge that helps to see such power structures, which decide what will happen to the forest: knowledge about lobbies, the politicians and big landowners who work with them, including the church, for instance. However, she also shows how these people and their ways of thinking shape the universities as well as the influential research; and how reporting in many parts of the media is shaped by information from those who often have their own economic interests. Sociologically speaking, they represent – as the working group is realising ever more clearly – a "clique" of relatively few men who control this area of the economy and the ideology of a whole country. But this power structure is supported by a middle class: we hear about the fact that the biggest owner of forest land, Sveaskog, actually belongs to all citizens as a state company, but still allows harmful practices. The Lutheran church, too, which still counts millions of Swedes as its members, plays a dubious role – as internal criticism has also pointed out (Ringberg et al. 2022) – given that huge areas of forest belong to it but are not sufficiently protected.

One thing further amplifies the sense of powerlessness among the young people and among us scientists: barely any of the eight parties in parliament want to change anything about these structures. The Social Democratic Party, which seems to be stuck in the past, is involved in this project of deforestation, together with the Greens, who legitimise the use of the forest as biomass, and the centrist neoliberal forces. The basic premise: nature is primarily private property. And the right to ownership is defined as being able to dispose of nature at will, to dominate it, even when this cannot be justified in terms of society as a whole, either economically or ecologically. All this could change.

And in fact, something does happen in society over these months. Making the power structures visible and drawing attention to untruths in relation to scientific discoveries does have an effect. The established newspaper *Dagens Nyheter* writes about emissions in relation to the forests, and about what all of this means for the intensification of the climate crisis (Urisman Otto in Thunberg 2022). Together with Scientists for Future, the young people agree on what must be guaranteed by politicians: they have to protect the forests, ban invasive species and monocultures as well as aggressive clearance methods; they have to support workers and guarantee their livelihood as well as their retraining in regenerative methods. Quite similar principles apply to protecting the world' oceans, for which a different global group is fighting – headed by the Fridays from Göteborg and Lysekil. They are also fighting for a new global agreement on biodiversity, which will be decided by the COP15 in Canada.

And still we ask ourselves: don't we need to see nature differently and define it differently in legal terms? I make notes for the approaching meeting with my colleagues. What are rights in the first place, in particular the right to own the forests and the soil, and to damage them? How are rights connected with the economy, politics, and democracy? Could we start there and find a bigger lever?

What becomes clear is that no analysis is possible without an understanding of intersectional power structures, meaning an understanding of how privileges and membership of specific groups (gender, ethnicity, class etc.) influence politics, the media and education. In more concrete terms, the question is therefore: how do we extend democracy so that this "substantial" dimension of power inequality can be dismantled and replaced with caring encounters at eye level; how can "formal" democratic processes such as elections and the functioning of institutions be reshaped in response to this?

And: can we ever ignore what we are or how we function? Can we ignore anthropology, a scientific picture of how we ourselves belong to nature; what it means to treat each other and nature in a caring way or a dominant way?

Humans as double creatures

All of this, too, is a subject of discussion for the activists in the woods. High up above the treetops, whose pinecones are being collected by the young people, birds are circling and flying over to Edsviken, the big lake in the north of Stockholm that borders directly on the forest, and opening up the space to the infinity of movement. We humans are such peculiar double creatures, I think then. On the one hand, we are so tied to the embodied perspective of our own position in the midst of the world, with our own specific experiences and our unique biographies, which shape our view – meaning that we can perceive the forest as an immeasurable gift, but also as threatening and uncanny. In that way, the centre of the world is always directly above us. The dome of the heavens, which we keep on shifting when we move, retains its central point above our heads, while touching the earth at the horizon.

And at the same time, we are the ones who can adopt a "view from nowhere", as the philosopher Thomas Nagel (1989) has said – an abstract, distanced view which is outlined by maps. We can extend the weather to the climate, we can develop technologies and altogether make the world calculable. Or at least try. This perspective discovers that trees take CO_2 from the air through photosynthesis and transform it into carbon, into "tree", and at the same time through this produce oxygen for us animals and humans. It can take account of geometry, physics, and mathematics. It can trace experience back to the firing of neurons in the brain, and analyse the life of trees in terms of ecosystem tasks and cycles of metabolism.

But there is also a third part of this, I often say in my lectures at the university, pointing to the Gestalt research of Maurice Merleau-Ponty (1908–1961); a third element for which we in our culture barely have a language, something we often overlook, but which is perhaps the most central of all; the dialectic dismantling of this opposition between the subjectivity of our embodied experience and the objectivity of technology, between the individual and the whole: namely connecting ourselves with the imagination and with empathy, with a shared spirit, a shared idea of being humane (Fopp 2016). Not because trees begin to walk, as in Tolkien's Lord of the Rings, but because we are tied to the vulnerability of embodied "being in the world", and can therefore transcend it, not as something that calculates, but as something that connects us.

I awake from these dreams and find myself back in the forest.

Back in the forest – "blah blah blah"

They all stand up and go on collecting pinecones. Some of them will travel to Italy. In fact, it is a preliminary meeting for the climate conference in Glasgow in November. What could be the most important message, now, in a late phase of the pandemic, with a view to the climate conference?

Everyone is talking about "There is no Planet B" and "Build Back Better": the idea that the gigantic investments which are being made by governments worldwide to get the economy going after the "standstill" should make societies more sustainable. They are still trying out the text, playing with all the phrases these politicians are constantly coming up with while doing nothing. Emissions are still rising. Most of the money which is supposed to be leading us out of the pandemic is still going to fossil products and to the banks that make them possible – to the outrage of the young activists (Kottasová et al. 2020). Greta will say in Milan: Blah blah blah! That is all blah blah blah.

Now, they sit together and eat carrots. Carrots have become important to the movement; they are a favourite food. The short poem which probably comes from Henrik Ibsen, "Little Carrot", has become a kind of motto; it is a hymn for the small ones who live in the shadow of the big ones. "Baby carrot: small. Ugly. Lives in the shadow of the carrot. Baby carrot." I admire the young activists for creating these moments of amusement amid the fight against time and against emissions. Because they have no planet B – even if Elon Musk and Jeff Bezos are going into space for the first time, releasing enormous emissions and dreaming of colonising Mars.

Here they sit, gathered in the woods to the north of Stockholm, doing everything they can to preserve a planet worth living on for everyone. For passers-by, they just look like ordinary walkers who are taking a short break in the woods and nibbling their carrots.

Agriculture – the EU's CAP ("Common Agricultural Policy")

From every country in Europe, the young FFF activists travel again and again to Strasbourg and Brussels and protest in front of the EU institutions. The Nordic governments should not be able to enforce their destructive forestry policies in the EU. If only many more would join, as People For Future, I think to myself. If there were thousands of us, especially older ones, we could change politics.

They gather in front of the EU parliament, together with their friends who are confronting the decisions made on agriculture, the "Common Agricultural Policy" (CAP).

For this group, it all begins with a warning tweet from Harriet Bradley, a specialist in agricultural policy. It is passed around the movement, and panic erupts. The danger to which the tweet is pointing: the EU will soon be proposing a policy for the next seven years, a crucial time, and this policy is still far too preferential towards large-scale enterprises. The policy will make the quality of the soil worse, cause huge emissions and animal suffering, and damage biodiversity (Pe'er et al. 2020). In these months, more and more reports are coming in of entire species going extinct – birds are affected, and insects in particular. And with them, the basis of agriculture. The "Living Planet Report 2020" talks of 60 % of all mammals, fish and reptiles having been wiped out since 1970. Bees are especially under threat.

After school, the group begins its almost daily work. They are sometimes joined by Leipzig researcher Guy Pe'er from Scientists For Future. He helps the young people in Zoom meetings. Which power structures are represented, and which interests are at stake? Soon, they understand better which of the lobbying groups prefer the biggest companies, so that many smaller farms are disappearing, in Sweden too; and the problematic role played by umbrella organisations, whose power games hardly take account of the climate and biodiversity crises (CorporateEurope 2021). And where are the trade unions? Climate and environmental protection and a secure future are some of the most important topics of the workers' movement, but the unions do not seem to be taking action (Bell 2020).

And once again the overarching questions: how can we anchor care for nature and other people in legal and economic terms on a deeper level? Because a similar pattern emerges which could already be seen in relation to the forests (for this pattern and a sustainable agriculture, see Pelluchon 2019). Many actors who are preserving the status quo are supported by a relatively small clique, dominated by men who combine lobbying, politics, private interests and research guided by economic profit, and who use the formal infrastructure of democracy to damage its substance, the "content" of democracy, meaning the provision of a sustainable basis for life for everyone, and encounters without domination.

And once again: why does the population not react? Isn't this about their future and about ensuring that the soil is not destroyed? But the young people are pretty much on their own. There are also very few European media that actually receive attention (euractiv.com is one of the only examples), and in national media the topic is not discussed enough. Understanding these power struggles would actually be more important, and more interesting, we say to ourselves, than any primetime series on TV.

That kicks off a discussion within the climate movement. Should the young activists talk about "stopping" EU agricultural policy or just about "changing" it? Should they be making policy proposals at all? The Swedes are sceptical. Isn't FFF a grassroots movement rather than a lobbying group or an NGO? In May 2020, more than ten European Fridays For Future groups publish a joint statement for a change in policy and thus for a different use of the 58 billion euros which are invested in agriculture annually by the EU (www.fridaysforfuture.org/change-the-cap). 3600 scientists protest at the same time (Pe'er et al. 2020) and support their demands, including the rapid replacement of animal with plant "products", and the replacement of an agriculture characterised by monocultures and pesticides with a regenerative approach. We need a transformation, they say, so broad that we can hardly keep track of every aspect of it, reshaping the way in which we interact with animals, the soil and nutrition (on the cultural background, see Foer 2019). But the most important aspect for them is that they stand behind the workers, regardless of whether they are currently working sustainably. The transformation has to be socially just, they say. And the protest has an effect: through the actions of the FFF activists, EU parliamentarians begin to rethink their position – and to change it. In particular, the social democratic block breaks apart and many of the MEPs endorse the arguments of the young activists and the scientists. But there are too few of them. In the end, the lobbying groups of the biggest companies and the associations with the most capital win out.

Here we can get a sense of an important "ingredient" for understanding democracy: evidently, purely formal democratic processes can be pushed through against social justice and thus against the "content" of democracy. That must change. And the organisation of the economy, with the financial sector at its core, seems to play the most important role in that.

The financial system

At the weekly global Zoom meetings of the FFF movement during the whole of the pandemic time, there are often a hundred young people from across the world taking part, so that together with their banners they don't all fit onto a single Zoom screen. The richer northern countries are in the process of getting vaccinated, while the countries of the Global South are barely receiving medicine.

Meanwhile, one insight becomes clearer to many of us during those months: one of the most influential factors in our relationship with each other and with nature is the way in which the economy is organised, in terms of its problematic centre, the finance sector, the banks, which make enormous profits in Sweden year after year, including and especially by trading in fossil fuels. In Sweden and worldwide, the rich get richer, partly because of this trading. There has not been such inequality since the global financial crisis one hundred years ago, according to the most important economic journalist in the country (Cervenka 2022).

Complementary to events in other sectors (forest, fossil industry, agriculture etc.), this is the factor which shapes everything, the core of a market organised along capitalist lines, obeying rules that could also be changed (see Raworth 2018; Göpel 2016). Thus, in this area too, a global FFF working group emerges which dedicates itself, together with academics, to precisely this financial system.

There are protest actions such as the one against the Standard Chartered Bank, which also sponsors FC Liverpool. They are often led by young people in the countries most affected by the crisis. They also try to show that it is not enough to follow the tendency of the big NGOs such as 350 and Climate Action Network, and just to ask the banks to "divest", meaning not to invest in fossil fuels, as most financial institutes have continued to do with billions of investment every year; just like pensions funds and universities.

No, say the young people, the financial system also has to become sustainable and really – in terms of content – democratic. And they receive support from many scientists who have joined together in Sweden to form the not-for-profit association Researchers' Desk (www.researchersdesk.se), a kind of Scientists For Future, initiated by Toya Westberg and some of her colleagues.

Some of these scientists point out the actual nature of our form of economy and draw attention to the fact that not all markets are capitalist, or have to be capitalist. Terms such as "investment" and "profit" are used misleadingly, when what is meant is returns on capital. Such insights could help to see the extent to which a relationship of domination is built into our society, between those who own capital and those who work in the companies that make these profits possible. Here the potential for change begins to emerge, leading in the direction of a just democracy, they say. Some researchers give a lecture for the young people on the structure of corporations, particularly banks, and their problematic legal definition, which basically forces them to pay attention to profit primarily and to exploit nature and humans – and they explain how we could replace them with more democratic forms of organisation and definitions of business goals, for instance in a postcapitalist economy (see the lectures on researchersdesk.se; see also Raworth 2018).

The Stockholmers join with specialists from "Fair Funding" and analyse the different banks in the city. The biggest look particularly bad when it comes to sustainability. However, some are based on a new model, and get by far the best ratings from the activists. So it really would be possible, we gradually begin to see, to organise the core of the economy through political regulations in order to reduce the incentive to destroy nature; and so that those who do have capital at their disposal no longer get richer without personal risk at the cost of the workers. In such a society, people would be able to meet each other on an equal footing, and power would be distributed. At the same time, one of the main motors behind these business models would disappear, since they automatically demand exponential growth (Schmelzer et al. 2022). There are numerous alternative models of common ownership, cooperation, or other approaches (see the chapter on the economy and Davos). They could – and if we don't want to exacerbate the climate and biodiversity crises, they must – become the legal norm through democratic regulations, according to researchers.

This could counteract the wrongs of the undemocratic situation we are currently in, in which owners of capital become ever richer without doing anything to earn their wealth. Countries such as Sweden have no tax on wealth or on inheritance (Cervenka 2022), which leads – along with the mechanisms whereby cultural capital is inherited in the education sector – to a situation in which power in society is pre-structured in a modern form of feudalism, unequal and unjust.

And once again the question: how do we formulate the basic problem, if formal democratic processes allow the wrong power structures to be defended – against the "substance" of democracy? Here, too, this is about a few huge actors controlled by a small sector of society mainly dominated by men, who are working with the same ideology as non-regenerative forestry and agriculture, and who are working together with the key powers behind the fossil industry.

The ideology of these actors still continues to dominate the economics training offered by universities (this has been criticised by the international university movement "Rethinking Economics", which argues that there has been a "dogmatic" distortion of scientific thinking: https://www.rethinkeconomics.org). And the broad masses of the population, including most of my colleagues at Stockholm University, may view these events sceptically, but do not speak up – as if there were no way of changing it or no alternatives. A further important insight is that the central banks could redefine their task and their structure and bring about sustainable transformation democratically (see Vogl 2021).

Is all of that actually legal, the activists ask themselves, and continue to collect pinecones. How can formal democratic regulations such as the definition of property law enable rich people to enrich themselves further by destroying what is needed for life?

This is about structural problems, it seems to me, which cannot be reduced to psychological phenomena such as greed. Similarly, it is not so much sociological categories such as "strata of society" that are relevant here, but rather political economic categories such as "class": it is not about what people want psychologically, or which part of society people belong to, but about the extent to which we reproduce the structures giving people more power at the expense of others. Structurally, the current definition of democracy seems to be undermining the actual preservation of democracy.

Legislation and legal cases

That is why another global group is working on the role of courts and the law. Can they stop this destruction of life through the courts; can they redefine the connections between property, freedom, and consideration for future generations?

Some of the Mynttorget group on the tree trunk are also reading up on this subject and starting court cases against their states and against the EU, because they are not responding to the climate crisis as a crisis. They are helped by specialists from Scientists For Future. But what can they base their cases on? On the declaration of the rights of the child? "You would have to point to concrete cases in which people have already been harmed now by the wrong policies," say some law professors at Stockholm University. Others refer to the right of children to grow up without existential angst. In the wake of the Dutch Urgenda suit, the young activists found projects such as "Aurora" in Sweden (supported by more than 1600 researchers), which prosecute the state. In Switzerland, it's the women of the "Klimaseniorinnen" ("Climate Seniors") who go to court (www.klimaseniorinnen.ch).

For many of them, it is also not possible to understand why the main owners and the upper management of the most important actors in the fossil society (banks which invest in fossil production; fossil corporations; forestry and agriculture which damage soil and forests etc.) are not taken to court more often, as a farmer in Peru did with RWE, the German energy corporation, because climate change was destroying the environment where he lived (Marusczyk 2017).

They are well aware that their production methods and profits have lethal consequences. Why should we accept that? The fabric of integrity, which we all share, is damaged by them, we could say.

A further FFF group is fighting to ensure that gas is not defined as sustainable by the EU. A sixth is focusing on mining economy and is fighting together with the Sami in the north of Scandinavia around Gállok against a new mine which is planned in the heart of reindeer country. A seventh is putting up resistance to problematic technologies such as "solar geoengineering". An eighth is forming a group together with climate activists in the trade unions. And of course, there is a huge working group worldwide to stop the fossil industry. Much of this is presented on the global Fridays For Future website. A huge, endlessly varied world of working projects has emerged, which everyone can join.

And still, I ask myself: what kind of structures are required by this struggle? How do law, politics, finance, the economy, and politics hang together? And what role is played by science? Why does hardly anyone listen to the researchers the young people are working together with? Why do task force experts have more effect when it comes to the corona pandemic than they do in relation to the climate crisis?

The Scientists For Future meet

The meeting with the European Scientists For Future finally begins. Around 40 researchers meet once a month for a quarter of a year. "There is a worry, not just among some of the young people," I say as an introduction to the discussions, "that we are losing too many of these struggles and trying to do too much at once. They are happy to work together with us researchers, but they are also getting burnt out." Most of the scientists agree. "And above all: in this kind of NGO-like work, they can't really express the grief and anger which they carry within them, when they see how bureaucratically the clearcutting of forests and the slaughter of millions of animals are being justified and continued." Agreement again. "Is this form of activism appropriate? Is there not a way, instead of only fighting the different heads of this hydra, to cut a single Gordian knot? Which scientific analysis can we offer on that?"

Over the next weeks, we keep on meeting, and set about dealing with the question of how we can approach the different struggles as part of one fight. Gradually, a picture becomes ever clearer to me, and a theory emerges which may not be shared by everyone, but which still seems to open up a new way to understand our difficult situation; and with that, possible ways out. There are three levels which seem apparent to me in these discussions. Few people have a problem with the first level. It consists of concrete policy proposals (investment in regenerative agricultural methods, bans on forest clearance, etc.); the second is the level of the economic, societal system (see Göpel, Hickel, Raworth etc.). But there is still a third level which I often find is missing from the discussion: the philosophy and worldview behind the definition of democracy.

Substantial and formal democracy – a new approach

I propose that we need a new understanding of the democratic framework of our shared life. And in that context, a fundamental distinction can play a central role.

If we want to change the situation, wouldn't it be helpful to distinguish between a "substantial" and a "formal" aspect of democracy, and to understand the relationship between the two – in politics and in the movements?

This distinction is not identical to the difference – defined in specialist literature – between "procedural" and "substantive" aspects of democracy (see e.g. Pansardi 2016; Jacobs/Shapiro 2013). "Substantive", in that context, relates more to the participation and influence of the population beyond elections, and thus still to aspects of democracy as a form of a collective decision-making process, not as content (for an affirmative and critical description of deliberative and participatory democracy, see Pelluchon 2019). And the common distinction between "input" and "throughput" also doesn't seem to do justice to the problem. This is not about process and result as aspects legitimising democratic decisions (see Schmidt 2013).

Substantial democracy means thinking about democracy in terms of the quality of relations and structures, as most people probably do intuitively. It corresponds to the demand or idea that all people can meet on an equal footing, free and of equal value, beyond personal and structural relationships of domination (and submission), and beyond the exploitation of nature and other humans; not forcing people to give up their connection with themselves, others and nature, but creating social spaces in which such contact becomes possible (see the chapter on education for a detailed description). It also entails that we all pay attention to ensuring that everyone has enough resources – or that they receive them unconditionally – in order to live a dignified life, and to be able to engage in political decisions, without overstepping the planetary limits, including for future generations. (In this sense, it has nothing to do with the imagination of a "people's will" in right-wing populism and the critique of formal democracy which is based on this idea.)

It is this substantial aspect which seems to be ignored in all these fights. Democracy is often only defined formally, with the focus not on the quality of relationships, but on the individual or the collective and the processes and rights affecting them; for example, the right to take part in elections or to own property (a classic text is Dahl 2015). There seems to be something wrong about the way we look at the relationship between these two sides.

Governments elected on a formally democratic basis pass laws which neglect the substantial idea of democracy, which destroy habitats, forests and soil, and exploit workers or avoid guaranteeing democracy in the workplace; they are limited by national borders, preferring parts of their own population at the expense of people in other parts of the world and thus making it impossible for people to interact and participate in a democratic way on an equal footing. This legitimises relationships of domination, and thus the opposite of substantial democracy.

Domination and care

This approach throws new light on a basic problem which has also been tackled by the New York philosopher Nancy Fraser (2022).

Late capitalism is not just an economic system, she says, but a societal system, which leads problematically to a limited view of democracy as an aspect of a pre-defined, small political sphere, artificially separated from the sphere of the economy. That is why so many people today have the feeling that they have no voice when it comes to the big questions of our time. Aren't the answers basically already predefined by this societal system?

It only seems to work because it exploits three areas: nature, which is seen as a "free resource" and as a place to leave "waste"; the riches of the people in the Global South which have been exploited throughout colonial history and through the current trade in raw materials, for instance; and the "reproductive work" which is often carried out unpaid by women, including childrearing and other care work. Only because these three forms of domination are ignored – that is, because we do not meet each other as equals – can this form of society exist. In my terms: the substantial aspects of democracy, the non-exploitative approach and the caring encounter on an equal footing including the distribution of power – all of those slide out of reach.

The outline of a solution

During the conversations with my colleagues, a problem comes up. A few of them say: you're mixing up two completely different aspects, apples and oranges; that is, the quality of relationships on the one hand, and structural aspects on the other. How can you make such a strange transition from a discussion of "encounters on an equal footing", from social interactions and how they are experienced, to political structures and economic systems?

But to me this seems to be the crucial point: we should be combining social psychology and structures in a new way, anchoring both in what I would call the "logic" of social and political relations: the logic of domination or of substantial democracy. With parallels to the work of such disparate thinkers as Hegel, Merleau-Ponty and von Redecker, this approach is about tracing a societal logic and making the phenomena of domination visible, not only in concrete relationships with one another, but also as solidified structures.

The basic phenomenon of domination can be studied on a small scale in terms of interpersonal (racist, antifeminist, heteronormative etc.) experience, on a medium scale in relation to nature, the river, a specific forest, or animals, and on a large scale in terms of political and economic structures. These structures, as well as our relationship to nature, must be imbued with the logic of substantial democracy.

With these ideas as a point of departure, I suggest that we might be able to understand better the desperate fights over individual political spheres – and why these fights are not enough. We might then be able to get a sense of the common features of forestry, agriculture, the financial sector and energy policy, which make it so difficult to push through policies that would correspond to the Paris Agreement and would therefore be democratically justified.

Often it is big companies that focus on economic growth through overexploitation of nature and workers. Regardless of whether these companies are private or state run, they are part of a market organised along capitalist lines; often run by men who belong to cliques with good connections to lobbying, the financial sector and political parties, and to the academic landscape. And all of this is justified in the name of democracy – understood in a narrow, formal sense.

How can these structures be changed through the idea of substantial democracy? A new framework could provide the sustainable guidelines within which all kinds of production should take place, so that enough is produced for everyone without going beyond the limits of the planet or destroying forests and soil. In this way, a kind of basis for a Marshall Plan could be created: with a strictly minimal CO_2 budget, an immediate moratorium on new fossil projects as well as the dismantling of existing ones; and the joint unconditional provision of renewable energy as well as the basic services which are necessary worldwide for a dignified life. This would enable freedom – understood democratically – meaning the scope we can make full use of without harming the lives of others (see the appendix for more specific policies).

This solution, the anchoring of a new political crisis framework including a plan for transformation, is therefore based on the concept of a substantial democracy. But the understanding of substantial democracy is not set in stone. It has to be discovered again and again and defined through formally democratic processes. This is a complex, fascinating interplay between substance and form, which continually redefines what it means to encounter one another without domination, as equal and free.

The spheres of law, family, education, economy, and politics might appear to be separate, constituting independent systems and defining what sociology calls modern, "differentiated" societies (Giddens 2021). This apparent separation is seemingly based on the urge for and the enabling of freedom. However, this system actually is much more bound together, and enables property relations, which support also a logic of domination. Von Redecker (2020) calls it "Sachherrschaft", an objectifying relation of domination. We can deal with animals and forests almost as we wish – restricted only by very limited legislation that always comes "after the fact", too late. In this respect, these sectors are not separate at all. The economy and private property (including forests, agriculture), the legal system, education and politics are all connected. They spell out and enforce this idea of freedom as licence to own and ability to use, often even if it is a destructive form of abuse. Whereby the real aim of it all, the protection of people, is turned into the opposite: the protection of those practices that do not adhere to the aims of substantial democracy.

The knowledge of how to build such substantial democratic, regenerative relationships with each other and with nature, on the other hand, is lacking in most of our society; we scientists agree on that. It is lacking at the university; it is – to a large extent – lacking in the education system that underlies agriculture, forestry, economics and so on (see the chapter on education for an alternative model). In this way, we have to lose all these battles.

In these months, we talk a lot with the indigenous people in Sweden, but also in other parts of the world, we understand better where our history has led us. What we need, some of the scientists suggest who come to see us in Mynttorget, are new Lockes and Jeffersons who would redefine freedom and integrity with dignity and human rights, now informed by indigenous thinking, eco-feminism and so on, thinkers who care for the environment and undo structures of domination. In short, we need an understanding of substantial democracy that permeates the definition of formal democracy. Not only in the form of constitutions that correct imbalances, but already in the definition of security, integrity and freedom.

That would be a possible definition of sustainability: when the formal aspects of democracy mirror the substance; and vice versa. It is then not a matter of undermining the modern distinction between ethics, economy, politics, and law; but of dialectically suspending it, preserving it, but transcending it into the dimension that connects them: precisely the preservation of the individual space of integrity and the common fabric of integrity. We could say: the space of integrity – unlike rights – cannot be granted or taken away from someone as an idea. It is there and can be violated. And we as a society have to come up with structures and concepts to protect it. Therefore, we should rethink rights. Rights in this sense are then not arbitrarily assigned laws as objective rights and entitlements as subjective ones, but more the result of a work of "making visible" what is already there, given to us as spaces of integrity.

But because incorporating substantial democracy represents a challenge to power relations and existing economic interests, such a framework and plan will never be implemented purely through formal processes such as elections, as the previous years have taught all of the activists. That is why we need our grassroots movements, the uprising of the "people". If we protest in the streets, we ought to do so under the banner of this new framework and the demand for substantial democracy, not only individual policies in specific sectors, and not only general slogans.

So we could shift the concept of this democratic encounter to centre stage, in universities, education, law, the economy and politics, which would then go well beyond the party political. But formal democracy, too, can be changed in this way: for instance, by introducing a chamber in parliaments which would take the form of permanent citizens' assemblies, advised by scientists, and taking account of future generations and young people. Or by adapting national constitutions or the UN charter (see the chapter about the first global strike).

The basic phenomenon

In my presentation to the Scientists For Future, I come back to the basic phenomenon, the exploration of what it means to be in contact and in exchange with one another – or domination, oppression, cutting ourselves off out of fear, out of compulsion; becoming passive, too. This central phenomenon seems to connect very different branches of science. When we understand that, we might hopefully be able to understand better how to be in the world in a sustainable way, what regenerative forestry, agriculture or finance is; and what creates the "armour" protecting the lobbying cliques (see Fopp 2016). Sustainability, in this approach, is not only the distribution of resources over several decades (Brundtland 1987), but also a particular relationship to the world, right now.

At Stockholm University, in our teacher training course, we set out to explore this "logic of domination – or being humane" by playing with the habitus of authoritarian characters when we write plays about them; about the grey men in *Momo* (by Michael Ende), and others. There are understandable mechanisms behind this position, this behaviour, this ideology. The violence does not become any better when we understand it. But a certain distance becomes possible for a few moments, including within the university landscape. We know that such "armour" on an individual level can often only be removed against resistance and at a societal level through movements which stand up and name oppressive behaviour as what it is and try to create the opposite: humanity which sees through domination and does not ignore it, does not just condemn it in moral terms, but looks at its impulses and transforms them into something better.

Ronia, the Robber's Daughter

And we try to approach this dimension of substantial democracy even in the theatre rooms. This time we are working with the book *Ronia, the Robber's Daughter*, by Astrid Lindgren (2018). The children of the warring bands of robbers, Ronia and Birk, who like each other, have run off into the woods and are sitting leaning against a tree trunk on the imaginary forest floor. I have divided the students into pairs. They have been banished from the castles of their fathers, who want to continue waging war against their neighbours. And now they are surviving in the forest. Improvise: what have you experienced outdoors today? How are your thoughts going back to the castle, to your parents, who are quarrelling? What do you want, what do you need?

All of them improvise simultaneously in every corner of the room, as is typical of drama education (see McAvoy 2022). No one is watching. It basically never happens that someone has nothing to say, quite the opposite, many of them want to stay in this fictional scene and in their roles, in the forest in the evening, and talk about those who make their lives difficult and how to get out of this situation. We all become present in the room, and at the same time absent – we disappear into shared imagination. But this is connected with nature, with our embodied selves, with our individual experiences, biographies, needs; in the play it is connected with hunger, with fear of the parents' generation, which clings to destructive behaviour and does not see nature; the fascination with the forest. We try to understand how we can forgive, look after each other, find our way out of the dark, surrounded by magic beings, trolls, and stories. Fear, for moments at least, becomes a natural part of the everyday, something that can be handled.

Here it is only a game. But reality looks very different for all the young people with whom the activists are in daily contact, and whose forests are being cut down. Many of them who are protecting the forests are brutally killed, in Brazil, for instance (McGrath 2022). This also becomes clear to the young people in their communication with their indigenous peers, who work together with them and chat with them online. Through deforestation, the focus on biomass production, and especially through global warming, their home is being radically changed. Its existence is increasingly under threat (on the global dimension, see Abate/Kronk 2013).

Does drama education and improvisation even achieve anything, I ask myself then. Is this just an incredibly privileged activity, here in a protected space for a middle class in the rich north? But we are discovering something fundamental. A reason for the destruction of the environment and logic behind agriculture is precisely the phenomenon of armour, of hardening, the development of authoritarian characters by patterns of muscle tensions, with their economic and philosophical ideologies, which so many hang on to across the world. And condemning this in a pure moral sense is not our only option; we can play with all these dominating impulses, getting into a realm beyond oppression and justice: of "being humane" (see the chapter on education).

In the end, the quarrelling robbers are reconciled with their children. In Lindgren's book, it is the children who have to take the first step, so that the older generation understand that their war is becoming a war against the young. The students are not much older than the Mynttorget group. Many understand the perspective of the children and young people. Slowly, they say goodbye to the imaginary world and go out into the real one.

The decisions of the German Constitutional Court and the EU: setbacks and progress

And it's moving on, after all – the real world. In the middle of the discussions in the Scientists For Future Zoom call, the sustainability researcher Felix Ekardt storms in (for his concept of sustainability, see Ekardt 2019). "Celebrate, we have to celebrate," he says, bringing the news that the German Constitutional Court has announced a trailblazing ruling, which is even discussed in the global media (Eddy 2021). The German government is ordered to make its climate policy stricter. This means that a charge is being upheld which Felix himself helped on its way years ago, and which was partly formulated by him.

Above all, he says, this is about the notion of freedom, and about the concept of intergenerationality. That is what is trailblazing. For the first time, the legally binding understanding of freedom has been clearly linked to the freedom of future generations, and through this, it is connected with the way in which we deal with the climate crisis (Calliess 2023).

At this time, some global activists begin to feel more hopeful in general. In another historic court case, the oil concern Shell is forced to take responsibility for the emissions caused by the use of products, meaning the burning of oil (Boffey 2021). The responsibility of business models is redefined and extended into the area known as "Scope 3": not only in relation to CO2 and the global warming caused by a business in its production process, or its suppliers, but also in relation to the gases released when the products are used, meaning the burning of petrol and so on. This is met with enormous relief. Could it mean that the courts can stop the fossil industry?

But the clouds only disappear for a few weeks. Then a new, even bigger storm begins to brew, with decisions coming in one after another which many people have been working towards for months or years. In the EU's taxonomy, gas and nuclear energy are classed as sustainable, against every scientific judgement of the situation (Hodgson 2022). Shell simply shifts its headquarters away from the Netherlands and avoids the court ruling; oil projects continue to be expanded worldwide. The head of sustainability at the British bank HSBC, symbolically for the financial sector, makes a similar speech to Malpass of the World Bank, in which he rejects the idea of paying attention to the climate crisis and calls on the financial sector to focus only on the next five years (Civillini 2023). The fossil industries make enormous profits, and the states invest the overwhelming majority of tax income in the fossil society, not in renewable energies. Even the oldest forests in Sweden are cut down, just like the rainforest, at record speed. The Swedish government continues to insist that the EU should relax its regulations on this. And the agricultural policy for the next seven years is softened slightly in the EU parliament, but still benefits the big non-sustainable companies which are exacerbating the climate crisis and the extinction of species (Boffey 2021/2). No real shift away from animal-based diets or at least from factory farming towards regenerative agriculture is in sight, which would be needed to obey the Paris Agreement.

In these months in 2021 and 2022, the fossil society fights back. And it hits those hardest who have done nothing to deserve it. But only slightly later, the historic international treaties arrive for the protection of the oceans and – at COP15 – biodiversity (UNEP 2022). The tireless activism of thousands of people worldwide has had an effect, especially because it has brought together different levels: university research and street activism, leadership by the most affected people worldwide and solidarity from tens of thousands of young people; work on clear transnational legal proposals (Greenfield 2023) and countless protest actions on the ground.

And finally, young climate strike activists such as Adélaïde Charlier are sitting in the European Parliament and presenting a "Beyond Growth" plan, together with hundreds of researchers (Evroux et al. 2023). Their joint project has clear contours: the demand for growth has to be given up as a goal and replaced by the "doughnut" approach (Raworth 2018). Basic services (not only education and health, but also transport, energy and accommodation) could be provided to all citizens unconditionally; production in workplaces has to be restructured democratically and organised in such a way that it does not rely on wearing out products and a circular economy becomes possible. Care work should be at the centre of the economy, as an aspect which can bring equality closer. In relation to the Global South, Europe should not only forgive national debt, but change its economic approach to prevent exploitative practices and structures. All of this would have been unthinkable only a few years ago, at the time when Greta made her speech in the same location. These ideas would have been rejected as extremist.

But the political parties are still not standing up for these ideas themselves; even left-wing and green parties are not doing so, at least not in most European countries. Nevertheless, in these hours, in these formally democratic spaces, the substance of democracy is being spelled out by an intergenerational movement which cannot be stopped.

People For Future and the "Theory of Change"

During these months, the older activists around the Mynttorget group put part of this program into practice, as People For Future. They organise themselves weekly, even daily, into a movement which is to be easily accessible to all (older than 28 years), but which should both stand up for systemic changes and push them forward, while not neglecting the struggle for the individual policy areas. In this way, the movement tries to tie together three levels: concrete policies to protect the forests and stop greenhouse gases; making a new political framework clear, which should be introduced globally and nationally in the form of CO_2 budgets and a moratorium on all new fossil infrastructure, with the help of a sustainable economic approach inspired by the doughnut economy and similar ideas; and the philosophical and legal changes in approach when it comes to democracy, so that the substance of encounters on an equal footing beyond domination is not damaged by the formal aspects but instead redefines and realigns those aspects.

Striking on Fridays remains the central focus for those who have the privilege to be able to do so. Many do not go to their workplaces, or at least take what they call an "emergency break". And they make this form of non-cooperation visible online. The conviction behind this: a new framework can only be established if a mass movement stands up for it and if for example hundreds of thousands gather on Fridays in solidarity with the young people and interrupt "business as usual".

A website is created for the People For Future, along with a Swedish manifesto. Andrea, Lena, Cilla, Christine, Jörg, Johan, Niclas, Jeannette, Jonas, Ann-Lis, Peter, Caroline, Eva, Anders, Shahin, Karin, Alexandre, Fabia, Valérie and many more invest their time in community work. The biggest project this year: demanding this new political framework, also with the help of a kind of scientifically supported Climate Task Force. A few go through the European models which already include similar practices, talk to those responsible, weigh the upsides and drawbacks of bringing science into politics, and sketch out a model which they then demand from the government in an open letter in the newspaper ETC (Herrera et al. 2021).

The pinecones and the global strike

Finally, the time has come. The global double strike in September and October 2021 is approaching, for which the young people have collected pinecones in the woods. The group prepares itself in the rooms of the Karolinska Institute, the medical university in Stockholm. The stage has to be organised, the march has to be planned and the speakers have to be chosen, as usual. Soon they will set off into the city and plaster it with posters.

On the Friday itself, they get up early – along with hundreds of thousands worldwide in different time zones – and distribute the pinecones on the ground in front of parliament: #UprootTheSystem is written there. Forest, agriculture, the fossil society and finance should be thought of in conjunction. If we want to stop emissions, we must see the full picture, they say, across the whole world.

When they then march through the city to the Vasapark, at around noon, accompanied by thousands of Stockholmers, the visitors from the MAPA countries which are most affected by the crises walk at the front: activists from the Philippines, Mexico, Argentina, Kenya, and so on. They are visiting their peers so that they can travel on together to the COP meeting in Glasgow. Their stories and their voices should be heard most clearly, the young activists say.

On the stage in Vasaparken, no more than a hundred metres away from the flat in which Astrid Lindgren came up with stories such as *Ronia, the Rob-*

ber's Daughter, not only the climate strike activists make speeches, but also the young activists from Black Lives Matter and from the Sami youth organisation Saminuorra. The young people are united in a generation of solidarity, with a transnational and intersectional focus, demanding truly fair sustainability. They stay for a long time in the park and enjoy the sunlight, which catches on the autumn leaves. Several thousand adults are here to support them, and to stand up as People For Future.

The manifesto

But still, there is hardly any reaction from politicians. Nothing but silence from Prime Minister Löfven and the Green Minister for the Environment Lövin – which is no different from their reaction to the huge marches before the pandemic.

And so the young people worry that these individual fights over the forest, over agriculture, over emissions and the financial sector will simply fizzle out. "Shouldn't we write down what is needed right now in terms of concrete policies, so that it is clear to everyone how far away we are from all that, but that it could also be changed?" they ask themselves in the autumn of 2021. "The next elections are coming up in one year. The parties are not really reacting to the things that make this a crisis. We can show everyone what would be a plan to make a good life possible for everyone." And for the first time since the founding of FFF in Sweden, the young people begin the project of a concrete "manifesto". What does their strike demand to uphold the Paris Agreement actually mean? They now want both: they don't want to lose sight of concrete changes to their world, the forest, the fields – and they still want to draw the focus to the big picture.

The big three topics are clear. The first is to take the crisis seriously as a crisis, which means drastically reducing emissions through CO2 budgets and making a plan for concrete measures to protect the forest and for regenerative agriculture (banning invasive species, moving away from monocultures and aggressive forest clearance, and so on), as well as globally putting a stop to all new fossil infrastructure and building a renewable infrastructure in its place. Secondly, all of that should happen on a socially and globally just basis; in terms of paying "fair shares", forgiving debt and paying reparations to the Global South, and in terms of the perspective on the rights of the indigenous population; and with the focus on the workers and the most vulnerable people in society. Thirdly, through this, democracy should be strengthened, young people should be involved in political decisions, and so should science; intergenerational justice should be ensured; media and education (school and university) should include a deeper knowledge of the crises and should be radically reformed, and participative democracy should be expanded.

At a press conference in Stockholm, Alde, Andreas, Agnes and Linna present the ideas which could give them a future; as well as to their peers in the whole of Sweden: to Claudia, Denise, Idun, Matilda, Karla, Falk, Eva, Valérie, Fabia, Erik, Sophia, Almut, Anton, Isabelle, Ell, Tindra, Chris, Taylor, Eira, Greta, Douglas, Janka, Simon, Samuel, Esmeralda, Nils, Raquel, Angus, Vega, Nora, Linnéa, Ebba, Elliot, Lilly, Lisa, Astrid, Hanna, Ozzy, Noëmi, Hjalmar, Sofia, Aron, Emmy, Maya, Leo, Judith, Ella, Filippa, Alex, Lydia, Alice, Hanna, Mina, Edit, Eira, Rocky, Minna, Melda, Astrid, and so many more. The text is published on the website of Fridays For Future Sverige (2022) and forms the background of the strike placards for the elections. It is aimed both at the rulers and – above all – at the whole population.

In this way, the unity of the struggles becomes visible. But in order to change politics, a mobilisation of the masses using disruptive methods – such as the non-cooperation and strikes on Fridays – seems to be needed. It is not enough to fight the many individual battles – and they can be fought by NGOs as well. Now, it is about expanding the movement as a disruptive grassroots movement. And how this is done should mirror the intended target of the political transformation: to establishing substantial democratic relations, and treating each other as equal and free, in a network which is open for everyone to join.

> News that the Thwaites Glacier shows cracks and fissures shocks the world (Vidal 2021). The layer of ice in front of the glacier is melting away from underneath, affected by warmer ocean waters. When this plug breaks away, there is nothing holding back the glacier ice itself. The Antarctic ice is melting so rapidly that researchers are saying it may already be too late to avoid a catastrophic rise in sea levels.
>
> The German Constitutional Court upholds the climate lawsuit and obliges the government to change its policies so that everyone's freedom is guaranteed, including the freedom of future generations.

In a landmark case, a court in the Hague orders Royal Dutch Shell to cut its global carbon emissions by 45 percent by the end of 2030 (Boffey 26.5.2021). Some months later, Shell moves its headquarters to London.

The EU defines both gas and nuclear energy as sustainable transition energies in the taxonomy (Rankin 2022).

Chapter 7: The Idea of Social Movements and the Journey to Glasgow – What Is the Right Way to Live?
November 2020 – December 2021: On a theory of democratic grassroots movements that can change the world

The two worlds

What does it mean to live together freely as equals? To create social spaces in which everyone can live a life in dignity? What is life about, and what are the movements about? How can they change history – and how should they be organised internally?

Some of the young people of Mynttorget stand near the huge stage, looking into the crowd. Tens of thousands of their peers are standing in front of them in the main square in Glasgow. It is November 2021 and still astonishingly warm in Scotland. Once again, all the countries have gathered their delegations for the climate conference. Two years have passed since the meeting in Madrid.

NGO employees who support the grassroots movements are also present. But what are these NGO workers even doing here? This is officially an event for grassroots movements, especially youth movements. And so, the question arises: how can we organise the relationship between movements and NGOs without a few employees disrupting processes of democratisation or even making them impossible, by intervening too much in the movements? The young people of Fridays For Future already have enough to do, trying to organise democracy among themselves and especially at a global level, I say to myself: who can appear here on the stage in front of the world media? Who represents the movement? Conflicts have come about in the last few days, important and necessary conflicts, including over the much too central role of the Global North.

Because at these COP meetings, which many young people describe as a kind of traumatic experience, as the world of adults in smart suits barely seems to take their future seriously, the question becomes especially important: how can we ensure that those people are heard who are most affected by the climate crisis and most vulnerable, in the Global South and in indigenous populations? They are barely allowed into the official halls of the conference, the "Blue Room", in contrast with the representatives of the fossil economy.

During these months, some of the privileged activists often refuse to give speeches when the UN and other organisations invite them to do so, and give other activists the chance instead. But then many organisers and media don't report on the speeches.

A few, mainly male employees from a few, mainly American or global NGOs (most NGOs help throughout these years in a way which the young people appreciate) have tried to put together a problematic kind of "Champions' League" of climate activism since summer 2019. Sometimes ignoring the movements themselves, they choose the "best" speakers and make others invisible in the process, often unintentionally. And usually, the ones who are chosen by the NGOs are young people who are already privileged and hold the most social and cultural capital, as well as being distinctly older than the 17-year-olds who continue to maintain the global structures and really participate in school strikes. This damages – from our perspective as activists in grassroots movements – the free and equal cooperation between young people. It is often justified with the argument that the Global South is being supported, although it indirectly means that employees in the Global North gain power. Given that I myself am in a similarly privileged position, this raises questions I have asked myself in the past few years: How can all possible actors help on an intergenerational basis, without intervening in the internal structures of the youth movement?

Behind the stage in Glasgow, I think to myself: if these children already have to be confronted with a world that's anything but democratically organised, where so many things are already shaped by privileges and the pressure to achieve, adults should at least respect the young people's internal democracy. At that moment, a new speech starts by the Global South activists and I wake up from these thoughts. In the audience, I can see so many of the Swedes who have travelled to Scotland. During the last months, they have succeeded in building strong structures as FFF Sweden, which is open to everyone under 26.

It is a counter model to the small elite groups of chosen ones: an autonomous grassroots movement of a hundred of children and young people.

But how does it work, and how does the whole movement work? In my university rooms I organise countless conversations with the FFF activists. What do you actually do when you're working on the climate movement, locally, nationally, and globally? A picture gradually emerges which shows not only how grassroots movements differ from associations, organisations such as NGOs, and political parties, but in a certain sense also what is really important in life.

Challenges for movements

Gradually, during the last year, the group of FFF activists has been renewed and doubled. The older ones from Mynttorget have stayed on, and everyone is happy to find that new people have turned up in many cities, including Stockholm, and have found their place: not only on the ground in front of parliament, but also in the chats online. It is the space in which the young people are connected, beyond all the concrete plans for strikes and without any pressure to perform. The main point: it's ok for people to admit that they don't know everything. For many of them, these FFF groups have become the best school you can imagine.

And now, so many of the members of these groups are standing in the audience in Glasgow, listening to their friends from across the world. They are talking about new oil pipelines being planned in their countries in the Global South, which are supported by European and American corporations (on these speeches, see Nakamura 2021).

And I am filled with a sense of despair. It can be seen in these months more generally, too: the structure of the grassroots movements, or of democracy in general, also has a vulnerable side. A few people from the outside, if they don't follow the "policies" for working together, can cause damage. A pattern becomes visible. There are a few principles which these directors of communications and strategists from NGOs keep applying again and again – at least, so it seems from my limited point of view. Most older organisers often help with good results by supporting the whole movement transparently at a global, national, or local level. In contrast with this, a few NGOs which are active globally have made it their mission to "support" the movements "from within". Their problematic strategy is expressed in various interconnected measures. As mentioned above, the NGO workers choose a few individual young activists and declare them "spokespeople", whom they want to transform into globally recognisable faces in the media. They claim that this is the only thing that works in the long term. The movement itself has no say – either in the question of whether there should be spokespeople, or in the question of who these should be. The differences in the organisational structure become particularly clear when we consider that the children are taking a risk and are striking, while the adults are doing their work and earning money from this situation.

Sometimes, some NGO employees even expose children to risky situations, without explaining the possible consequences. They then end up facing governments and the public and being flooded with hate and threats. Finally, a few privileged young people are often grouped into small elites for which separate channels are made available and meetings are arranged which others know nothing about – which sometimes hinders the young activists from building democratic channels of communication and decision structures. One effect of this dynamic is also that problematic power relations arise within the movement, between a few adults and this "elite group". It may be the case that officially in these groups no "formal" decisions are reached, but informal decisions are often taken, for instance regarding "narratives" which then shape the whole strategy. And all of this nourishes a way of thinking which leads away from the children's rebellious school strike and focuses on young adults who are involved in professional "campaigns". Children who have helped to build up the movement for years are ignored, and end up withdrawing.

This particularly affects those who cannot express themselves perfectly and who can perhaps not take as much stress, I think to myself, feeling the winter air in Scotland. And it does – that is the insight I've reached over these years – have consequences for world history. The basic structures of the movement become weaker. Of course, one can also argue: the NGO workers are doing excellent and effective work, and they only want what's best. That seems obvious. For some of the young people, especially in the Global South, the support of the NGOs is crucial, giving them the time, space and other resources to be activists and to reach a much bigger audience. And still, in Glasgow I can't help thinking: that also causes destruction; the idea of democracy itself is being destroyed, the very idea that we are all equally valuable and that young people in particular have the right to shape their world and their movement.

How to change history

But what is a grassroots movement? In the interviews with FFF activists which I conduct at Stockholm University, I try to understand the necessary ingredients. History shows: to change politics, we need grassroots democratic movements which are structured differently from associations, political parties or NGOs, but which can be supported by them. The second necessity: disruption, meaning not just demonstrations or "campaigns" but civil disobedience, strikes, and blockades. This combination of democratic popular movement (not small groups organised top-down) and disruption is explosive (Chenoweth in Thunberg 2022). Such movements were at the centre of most or almost all progressive social and political changes: the workers' movements, women's movements, and civil rights movements, XR, FFF and Black Lives Matter (and they are often not discussed enough in "European" theories of transformation, such as Göpel 2022 and Rosa 2020; detailed counterexamples are Celikates 2016 and von Redecker 2021).

But most people, I suspect during these months, do not really know what such democratic movements are (movements for adults, that is: XR, People For Future, etc.) or how they work – and perhaps that is why they don't join in the first place.

The logic of substantial democracy – what is a grassroots movement?

A large part of the research into grassroots movements seems to ignore two points, or barely to emphasise them. These become clear from the example of the FFF and XR groups: on the one hand, what I call the "non-instrumental" convivial logic of the relationships on which grassroots movements rely. And on the other, the fact that they are really something different from just a part of a "civil society", as mainstream sociology claims. This dimension seems to be much clearer in theories and reports from the Black feminist movements and from BIPOC communities, and to be captured better conceptually and in terms of social theory (Springer 2005; Rodriguez 1998; Garza 2020; Keisha-Khan Y. Perry 2016; Gomez et al. 2011).

Thus, Gene Sharp (1973), Graeber (2014) and Engler and Engler (2017), standard works on activism and nonviolent resistance, do explore how important movements are to progressive change in human history: with the discovery that we can connect nonviolent direct action with the organisation of mass movements. But the internal organisation of these movements is only discussed very rarely, or only in technical terms relating to decision processes ("Self-Organizing Systems"; "Holacracy" etc.: Robertson 2016). This means that the core, the idea of non-instrumental relationships, is barely mentioned. Perhaps we could think about sociology and theories of democracy differently, starting from these movement structures. This could change the concept of democracy in the social sciences and humanities, which still don't pay enough attention to the crucial "substantial" aspects.

So what distinguishes the grassroots organisations that change history from other forms of organisation? In the research there is a good understanding of the fact that they are organised democratically on a grassroots basis, not hierarchically (with all the problems of informal power relations), although they address the masses, the broad population, which is invited to participate politically. They are devoted to a cause, something which must change, and they often tie this clearly to a political demand (such as women's suffrage). Additionally, this involvement of "ordinary" people takes the form of community building and organisation, often meaning that methods of strictly peaceful civil disobedience are applied, which at the margins of the movement may shift to forms of sabotage (Malm 2021) – including in the civil rights movement and among the suffragettes – but are never supposed to harm people but to respect the dignity of everyone.

But what is first needed to make all this possible, I realise increasingly in my studies, is a different "social logic": the logic of "non-instrumental" relationships, to use Max Horkheimer's central concept from his Critical Theory (2013). This means not wanting to gain an advantage from other people; it means not using them, but affirming each other, caring. No one is profiting from the work of others; all of them can show themselves transparently in public and don't have to hide themselves as the people involved in the central organisation. In addition to that, no one has to prove successes to funders, as is the case for NGOs, which creates pressure to perform and even something like ableism. Being imperfect is fine, being ill is okay, and in fact it's okay to be lying in bed at home, as in the opening of Astrid Lindgren's *The Brothers Lionheart* – by listening, you can still be an important part of the movement, because someone is keeping you updated and telling you what's going on. This kind of community building can also become what Spade (2021), von Redecker (2021), and Garza (2020) argue is the centre of really transformative movements: mutual aid, helping each other, meaning that the living situation of the people involved is at the heart of the movement, along with the material and existential worries and resources which are shared. All of this describes the ideal case; in real life, platforms, chats, and central organisational places might last for one, two or three years. Then they have to be replaced with new inventions.

However, that does not mean that in grassroots movements we are present as private individuals, as some people might believe. Quite the opposite: someone who is there as a private individual will damage the movement, sooner or later. Because it is only when we operate as part of the movement (in that sense not dissimilarly to members of an association or a democratic workplace) can we see everyone as equal; not preferring some people and only interacting with them, or putting our own interests first, creating informal centres of power.

Challenges in Glasgow – unequal, but still equal

"Watch out, it's starting!" The crowd in Glasgow applauds when a new activist climbs onto the stage. And there they stand, talking to their friends from across the world, the Swedish group. They send pictures and messages to those who had to stay at home.

Across the world, activists are trying to create similar structures, confronting parallel challenges when it comes to building a global democracy. The working groups are useful when it comes to finding new global strike dates or agreeing on a hashtag, but there is still a lack of structures that would make it easier to get involved globally, some activists say. And at the same time, there is the challenge that those young people who come from the regions, classes and sectors of society that are most severely affected ought to take on a leading role, but without just having the responsibility foisted on them by those in the Global North. That is what most of the conversations in the corridors of the COP are about. How can we organise ourselves as equals when some are so privileged?

How do you do that, I ask the young activists in our conversations about the development of Fridays For Future in Sweden: how do you try to meet each other as equals?

The development of Fridays For Future

They have created a code of conduct to structure the way they interact, as well as guidelines for the content of their cause, which they publish as a manifesto.

The crux of all this is the willingness to respect one another and set off on a path of learning and deepening democracy, in which people are prepared to question their own standpoints and help each other to understand themselves and others better.

They create democratic monthly meetings and "rotate" these, choosing different "facilitators" for each meeting, and establishing a clear agenda beforehand, along with ways to make decisions. Basically, they are consensus-oriented and only vote in an emergency, in which case a proposal needs 80 % agreement within one or two days. In this way, they avoid the process of fighting for majorities, which is so crucial for political parties and associations and for formal democracy in general.

These unifying organisational solutions are strengthened further by the strikes as a common form of action. The experience of the strikes, in which the young activists obstruct the status quo of the fossil society as equals, draws them together even more. Most of them describe how the weekly strikes, this form of civil disobedience, contain both a feeling of power, since they no longer just have to watch while their world is destroyed, and also a sense of exposure, which can be connected with hate from passers-by. This also leads to grief – and to the danger of burning out if that grief is not expressed and met with support.

In this context, their invention is all the more central: they have the chats, in which they can all participate as equals, and the physical places such as Mynttorget where they can express grief and fear, too. This combines the logic of social relationships with specific aspects of FFF's organisation: the young people do not simply organise themselves decentrally, as Extinction Rebellion does. In XR there is no real central meeting point, but at most allotted rules and mandates, through what is known as the holacracy or SoS model (Robertson 2016; Extinction Rebellion 2023). This is what makes a crucial contribution to the strength of the movement, and works: a permanent exchange on an equal footing in a central location, which is also free of the pressure to be right or to excel. Without such a central place (ideally both virtual and physical), movements seem not to work, contrary to all theories of decentralised movements.

FFF and People For Future are in this sense a very specific form of social movement, because they combine the open democratic structure of a decentralised set-up with organisational solutions (shared meeting points) which enable much more direct and non-instrumental relationships with everyone. That solves (at least to a degree) the problem of other movements that in spite of the decentralised approach the same people generally decide what is to happen – the people who are the loudest and who have the most resources, often middle and upper middle class people (see Bourdieu 2010 for the dangers of classism within social groups which create informal ways of communicating – by mentioning some specific culture and showing a certain "habitus" – which allow only a few to feel truly included).

I look across at the apple trees in front of the university building and suddenly work out that this is actually about democratisation not just in terms of content, but also in terms of organisation. Sometimes outsiders understand this to mean that everyone should do everything, as if there shouldn't be any roles or mandates. But that is not the case. And it is also not about levelling down. It seems absurd to stop those people from making speeches who are the best at doing so, in the name of democracy; the point is that they shouldn't be the only ones who are visible. So it is about making sure everyone is involved without one person deciding how the others should behave; the older ones, in particular, should not be deciding that, since there is a power imbalance between them and the younger ones. That is why the independence and autonomy of the movement as a youth movement is so important to me. Instead, as People For Future for activists over 26, we build our own complementary structures.

How organisations can work together with children – new rules

A few of the older ones then intervene in relation to the role of the NGO employees. And with time, the bosses of these few organisations do see that something is going wrong in terms of the power relations. Many of them don't even have rules defining how their employees should interact with children, which ought to be standard and which is the case for most social organisations. The result of these interventions is a document which I write at the university. It is anchored and accepted in meetings with those responsible for the biggest global climate networks.

It is called "Guidelines for cooperation between generations (NGOs and youth movements)," and it states that the movement must belong first and foremost to the children and young people, and to all of them, democratically – they have a right to self-organisation. Secondly, power relations must change, so that NGO workers communicate with the movement as a whole and do not intervene in the structures, splitting off groups and forming elites, and making the children dependent on them in an unhealthy way. Thirdly, children's attention must always be drawn clearly to the dangers of current and possible future situations, and they must be given information so that they can make informed decisions themselves. And fourthly, bringing everything together: the welfare of the children and their own position as political subjects must always have priority; adults must put this first.

During these years of intergenerational cooperation in the climate movements, it is astonishing how many adults do not want to commit to such formulations, talking of "ageism" and claiming that they are being discriminated against when they are asked to give young people precedence. In Sweden, already early on, since the 1970s, the playwright and theatre director Suzanne Osten (2009) was drawing attention to the unequal power relations between children and adults, and formulated this as an imperative for the whole of society: to make these very power relations visible, and respond to them in such a way that young people are not oppressed but are taken seriously as subjects. Roger Hart (1992) makes similar points for UNICEF in his text "Children's Participation – From Tokenism to Citizenship".

More than "civil society"

However, in a certain sense, through this limitation on the power of NGOs, an even bigger issue becomes obvious. Because what the young people of FFF and the adults of Extinction Rebellion have built up in the previous years is not at all appropriately described as "civil society", a term which applies more to organisations and associations. The concept of civil society and the theories behind it seem to belong to and contribute to an ideology which trivialises protest and weakens it into a friendly "deliberative discourse", to use Jürgen Habermas' term (see Chappell 2012).

But doesn't the term civil society – and often theories of "global governance" in relation to particular actors in sustainability transformation (see e.g. Linnér/Wibeck 2019) – come from a way of thinking that takes the sting out of these very movements by dividing society into four sectors – the public sector, the market, the private sector and the sector of civil society, which is more about ideas than about material concerns? And then, in such a "friendly" way, even gives these "ideas people" a seat at the table of "stakeholders", so that there can be a discussion at a "round table"?

But do movements such as FFF and XR not want to change the power relations, not just bring their arguments into existing discourses of those in power? So mainstream research problematically contributes to a process which endeavours to weaken the movements through its manner of describing them.

This seems to be a crucial difference to an (often right-wing) populistic movement that tries to speak for a fictitiously created "will of the people". In contrast, the democratic grassroots movements (even historically) are not fighting for the specific interests of one group in society (even if it looks like that on the surface), but for the inclusion of one group in the realm of the dignity of all people, the equality and freedom of all, beyond borders, ethnicity, gender etc. This is an aspect which gives the movement its democratic legitimacy. It is not about pretending to understand the "will of the people", or about asserting particular interests, but about creating real equity, justice, and humanity for all.

That is why it is so important, I think, that these groups use means which guarantee everyone's dignity, even the one of the persons who disagree. The danger is real to hurt the common fabric of dignity in the name of justice. In this sense, every movement needs – if one doesn't want to rely on the lucky constellation of people who have a strong informal power position and use it to include everyone and go beyond all forms of domination – a permanent learning process how to create such real substantial democracy, an open space for different analyses, real inclusion; learning, how to stop the concentration of power in the hands of few (see the chapter on education).

The "blue zone" and the negotiations in the COP rooms

After the protest march through Glasgow, activists of all ages sit together for the next days inside and outside the COP rooms. The local organisers have done an enormous amount of work, over many months. They not only organise the strike with tens of thousands of children and young people. They create a meeting point in the middle of the city where they can all gather: Dylan, Saoi, Sandy, and so many more. Some of the Swedish activists have met already in Brussels in March and in Lausanne in August 2019; all of them form the group of activists who keep in contact the whole time, organising the movement day after day, month after month, year after year, upholding the common structures and debates. Together with Mitzi, Marja, Sommer, Eric, Arshak, Annika, Theo, Patsy, Ianthe, Erik, Yusuf, and so many more from all five continents – who keep in contact with each other and many people on Mynttorget over all these years.

Often, during these two weeks of discussions in the Cryosphere Pavilion, the activists end up in the middle of what is known as the "blue zone" at the conference; a space which is cordoned off and guarded, in which the crucial negotiations also take place. This is where research on the melting ice is presented, and it is also where (far too few) indigenous activists from countries such as Canada report on the devastating consequences of the crisis for their livelihoods.

How do the young people actually feel here, in these COP spaces: included or excluded? How do they experience what is seemingly the most important meeting on the climate crisis, with their specific intersectional and regional backgrounds? Isabelle from Mynttorget is not only here as a strike activist, but also as a researcher. She began going on strike when she was in high school, and now she is finishing her studies. She has made it the goal of her final project at Stockholm University to gain a better understanding of her fellow activists globally.

While she records interviews, I walk over to the official negotiation room. They are sitting there together, the official delegates of all nations. They have the mandate to respond to the crisis with political decisions. They could theoretically – in conversation with governments – agree on real solutions, a crisis plan to stop emissions, keep fuel in the ground and protect forests, animals, and the soil.

I sit down towards the back of the huge plenary hall. One country after another takes the microphone. One after another, they describe their national situations. Proposals are presented for changes to a joint document.

After I've heard three delegations, at most, I begin to get restless. Once again, the speaker begins to say that he – the room is dominated by men in suits – agrees with all the others and wants to commit to a "net zero goal for 2050", "as set out in the Paris Agreement". There's no point in even talking about that, I think to myself. And after half an hour I understand the strategy of almost all governments across the world.

Rather than focusing on the 1.5-degree limit in the Paris Agreement (or even on the "well below 2 degrees" target) or agreeing on concrete measures, they behave as if an abstract goal like "net zero in 2050" is enough for the situation. But the crux of this is the absolute quantity of emissions; naming a year does not define this at all.

First of all, countries ought to be aiming for quite different zero emissions goals, depending on their wealth and infrastructure, I say to myself. This global solidarity is set out in the Paris Agreement. And secondly, the whole CO_2 budget for this limit will already have been used up in a few years, or in six or seven years, to be precise (Anderson et al. 2020) if emissions continue to remain the same. Used up forever.

The speeches of these ministers on a 2050 goal in this huge hall seem to me to make a mockery of the young people sitting in the next rooms. And they are not even aiming for "almost" zero emissions, but for "net zero", a formulation which allows compensations and "offsetting", including enormous loopholes for measures such as carbon capture and storage, which are often impracticable or unjust, and which barely prevent emissions (Skelton et al. 2020).

The ministers from the two countries leading the negotiations collect the proposals. Then they withdraw, together with a few select representatives from groups of countries. Finally, they present a new text which will again be discussed in the plenary meeting – and so on, until there is a finished document.

The interesting thing about this afternoon are the descriptions of the concrete changes to the climate in all the different countries. One minister after another describes a nightmare: what is happening in Peru, in Australia, and so on. I become increasingly furious, and I want to yell: what's the point of all this? You ought to... But what? Stop the emissions, with clear crisis plans for the next five years which are transparent and binding, and verifiable. Instead, the ministers continue their speeches on "net zero 2050" in their own countries well into the evening. Global justice and democracy are barely discussed at all. As the climate Twitter community keeps reminding us: emissions are still rising across the world, in spite of the 26 COP summits.

Following pressure from India – and in an undemocratic move at the last second under the direction of British politician Alok Sharma – the final document does not even include the phrase "phase out coal", but only "phase down coal", as if the world could just keep going on taking coal out of the ground and burning it. For the first time, and that shows the problems of these meetings, a fossil fuel is mentioned officially in the context of the COP. In the Paris Agreement, coal, oil, and gas are not discussed at all, even though they are at the core of the problem; the power of the corporations is so strong – or at least, this is one explanation – and the fear of the politicians is so great, or else their self-interest. And this goes for petrostates such as Saudi Arabia and Russia, just as much as it does for Germany's coal industry and for the Swiss financial sector. On Swedish public radio, the science reporter points out rightly that the conference seems to fall into two parts (Sverigesradio 2022): the self-promotion of the countries, and the sober observations by activists and scientists, who are practically in despair.

Accordingly, the young people mainly meet their friends from across the world outside the COP site and make this failure clear at "rallies" – spontaneous demonstrations; and they join and support the workers' strike which happens during these days. They give speeches and chant: "You can shove your climate crisis up your arse" (Nicholson 2021)! And they plan the next protest actions, at least looking back with good memories of the huge strike on Friday in the lanes of Glasgow with tens of thousands of young people; including many schoolchildren at their first demonstration.

And while the young people are already setting out, the older ones are still working on the movements which can build up the "people's power". In the evenings, we meet up with people in the pubs along the river Clyde, including trade unionists from the international umbrella organisation ITUC, and many others who are building grassroots movements, like the activists behind the Momentum movement in England, which anchors the idea of climate justice right in the local communities which are affected, and has renewed the workers' movement. How do you do that, I ask again and again: how do you organise locally and globally for this other, democratic voice of the populations?

Back from Glasgow - on questions about the class society

We travel back on the train to Mynttorget. At Stockholm University, I have to report back to my colleagues. How did it go for the young people and for us researchers? How is the global community reacting to the crisis?

The colloquium of my colleague Linnéa begins at our Department of Child and Youth Studies. "Critical Youth Studies" is the name of the event. And suddenly, as the discussion develops, I realise again what seems to me to be the central aspect of movements and the real challenge for a rapid transformation of society. Linnéa is leading the meeting. She has written some of the most important books about the children in Swedish society who live in precarious circumstances and are affected by poverty and violence, and about the policies which could prevent this (Bruno/Becevic 2020).

She chooses a rather unusual way to begin the seminar, and suggests that if we want we should discuss our own "subject position". A subject position is – in the tradition of theorists such as Foucault – the socially and discursively defined interpretation of our position in society, particularly in terms of the interconnected dimensions of discrimination and privilege, such as gender, class, and so on. But today, the focus is on the class we come from and in which we live, and not on gender or ethnicity, which are so often discussed in the humanities and social sciences.

Many educators do speak about these dimensions sometimes in their lectures, but mainly just as an object of research, as something we should be thinking about during an analysis – not as part of our lives at the universities. And it is the same, I think to myself privately, it is exactly the same for us activists. Our positions in terms of class and origin seems almost to be taboo. In the seminar, many people hesitate. Something is happening here which goes against the norm. Usually, in the way that we speak and behave, we behave as if all of us belong to the same (upper) middle class. But how can we talk openly and sensitively about class, about belonging to a class and about oppression, without forcing anyone to say something they don't want to reveal? Without creating shame? Without excluding anyone – on the contrary – with the intention of drawing attention to processes of democratisation and the structures behind them?

Many of the adult activists themselves live on a very low income and in insecure, temporary work contracts, or else they are ill or unemployed. Travelling, for instance (to events such as the COP), is often only an exception made possible by support, and this goes even more for our colleagues from the countries most affected by the climate crisis. This makes it even clearer how even environmental NGOs differ from grassroots movements: a few NGO employees, though certainly not all, come from the middle class, fly around the world from meeting to meeting, drive their cars around and join cliques with similar people in which the boundaries between private friendships and economic interests, jobs and business connections become unclear – not so different from the cliques which cling on to non-sustainable forestry and agriculture. And at the universities it's not so different either. Many things are shaped by instrumental relationships, in relation to possible joint publications or jobs, for instance (see McGeown/Barry 2023). Our own position in society is not discussed very readily. And educational justice is generally lacking: it is mainly the same middle class which reproduces itself through the forms of education and the structures (Warren 2014).

The alienation in Glasgow is, I think to myself, not only an estrangement between the young people and what is happening on the COP site, but also between those who are committed to noticing intersectional injustice, including most of the young people, and those – including the delegates, but also the environmental lobbyists – who do not talk about this dimension. The topic is still mainly taboo – at the university, too. When someone like my colleague then begins a seminar by offering to discuss our position in society, most people remain silent. They look out of the window. Even the ones whose research focuses on people in precarious positions and on critiques of capitalism are not sure how to deal with such a situation.

Perhaps – that is my suspicion – this is rooted in the fear that we will suddenly have to step forward as private individuals and offer help to others, or that roles will get mixed up. Or that we will have to take responsibility for structural injustice and blame for the suffering around us.

But what is it really about? In the movement, many people say, "Yes, but we are all fighting to protect the environment and the climate. Everything else has nothing to do with that; in best case, it's private. We all have our worries and problems, let's focus on the most important thing." But that creates a problem. We cannot look past the structural, socially produced premises of these concrete life circumstances; the structures exist and are part of them. And these circumstances, as well as the poverty or fear which result from them, are connected with class origins, with the perception of skin colour, with questions of gender, with exclusion mechanisms in the existing, purely formal concept of democracy. And these are what make the privileged position of the upper and middle classes possible, locally and globally. Ignoring them is not an option, I say to myself. Because the left-wing and green government is not building any affordable housing, those who own houses and flats are becoming richer so much more quickly at the expense of those who have to pay rent. Because politicians are not ensuring, institutionally and through changes in teaching methods, that workers' children can study, white middle-class people have an advantage. And so on. Poverty and precarious situations might seem to be a private issue, but they are structurally necessary elements of a majority society which takes advantage of the status quo, the researchers around us show. And this only reflects, according to Nancy Fraser (2022), precisely what defines the basis of the climate crisis: that the modern economic system has always been built on this form of domination, on the exploitation of the precarious situation of the working class, of unpaid reproductive work carried out by women; on the exploitation of nature; and on the exploitation of the Global South. This is not collateral damage – according to this theory – but the conditions which make it possible for some people to be well off while others aren't.

At the colloquium, when I talk about the trip to Glasgow and the movement, this becomes clear to me: how easy it is to ignore this deeper understanding of encounters as free and equal people. Everyone quickly agrees with the demand for democratisation – but really reshaping relationships in concrete terms within structures is another matter. If this is to change, we have to start from within, in our own surroundings, at work or in activism, I conclude at this meeting as a kind of research result. We need new centres for sustainable, substantial democracy, everywhere, in universities, movements, and cities.

In this sense, I do not understand the "Theory of Change" which claims that only a small upper class of oil barons and financial sharks are responsible for the climate crisis and that protests should be addressed to them – the "1 %" – even if it is true that they are responsible for a particularly large proportion of emissions and structures. It is also the middle class, and particularly the upper middle class, which maintains the often subtle structures which both destroy nature and build up privileges at the expense of poorer classes. That could be why "solving" the climate crisis is such a complicated undertaking. Two steps seem to build on each other: so that the upper ten or three percent of the population can be addressed as those who produce the most emissions, have power over the fossil society and prevent structural changes that would create a more sustainable society (through media, corporations, lobbying etc.), the fifty percent of people who belong to the middle class and upper middle class would first have to admit that their own relatively wealthy situation is based on the fact that the rest, the other half, are "kept down", dominated.

But this is exactly what is not happening. And for this to be covered up and suppressed, aspects such as our own subject positions within society have to be taboo at universities. "We are all equal, after all." But that is not true; some have advantages at the expense of others; and there shouldn't be (political, economic and societal) structures which create different classes after all, if we want to live as equals, in democratic relations, one could argue. This is not just about classism: that middle and upper middle class people speak and act in ways that exclude working class people from movements, for instance (Bourdieu 2010). It is about the economic structures which produce classes – and the interdependent sustainability crises (Fraser 2022). And in grassroots movements, this whole topic becomes especially striking. In a certain sense, they break through this taboo. In them, people can meet on an equal footing, because everyone can admit and deal with the fact that the democratic structure of "meeting each other as equals" is not yet a given. (In this sense, the demand for basic income and basic services – see appendix – aims not only to counter injustice in terms of gender and class, but also to enable more people to become active in movements and political, democratic work without being chosen and paid by organisations.)

What distinguishes transformative grassroots movements – a new theory (on social logic, organisation, and communication)

With this perspective in mind, we can revise the existing theories about what distinguishes grassroots movements, movements that have changed history and which are the only means known to us of potentially changing history in a democratic direction; from liberation movements to the civil rights movement around Rosa Parks and Martin Luther King, from the suffragettes to Black Lives Matter, as well as the newer climate justice movements.

The core, I propose, can be described as a threefold process of democratisation.

The first level is that of "inner social logic": contrary to the theory of Chenoweth (2021), Sharp (1973) and Engler and Engler (2017), it is not only about combining three aspects such as "direct action", mobilising a politically participatory mass movement, and nonviolent resistance, although this analysis is already fascinating.

This is about the fundamental question of how and to what extent these movements internally allow and encourage grassroots democracy. In this context, democracy means more than consensus orientation and a lack of hierarchical structures; it means real encounters on an equal footing, radical inclusion and affirmation; seeing through micro-transactions of domination and establishing relations beyond them. This is their secret centre and in a way the core of their strength. It has probably also been underplayed in research because researchers often don't really have an insight into the everyday democratic challenges which are connected with the building of movements. The key strength of democratisation means, more specifically, that FFF and XR were successful precisely when they acted as grassroots democratic movements and gave the public the opportunity to join them as equals – so that every child, in every village, had the sense of being able to contribute and "own" the movement, not being disturbed by NGOs working behind the scenes and preferring some "chosen" young people.

The second level is that of organisation. The general level of the "will to democracy" is reflected on this second level: how are the movements structured and organised internally? Decentrally, like Extinction Rebellion, with holacratic elements through which roles and mandates are distributed? Or through other non-hierarchical models which distribute power and at the same time regulate how responsibility can be taken on? Because this is the challenge: how do we ensure that particular work processes can be approached in a structured way? Without every task either being doubled up or forgotten? If there are no bosses to hand out tasks, supervise them and coordinate them, how can this lead to anything but improvised chaos? More and more organisations turn, at this time, to the model with roles and mandates (Robertson 2016): by defining roles with particular mandates (role: "responsibility for press contact"; mandate: "can write their own texts"), work processes are organised transparently for all, without this being tied to bosses or to particular people at all. Roles can be switched; someone can take on two different roles and so on, as long as this is communicated openly.

But something seems to be missing from these models and from the organisational theories behind them: namely the idea of a shared "place of exchange", like the Mynttorget or the chats, which exist beyond specific achievements, mandates, or roles, and enables something like the spread of information and a playful way to meet one another as equals. All models of holacracy do assume that central meetings will take place, but these only serve to decide and redefine roles and mandates, meaning that they are purely work meetings. A shared democratic "playground" is something different. If it is missing, it is difficult for people to take care of each other, because there is no concrete understanding of the challenges.

In many widespread theories of organisation (including those discussing parties, trade unions, and corporations), there is often a lack of understanding for these deeper processes of intersectional democratisation, which are emphasised by authors such as Garza (2020) and Spade (2021).

The final, third level on which the process of democratisation can emerge as the core of grassroots movements is that of communication: leading and conducting meetings, creating a "code of conduct", and structuring consensus-orientated decision processes. What still tends to be ignored in that context seems to be the aspect of democratic leadership (see the chapter about education). It is about the fact that creating democratic spaces is active work, leading to radical inclusion; making everyone's dignity visible. It does not just mean remembering principles, but actual leadership, encouraging people, freeing up resources which will mean that space and time can be allocated fairly, and stopping domination; distributing power, especially as the ones who possess the most informal power, as well as resources of all sorts, including cultural and economic capital. As a small example, the older ones can show that different approaches to the world's problems are possible (see the chapter on education for a detailed description of democratic leadership).

To summarise, the hypothesis is as follows: if all these three levels of democratisation are combined, that of social logic (non-instrumental relationships beyond domination), organisation (creating shared spaces), and communication (democratic leadership), then a quite unique energy can unfold and change the world. This is to some extent what happened when XR and FFF emerged in autumn 2018, and since then it has been the daily task for those who work in them or in similar grassroots movements. Substantial democracy should be created, internally and externally. That provides the compass, motivation, and strength.

Fear and (informal) power

So it is not enough just to point out what is wrong about informal or structural relations of domination. That is only the beginning of the process of getting out of them and bringing about democratisation. And that is where an insidious sociopsychological mechanism comes in. The same goes for a family dominated by a "good" "Pater familias", and for an economy which structured by "caring" capitalists who "create" jobs, as well as for "caring" NGO workers who prevent democracy in grassroots movements by forming elites and exercising power themselves (even by establishing a "tyranny of the good" so that everyone has to think alike).

For those who are closest to the people with more power, it is often about important personal relationships based on trust. In that context, to suddenly say, "That is a form of well-meaning, but still subtly violent domination; please leave the organisation to all of us as equal members," is difficult, often almost impossible. And for those who are furthest away from the people in informal power positions, there is the threat of being stigmatised: suddenly you are criticising a whole system if you demand non-instrumental relationships, and the building of a broad bottom-up, substantial democratic learning community.

For that reason, in my research, it seems particularly important that there should be explicit democratic leadership in all groups (and centres of knowledge where everyone can explore the principles of seeing through domination and create humane relations, making the dignity of everyone visible), whether in the family, at school and university, in society as a whole or in social movements.

Can we delegate climate activism?

The definition of these democratic grassroots movements then becomes clearer somewhat later in comparison with the other groups emerging, particularly in Europe, and gluing themselves to the streets. They all belong to a loosely connected network called A22. What many people simply see as a new kind of climate activism, a kind of continuation or complement to XR and FFF, turns out to be something quite different from the standpoint outlined here. It does not connect the two components which are so crucial for historic change: broad people's movements which are democratically organised and accessible to all, and disruption.

That is ignored by the media debate – and also by many debates in the movements themselves. Criticism is quickly focused on the form of protest action – people gluing themselves to the ground (even if I have difficulty understand how methods which exclude most people from engaging, disturb the working class, and potentially lead to violence should be effective). But the problem, measured against the yardstick of democratisation, seems also to be one of internal organisation. These are small groups, often – not always – organised "top down". The theorists and practitioners around the A22 network ("Just Stop Oil", "Last Generation", and so on), partly financed by the Climate Emergency Fund, have quite a different understanding of movements. Some of them have left the XR and "For Future" movements because they don't believe that the difficult work of democratic grassroots processes is effective. Crucially, they are also distinguished by the fact that large sums are being invested in recruiting and paying individual activists (Milman 2022). I keep on thinking: if they would only support the grassroots movements, their actions would have more benefits. Because their cause and the knowledge behind it is the same as in the grassroots movements. They do put the crisis on the agenda; and with great urgency. In that sense, they deserve solidarity. But how are we going to change the political approach in our societies just through small disruptive action groups? They can make a contribution, but without popular mass movements such as FFF, XR and Scientists For Future, the project seems hopeless; and lacking the focus on substantial democratisation which can be the core of the movements and the politics they fight for.

Conversely, there are also those who want to go in the opposite direction; just as problematic, it seems to me. For example, when Sven Hillekamp (2023) declares before a For Future general meeting that FFF is the opposite of A22 because the young people do not involve themselves in civil disobedience, that is also not true. FFF and XR were successful because they are both: they are grassroots movements, and they are disruptive. The school strike was an act of civil disobedience. If there had not been children who were required to attend school and refused to do so, coming back to Mynttorget week after week, society would hardly have reacted. It is by acting against the law – but with legitimacy – that the whole movement becomes a topic of conversation and gains force.

That is why I sometimes criticise the For Future groups of older activists who only support the young people and do not take action themselves to show non-cooperation with the fossil society, or at least take an "Emergency Break" on Fridays, even if it's just for an hour – if they are in a situation which allows it. They are pushing the movement into the realm of harmless activity within civil society. In London, hundreds of thousands demonstrate in these months in front of parliament. "The Big One" is the name of the first non-disruptive action by XR – which is entirely ignored by the media and by politicians, and by the people behind the fossil industry and the financial system. But at the same time, it seems that professionally organised top-down actions by a few people do not really create social change either.

And so, the fundamental question arises: which methods and forms of action would be fit for popular movements, disruptive but not exclusive; methods which do not reinforce existing privileges; ensuring that those who are most affected by the crises and dominated by the people in power are able to take the lead? Are there forms of non-cooperation, for example? It should be possible for all concerned people to join in these forms of action – and still put an indefinite stop to "business as usual". Stopping on a Friday works as a signal: the approach of grassroots movements consists in ensuring that inner and outer aspects of the movements correspond to each other. What we are fighting for, intersectional, substantial democratisation of society and a sustainable life, must be reflected in the structure of the movement and in its collective actions.

The new year begins

After the trip to Glasgow, a new time begins, for the global grassroots movement, a time of mobilisation. Information meetings are organised. The UN climate conference is coming up in Stockholm, and so are the first elections in four years, since the beginning of the strike.

While some members of the Swedish movement are commenting on the election campaigns by the political parties, others are preparing for their friends from Brazil, the Philippines, and Uganda to visit Stockholm and thus also realise global grassroots democracy.

> New guidelines for cooperation between generations (NGOs and youth movements) – developed at the Department of Child and Youth Studies at Stockholm University: Firstly, the movement must belong to the children and young people – to all of them, democratically. Secondly, power relations must change, meaning that NGO workers must communicate with the movement as a whole and not intervene in structures, splitting up groups and forming elites, and thus making the children dependent on them in unhealthy ways. Thirdly, children's attention must always be drawn clearly to the dangers of a situation, and they must be given information so that they can make informed decisions themselves. And fourthly, bringing all this together: the children's welfare and their own position as political subjects must always have priority; adults have to put this first.

Chapter 8: The War, Fuel, and the Global Social Contract
January – August 2022: Towards a new world order to lead us out of the crises

The war and the new world order

Then the world order changes.
 A few weeks have passed since our trip to Glasgow. Christmas is over, and the UN environment conference in Stockholm is approaching, along with the first Swedish elections since the founding of Fridays For Future four years ago.
 A group of activists climbs up the hill over Lake Mälaren. It is late winter 2022. They look around them, down at the lake which surrounds the islands of Stockholm, and then they continue, down into the crowd which has gathered in front of the Russian embassy. Many of them painted blue and yellow signs the previous evening, and planned to go together to demonstrate their solidarity with Ukraine. Because on the 24th of February, the Russian government launched an attack on the Ukraine, crossing the borders and beginning a terrible war – and now it is trying to mark out new territory, against international law.
 How does it feel for these mainly young people, I wonder, when they have just survived the corona time and are finally allowed to leave their flats? How can we make a space in which they feel safe? How can we think up a transition to a new, safe world order together with them, and make it a reality? Because a new geopolitical order is emerging, but so is a conflict between worldviews.
 For many of them, the final year of school is nearing its end. Adult life is beginning. Many were 15 or 16 years old at the start of the strike, and now they are 19 or 20. New tasks are waiting, new flats, if they find them, new relationships, and studies. The pandemic is no longer so oppressively palpable in Eu-

rope; the climate crisis is all the more so. Temperatures in the Arctic are rising to unheard-of levels. Suddenly normal temperatures are exceeded by 30 degrees (Harvey 2022). The news agency Reuters announces on all channels that 2021 was the year with the highest levels of CO2 emissions in human history (Twidale/Chestney 2022). But they would have to sink by more than ten percent annually, at least in Europe. They are rising because coal, oil, and gas are still being extracted and burned. And all of this is connected, as soon becomes clear.

That is also the big question in all the newspaper articles about the world situation: the question of how to ensure security for everyone. Amitav Ghosh (2022) sees colonial history as the root cause of both the climate crisis and the war. In activist circles, too, connections are discovered between the two crises. The war appears to be about nationalism, authoritarian patriarchal attitudes, and the logic of energy production – but specifically also about fossil fuels and thus the causes of the climate crisis (Milman 2022). It is also, I think to myself based on my research, about the violation of spaces of integrity and of relationships which create real freedom; it is about clinging to a dominant, violent ideology.

In this sense, a different struggle suddenly emerges: the struggle over worldviews and philosophies which stick to what is dominant – and those which would make sustainability and democracy possible. In front of the Russian embassy, memories surface of the security conference in Munich in winter 2018–19, when Chancellor Angela Merkel commented on Fridays For Future for the first time. In the improvised "Q and A" section, she seemed to claim that Russian trolls were involved in the rise of the youth movement, and in a context in which she also praised the Nord Stream pipeline to Russia, the German car industry and borderless trade in general as a peacekeeping measure. And now we can read in the newspapers: holding on to the fossil society which drives climate change has not only made so many countries and their governments dependent on Russia, but is now also financing the war. The Russian gas and oil oligarchs and the regime are revealed to be closely associated with European states. An almost unimaginable estimated 80 percent of the Russian trade in raw materials takes place through Swiss banks (Parliament 2022). Material is transformed into wealth that has no physical location – an enormously lucrative business. Switzerland will then also initially reject the EU sanctions against Russia, and will only later react to protests by the international community (Pfaff 2022).

What Scientists For Future have been pointing out for years: this dependence also highlights the hesitancy of European countries to develop renewable energy, especially solar and wind power, with the necessary electricity grid. Wind and sun are free, they don't obey territorial borders and they can also be turned into energy and shared wealth on a small scale, decentrally, which means that they don't fit into the thinking, the worldview or the political and economic order which has tied together the governments of Europe and Russia in a joint project: that of corporations which transform nature into burnable commodities, through which individuals become rich in a peculiar mixture of maintaining territorial national borders, including the marking out of property, and the borderless movement of goods and money between Moscow, Zurich and London. And now, the poorest and most vulnerable across the world are paying the price for the fact that many have held onto the trade in fossil fuels and cooperated with those who do not obey international law.

Solidarity

Now, in these February days, it seems almost cynical to spend time analysing energy policies. The priority should be the people in the cities which have been bombed, and the people who are fleeing; and also our own dismay and fear. Combined with the climate crisis, the war is soon affecting the whole world. The horn of Africa is suffering from the most terrible famines (Unicef 2022). The shared fabric of integrity is literally falling apart. How can it be woven back together, is then the question, when these problems are discussed as a whole.

Only a few days later, on a Wednesday evening, the climate activists stand in Sergels torg, the main square in Stockholm, and take part in the solidarity demonstration. Those who have fled describe the horrific crimes of the war. They are real people with names, speaking to the crowd as the whole environmental movement stands in the square in front of the "Kulturhuset", the huge cultural centre which forms something like the central point in the city. The Ukrainian ambassador nods at the young people, joins them and thanks them for being there. What a good thing they have the movement and are in contact with each other, I think to myself. And: how can we comfort each other? How can we talk about this war? How can we even think about security and safety and communicate those ideas?

Ideas about a new order

During these days, researchers such as Julia Steinberger, with whom the young people are cooperating, open up a new perspective on a possible more secure order. They point out that there are actually enough resources for everyone (Millward-Hopkins et al. 2020, Hickel et al. 2022). Looking at the world from a distance, at the situation now and in thirty years – if we do react to the climate crisis – they see that it is possible to organise our shared life in such a way that everyone has enough to live a good life, without breaking the limits of the planet with an enormous "throughput", wearing out materials and using up energy. It just doesn't work – they show – with the way in which clothes, food and so on are currently produced, or with how their production and distribution are organised in economic and political terms. Geopolitics must be conceptualised together with a new economy, they say.

This groundbreaking research seems like good news. The wrong idea of global coexistence has been institutionalised, and that idea is coming to an end – or else we have to ensure that it does, I think to myself, as the activists, young and old, climb the snowy hill and look out at the city and into their future.

Why, for example, does the UN just sit by while one of the members of the Security Council is so clearly contravening Articles 1 and 2 of the Charter, governing the protection of territorial integrity? Why does the global community not react in such a way as to ensure peace, and what could that reaction look like? How could all of this relate to a new concept of security or "stewardship", which would include respect for autonomy as well as keeping nature intact, and distributing "resources" as something shared, as part of the commons (see Dixson-Declève et al. 2022)?

And why do no politicians go ahead and begin to create new global rules, as was at least attempted after the Second World War, with the Declaration of Human Rights and the UN Charter? It would be possible for politicians to discuss the war and the energy crisis, but only if this new global framework, which is political and democratic in the broadest sense, is debated at the same time. Instead, many seem to be cobbling together quick solutions for how new gas, new coal, and new oil can be extracted if Russian fuels are soon no longer available.

How can we make it possible to shape the world on the basis of transnational justice, and at the same time respond to the environmental crises?

At exactly this moment, the world is looking to Sweden when it comes to the environment. Fifty years after the first international environmental conference of all, opened by Olof Palme, the anniversary conference Stockholm+50 is coming up. Many people are wondering how the idea of a new post-fossil world order can be advanced in concrete terms. What should really happen when the governments of all countries meet? How might we be able to present a new global social contract? In the coming weeks, we all learn a lot: about ways in which the UN might be reformed, but also about ways of building a radically new, global community.

The Fossil Fuel Treaty group and the social contract

Every Thursday evening, before the weekly Friday strike, the older adults in the People For Future group meet up online. Some of them belong to Scientists For Future, others to Parents For Future, and others to Artists For Future, while others don't belong to any particular subgroup.

Many of them work closely with the "Fossil Fuel Treaty" group, the group I stumbled on back in Madrid during the COP conference, which is guided by the model of the nuclear disarmament treaty. It is a team of researchers, politicians and activists who want to establish an international treaty for a just end to the fossil society (www.fossilfueltreaty.org). Would that be a solution for the new global security policy?

We have to expand the Paris Agreement, they argue, because it says nothing about the background of the crisis, the burning of oil, gas, and coal. Many of them want to come to the conference in Stockholm. The goal: cities and countries should join the new treaty. This treaty should first of all bring about an immediate moratorium on all new fossil infrastructure. There can be no new oil drilling, no new pipelines – and no financing of such projects. The next focus is the dramatic annual downscaling of fossil infrastructure, including the expansion of renewable structures worldwide (with the UNEP Production Gap Report as a guide). And these two points should, thirdly, be connected with a fair global transformation (from finance to a new economic order between the Global South and North), through which workers in the fossil sector should also be helped (see Appendix).

A great deal has happened since Madrid, when I sat in a dark theatre with about twenty researchers and activists. I kept in touch with the people involved, and my own research often overlapped with their work. Many city governments have officially joined the call for a treaty: Barcelona, Vancouver, Paris. And more than one hundred Nobel Prize winners have joined, too (Garric 2022).

Many of the young people are campaigning for a new social contract worldwide, together with Brenna Two Bears and other representatives of indigenous populations. Loukina from Switzerland – part of the steering committee – is pushing the developments forward in Swiss cities, together with a small group. So, in next to no time, a very active international organisation has emerged under the direction of Tzeporah Berman from Canada – with a communications team, research, and representatives in all continents, including specialists in questions of global justice.

Many are working specifically on the "Marshall Plan" policy which would be necessary: from policies in all sectors (such as establishing local renewable energy systems owned by communities themselves), to proposals for systemic change, including global, unconditional basic services, a circular economy, debt forgiveness for MAPA regions, or a basic income (Hällström 2021). Environmental organisations across the world, such as Friends of the Earth and the For Future groups, adopt the idea of a Treaty Initiative and form local groups. Soon, behind the scenes, diplomatic paths lead to more city governments being convinced, as well as countries such as Vanuatu. They publicly demand a treaty and join the project for a new global order. And now the project might reach a bigger breakthrough in Stockholm.

But at the meetings, one challenge keeps on being discussed. It is not enough if individual cities and countries are won over to this new order beyond the fossil society. Why should the governments of oil states suddenly cooperate and transform their economies? That is hardly likely to happen, I argue, unless we simultaneously create global pressure and worldwide solidarity from the streets.

So a plan is formed. We need a dual approach: the grassroots movements should be expanded worldwide, with the help of FFF, XR and PeopleFF. They can create pressure for this new order, partly through civil disobedience, and they can push for system change, while the Treaty group works with civil society and from within at a political level to organise binding legal changes. The idea behind both approaches: we have to make this notion of a shared humanity and a shared humane attitude visible in the first place – a humanity which lives on one planet and actually has enough resources, if we organise them carefully and justly, and are fair in our dealings with each other.

Sometimes during these weeks, I think about the website on which – in the year before the school strike – I naively outlined and called for a conference, somewhat different from the Treaty group. Because the global community could put an end to the worst fears in one day. It could agree what quantity of coal, oil, and gas should be extracted where (and what should be left in the ground), so that we can be sure that we have control of the quantity which will not lead to more than 1.5 degrees of global warming. Then, practically overnight, the main danger would be averted. That would be possible. Such a decision can be imagined. But we don't have the organised global community, the order that would be necessary to make such a decision. Not yet.

A new scenario – the four secret rooms

While we older ones work on this plan, a year of strike preparations and marches begins for the young people. Strike activists across the world have agreed on the 25th of March as the next global day. This will be followed by the strike on the 3rd of June during the UN environmental conference. And the Friday directly before the national elections on the 9th of September is also planned in for the Swedish activists.

That is how the plan looks for the climate activists. In addition, as the "Aurora" association, they want to prosecute the state for doing too little against the crisis, and for not behaving fairly in a global context. And similar things are happening across the world, in all local FFF groups, in metropolises and tiny villages. In the regions most affected by the climate crisis, other protest forms are often sought and preferred, because strikes are often too dangerous. From there, many of the young people make their way to the conference, to visit the Stockholmers whom they know so well from the chats, and who have spent months preparing for the days in June.

What ought to happen during the three-day conference? We need a scenario for what ought to happen within and outside the trade fair site in the south of Stockholm. I set out to write a scene that aims to make the problem clear. Couldn't the cultural institutions of the city focus on this – on what ought to come out of such a global, new social contract – and present the results during the conference, also as a reaction to the war? An email with my scenario soon lands in many inboxes. The Royal Dramatic Theatre reacts with interest. It is located in the centre of Stockholm, almost within eyeshot of Mynttorget. In a conversation with two interested dramaturgs and directors, we decide to form a group of theatre people who will take up these questions and develop projects. We need a new story, without drifting into propaganda or instrumentalising art.

For the meeting, I have sketched out the following basic scene: passers-by go past four rooms and hear conversations and debates arising around this new social contract. In the first room, a transformation plan including CO_2 budgets is being formed, and reductions in emissions are being broken down, across the system and in all sectors; in the second, the stopping of the fossil industry and the regeneration of forests and soil beyond the production of animal protein is being determined, inspired by what the Treaty group has been working on every day; in the third, the discussion is about providing for everyone's basic needs and organising fair sources of funding, so that everyone across the world has enough resources and infrastructure for a dignified life beyond poverty and hunger, including cooperation on financing a renewable energy system (Jacobson 2019; Teske 2019).

But the most important room is the fourth one, to which the people in the other rooms are constantly hurrying. There, the questions of global justice are discussed, questions of an intersectional analysis of structures of domination, the deepening of democratic processes, and the legal restructuring of our current order, including that of the UN: how the charter could be reshaped and peace could be secured; how reparations and debt forgiveness could be organised, as well as a fair reorganisation of the flow of resources between the Global North and South; how to distribute the CO_2 budget fairly, as well as help for workers in the fossil sector; in short, how the crisis plan should look which will make our global society sustainably and democratically fair. The fictional passers-by listen, are drawn into the debates and set out the conditions which should characterise this new social contract.

And that is really what the members of the Fossil Fuel Treaty Group are aiming for. Soon, a first group of countries is to make a start and establish a prototype for this new cooperative world order. This is about a kind of global democracy which will extend the model of compromise between national and often nationalist governments – as well as circumventing and replacing that model.

What about the UN Security Council?

At this moment, a new group turns up which says: aren't there already instruments to shape global politics in this way? This is all taking far too long. Don't we already have the Charter, the General Assembly, and the Security Council of the UN? Why should we now suddenly develop a completely new form of international cooperation for safety and security?

During these weeks, some Scientists For Future get in touch who have been researching this question in the context of Security Studies. They organise a meeting at the Marc Bloch Centre, an institute for social studies and the humanities at the Humboldt University, with the title, "The Climatization of International Peace and Security". How can we approach the climate crisis, the war and security for young people together, through a reform of the UN? This question is raised, among others, by the researchers Judith Hardt, Anne Dienelt, and Adrien Estève. They are also in contact with Louis Kotzé (2022), who has a published key research on this question, as well as with researchers and decision makers in all the countries represented in the Security Council. But does it even make sense to hope for a solution to this question from the very top of the UN, or to work on such a solution? Which texts and which parts of the law would have to change, and how?

Apparently in the context of the work of the Security Council there is a "draft", a sketch of a redefinition of what should be understood as security, the researchers report, so that the climate crisis is better integrated (this is also a subject of research for the Secretary-General's High-Level Advisory Board on Effective Multilateralism). However: the Security Council is not a democratic reflection of the world population. And this sketch seems to be full of holes. The causes of the crises, for example, are entirely ignored. There is no mention of climate justice, and only the effects of the crises, such as migration, are discussed. The rights of future generations and of young people are deliberately not mentioned. The focus is almost entirely one-sidedly on the security of states, and barely on the security of people or nature beyond state organisations, at most when populations are threatened by regimes so that humanitarian interventions might be necessary. The concept of security itself is understood in a truncated, military and "masculine" way, according to some, so that the existential dimension of help with poverty, drought, floods, and social conflicts is missing (see Hardt et al. 2023).

At least the researchers also say: while there may not be any legal framework within the UN to create security in relation to the climate and biodiversity, there certainly would be potential for this, through resolutions, for instance; through changes to the Charter, or through the invention of new institutions such as a new court to supervise the implementation of the Paris Agreement – for which the Paris Agreement would have to be expanded. There is no legal basis per se which would place limits on this; there is only a lack of political will to think beyond conflicts between nation states.

A reform of the security council's decision process is also not in sight, but it is possible. It could prioritise protecting people and nature, not only states, perhaps through the softening of the veto power and a more democratic representation of the populations of all countries. It is also unlikely that the General Assembly will be equipped with more rights, according to the prognosis; after all, it consists of governments which are often nationalistically motivated, and which often precisely do not see themselves as the joint representation of one whole human race. And even more dangerously: behind all these processes loom think-tanks and private economic actors which want to present the environmental crises as if they could be solved by private military means, according to Adrien Estève. These problems are also reflected in the other UN organisations which are directly responsible for dealing with the climate crises, including the UNFCCC, the framework for the COP meetings where the nations cannot even agree to name fossil fuels as the problem.

That is why the Fossil Fuel Treaty Group was formed – as a kind of answer to the failure of officially established international cooperation. Because a top-down process from the Security Council, the UN General Assembly and the COP meetings is not going to lead to any binding guarantees of security for human beings or for the planet, a new social contract is needed: so that the towns, regions, and then individual countries can try "from below" to join together gradually, with help from civil society and from mass movements. Despite this, the Scientists argue in Berlin, it clearly makes sense to push for the radical democratic reform of the UN. Again and again, they argue against the Treaty Group: why should gas and oil states ultimately join, if all of this is organised voluntarily "from below"? Isn't the same thing happening here as with the treaty for nuclear disarmament, meaning that those who ought to be the first to make a move, because they own such weapons, are the ones who don't participate? The Treaty Group disagrees. They insist that the UN is not acting, and that moral pressure needs to be built up so that oil states are seen as rogue states. So the arguments go back and forth.

And soon it is possible to see that in June, during the environmental conference, neither the Fossil Treaty Group, nor the proposals to reshape the UN will be able to take on an official role. All such applications, which spend months going through the official channels, are rejected in the final weeks before the conference by the representatives of the Swedish government, the host of the conference.

Thus, the global FFF group which is preparing for the conference in the chats only finds its position confirmed: a third way is needed, global cooperation from below in the form of a disruptive mass movement. What FFF and XR have built up in the previous years must be expanded further and broadened: as a form of organised people power. And so, they plan the march through the city, including a program of events on stage, anchoring this plan in the worldwide chats – as a small step in the multiyear strategy to strengthen an uprising of quite ordinary people from below; of schoolchildren and all concerned adults, led by those who are most affected.

The idea of a second chamber – cosmopolitanism and global democracy

One question remains unanswered. These approaches still adhere to a model that is mainly based on nation states with their nationalist governments. What if we need to anchor democracy in transnational thinking?

Taking their cue from this idea of bottom-up global democracy, other activists are also searching for ways in which the basis could be changed, which other approaches leave untouched: how citizenship and democracy can be expanded and defined worldwide; how borders and spaces of freedom, ownership and the stewardship of nature can be redefined on the basis of cosmopolitanism (see www.globalassembly.org; for a philosophical debate on cosmopolitanism, see Hooft's (2009) discussion of classical analytical texts by Scheffler, Rawls, Nussbaum and others).

An element of this is the question of how decision processes can be changed at a global level; for example, by creating a "global assembly" of representative but ordinary people. Some say that this could parallel the UN General Assembly formed by heads of state, as a "second chamber", while others say that it could replace it. Some like the idea that the ordinary parties which compete locally, nationally and (for example) at EU level should also come up with global policies, so that a global Green New Deal could be established, for instance (Taylor 2021). But many dread the prospect of a Star Wars-inspired world government which would trump the national and local levels – most focus more on a grassroots democratic path such as that of the global assembly.

The idea of a "second chamber" becomes increasingly influential in the movements and takes on a central role in democracy research (Pelluchon 2019): what if we invent a second chamber which could also become part of global, local and national parliaments? It would be composed of people chosen at random but representatively, who would be guided by scientific climate and justice experts and advocates for children, for nature, for non-humans and for future generations, and would develop plans and reach decisions – which cannot simply be ignored, as currently often seems to be the case with such citizens' councils (for a critical analysis, see Machin 2023). This would mean that there would be a kind of guarantee that planetary and human limits as well as the dignity of all people would be taken into account – a democracy "task force" chamber, for which the grassroots movements could stand up until it is established at all political levels. The substantially democratic idea of the second chamber would give the demands of the climate movements for citizens' assemblies (Extinction Rebellion; Last Generation; Global Assembly; Occupy; etc.) a permanent democratic form.

All of that often sounds utopian to the researchers in Security Studies, who point out that such processes of transforming democratic infrastructure take an incredibly long time – and even then, there is the problem that in a certain sense these assemblies simply reflect unjust power relations, such as between classes, rather than dismantling them. Workers' movements have pointed this out (Bell 2020).

Geoengineering: the balloon stays on the ground

From one day to the next, this struggle for a new world order and different worldviews, which had seemed so abstract, suddenly becomes very concrete.

Once again, the Fridays and People For Future are on a Zoom call, this time together with representatives of many environmental organisations in Sweden. Researchers linked to Harvard professor David Keith want to use the space station in Kiruna in northern Sweden, of all things, to conduct the world's most important experiment in "solar geoengineering". A balloon is to be sent high into the sky. It could theoretically distribute sulphur particles, later, as a possible step in a gigantic experiment. The sun would be dimmed for years – or forever – and thus the temperature on earth would be reduced; as well as droughts and floods. Some scientists see such an intervention in nature as unavoidable if emissions cannot immediately be stopped – otherwise, billions of people will be exposed to lethal heat.

Here, different worldviews collide, I think to myself on the Zoom call. Could this be the biggest debate of our time – whether we should continue to insist on using a specific form of technology to intervene, subjugating nature and concentrating power in the hands of a few, or whether we should try to find our way to a new form of cooperation between nature and technology, by deepening democracy globally?

The experiment seems to be the counter-model to democratic cooperation between nature and technology – in a crude form which brings together economic interests and research. Individual researchers have an interest in profiting from these projects, and many private universities often are closely associated with the fossil industry.

The opposite philosophical model would focus on a sustainable democratisation of the energy sector and the economy, on fair ways to put a stop to emissions, and on making essential resources available to everyone equally (on this, see Dixson-Declève et al. 2022). Instead, researchers, think-tanks with economic interests, and whole scientific academies such as the American "National Academies of Sciences, Engineering, and Medicine" (NASEM) as well as well-known journals, are demanding that the global community research how to obscure the sun with small particles for ever (Voosen 2021). This might mean that we never see blue sky again.

At the meetings of the environmental movements, leading researchers gather and exchange the most controversial arguments. After thorough investigations, some of them decide to stop this technological quick fix process. That is why they are now all on a Zoom call, thinking about how the balloon experiment can be prevented. Should they travel to the space station in the forests of northern Sweden and block the way?

In some activist contexts, doubts spring up: what is really so bad about this experiment, some people ask. It is just an experiment, after all. And if it is realised, it can be used as an emergency brake when the earth is burning and billions are dying. Isn't it better never to see blue skies again than to make billions of people go without water?

The specialists point that a widespread use of geo-engineering, towards which this experiment would be a first step, faces massive problems in terms of ethics, politics, physics, and tactics (on these arguments, see Biermann et al. 2022; Hällström in Thunberg 2022): as a technological quick-fix, geoengineering would draw decision-makers' focus away from what should be the main priority: avoiding carbon emissions. Reliance on geoengineering means also that our dependence on technology could never be stopped: a sudden heat shock would follow any interruption in the artificial regulation of climate. Geoengineering would thus make humanity's future dependant on technologies that would likely be controlled by a few powerful corporations. Besides, keeping the infrastructure necessary for continuous geo-engineering seems entirely impossible in global political terms. Finally, unpredictable consequences of geoengineering could be negative for various parts of the world. For example, the monsoon rains in India could suddenly change their pattern. Who decides on which risks should be taken, by whom? The UN Security Council?

Another problem: the acidification of the oceans and many other things will not be affected at all by this "solution". Whereas those could be solved by a radical reduction in greenhouse gas emissions, by stopping the burning of fossil fuels. "Wouldn't that be a real solution?" an activist asks the world-famous physicist Michael Mann, who developed the hockey stick graph which illustrates global warming and CO2 increases (CIEL 2021).

What we could instead think up and realise, I think then with my research in the background, would be a humane technology that would be compatible with nature, interacting with it gently, protecting ecosystems (see Vetter 2022); and a basic attitude which anticipates problems and acts preventatively. And focuses on democracy, unlike the suggestions of Bill Gates (2021), David Keith, Elon Musk, and Mark Zuckerberg, whose purely technological approaches without real consideration for democracy shape the climate debate during these months. Many activists are very concerned that mainstream society will suddenly be convinced by their attitude and that the world really will drift in the direction of geoengineering; that all of this will be normalised. (And in these months, most of them shift entirely to open source tools, away from the capitalist-organised tech platforms Twitter, Meta, Alphabet, etc.)

Finally, this most renowned of geoengineering projects is successfully stopped. The process is led by a few organisations of indigenous populations in northern Europe. Led by Åsa Larsson Blind, they write an open letter to Harvard University – or rather, to the ethics commission which is supposed to be monitoring the project (Goering 2021). This startles the commission, and the project is postponed indefinitely. The earth thus seems to be more than just a territory to be controlled – it is a place for life.

The five dimensions of convivialism

At the university, I brood on the previous weeks: the worldview behind "geo-engineering"; the relationship between technology and nature; the war and Security Studies; the crises and the scene I wrote, which was supposed to propose a new social contract.

I try to develop the perspective which still often seems to be missing, and which only becomes apparent to me at all in conversations with the young people and with indigenous activists and thinkers. It is a kind of compilation of collective knowledge, or joint research. Central to this is the idea that we should think about states and geopolitics by connecting this thinking with the studies of what people regard as a "good life"; what enables them to live and be in contact with themselves, with nature and with others, without shutting down because of fear, poverty, hunger, or violence. The social and political spaces which make this possible are the ones which enable substantial democracy, along with the protection of the essentials of life beyond power relations.

With that, we end up at a different point of departure from most of the theories of "global governance", which argue that we have to leave aside the sociopsychological and anthropological and ecological conditions of life, either for reasons relating to the political "reality" or for reasons of "liberal" caution about colonial universalisations of values.

But anyone who argues in this way is arguing from a purely formal understanding of democracy, and neglecting crucial aspects of the question of how a political order can be scientifically justified at all. Can we just "elect" certain worldviews and ideas about global cooperation, or do investigations of reality offer us a framework to measure this? What is the point of a science of democracy? How should we measure a successful way of living, globally? And if we have found the framework, what form of society results from that?

I try to record the whole picture in a kind of table. First of all, the form of society can be called "convivialism", based on a French research tradition (Adloff/Leggewie 2014). On the structural level, it is characterised by the dismantling of relationships of domination. This means that it can be combined with a post-growth economy which does not force us into an exponential throughput of material and energy, but instead means treating what there is or what we create in a caring, regenerative manner. There should be enough resources for everyone, without the underlying ecosystem being exhausted.

Secondly, a shared project for humanity becomes visible through this (the repairing and weaving of a common fabric of integrity), which connects us with each other historically: with all interactions that damage us or make it possible to develop a space of integrity. We can work on this fabric: for instance, the Global North can cancel the Global South's unfair debts, and the economy can be reshaped so that it is no longer based on exploitation. Thirdly, the form of society and this fabric are thus connected with the project of helping each other develop social and individual spaces of integrity (education, health, etc.). Fourthly, through this, exploring and creating "connectedness" becomes central; meaning gentle contact and democratic exchange on an equal footing, as research in sociopsychology and neurophysiology has discovered and could discover further (Stern 1985; Immordio-Yang 2015; see the chapter on reorganising universities). And fifthly, this can then be described on a level which combines all these aspects and aims to capture them: humanity and integrity as a replacement for dominant relationships with the self and the world. From there we can develop a new thinking regarding property, rights, nature, and making democracy possible in political structures, including transnational cooperation.

Thus, at the institute, we incorporate the question of sustainable coexistence, by endeavouring to connect the smallest scale with the biggest: the way in which we can successfully meet each other through substantial democracy, as parents and children, or as people in educational institutions by ensuring that we do not force people to lose contact with themselves and others, and by fostering a "humane energy" (Fopp 2015) – and the way in which cooperation between states or supranational democracy can be organised. Seen like this, institutionalised political structures at a local, national, and global level would have the task, as "formal" structures, of enabling substantial democracy, meaning encounters on an equal footing, in freedom and equality.

If we look at the theories in the realm of political science, international relations, or global governance, some of them claim to be descriptive (realism, liberalism), while others offer a normative compass. This approach here does the latter, attempting to combine intersectional approaches (queer feminism, postsocialism, postcolonialism) with ecological perspectives (theories of sustainability, posthumanism etc.) and to find common ground. If we understand what allows us to open up a space of integrity for each other and develop it by organising our lives – using rules and resources – in such a way that this is possible for everyone, then we have a compass which also helps to determine how the higher levels should be arranged, such as the organisation of social spaces and the global political order. The proposal is therefore: the compass is already provided by the way in which we humans work, because we are capable of losing democratic contact with ourselves, of breaking it off or restoring it.

Nature as property or as "commons"

But how to make this all happen? Not in an utopian world, but here and now? Many of the young activists are working with a proposal from the Earth4All group which takes a similar line. They want to redefine how we treat nature globally, and thus create security and a new global order. Nature should not be treated primarily as private property, but as the "commons", something we have to look after together (Dixson-Declève et al. 2022).

The Arctic, the rainforests, the permafrost regions, all these sensitive ecosystems which are approaching devastating tipping points so rapidly, ought to be assigned to all of us or to no one. Anyone who "uses" them should pay a contribution which goes into a fund. From that – some of us continue the thought experiment – it would be possible to pay for global basic services (living space, transport, education, health) or for a basic income (Bidadanure 2019). This would help fight poverty and hunger, according to this approach. This is compatible with the core idea of the Treaty Group and with the cause we are fighting for in the global movements.

Two interpretations

But when considering this proposal, it seems to me that there are two ways of interpreting the basic idea which aims for "cooperation between all of humanity on one planet". The point of departure seems to be the same in both cases. The climate crisis and satisfying everyone's basic needs should move to the centre of global cooperation – and this should take place through the redefinition of nature and how we provide resources to each other and distribute them.

The basic idea seems plausible to me: the big ecosystems should be defined as commons, and in that sense the approach would be close to some indigenous traditions which have long inspired many activists. But the form this takes in political and economic terms still seems problematic. In their approach to the commons, Dixson, Göpel, Gaffney, and Rockström take their cue from Elinor Oström's theories and start from the assumption that this is mainly about limiting the use of shared resources. However, this means leaving aside the actual concepts of nature (as anything other than a "resource") and of property (in John Locke's tradition, as something we can dispose of and abuse), as well as the concept of integrity. It is mainly about restricting exploitation and damage.

A different – partly complementary and partly competing – approach would be to take a "holistic" and "structurally relational" view (Shiva 2020). This would mean asking how we can prioritise the regenerative treatment of nature, so that all people and non-humans can live a dignified life together beyond relationships of domination: this would entail providing enough resources for all. Rather than primarily starting from the limits that are to be placed on whichever capitalist markets and forces, we would focus on the fact that nature does not primarily belong to anyone and that it can be preserved for everyone by everyone and treated regeneratively. That involves focusing on our relationships with each other and with nature (and not only how nature is separated from us and commodified), and on what a more sustainable relationship with nature would be.

Or in terms of formal and substantial democracy: this is once again a complementary approach to the individualistic non-relational logic of formal democracy, as reflected externally in property law and election procedures. The new, relational logic of substantial democracy prioritises a caring "metabolism" and a democratic process of exchange. All the terms we use can be adapted to this thought: this is no longer about a capitalist market (1) in which users of resources (2) ought to be limited by boundaries (3); it is not an appeal (4) to collective responsibility, to "stewardship of ecosystems", and it is not only about sanctions for "free-riding" (5) – but instead it is about the regenerative organisation of non-dominant exchange, a productive kind of metabolism which allows full contact in a framework in which everyone's needs are met. From this point of view, forests and so on don't belong to anyone, not even to themselves (contrary to Wesche 2023). They are not "things" which we all "share as resources", but something different; they are the environment in which we live together and which feeds us, if we don't dominate it but instead help to regenerate it.

Legally and politically, parts of these two models of nature as "the commons" (the model of Earth4All and the one sketched here) are complementary and can be realised immediately. However, other parts are mutually exclusive, especially when it comes to the underlying political economy (see Appendix). Still, both are based on a new, different view of the earth as the home of an interconnected humanity.

The poster, the police officer, and the UN conference

And so, early summer arrives. The Russian regime continues its war. Report after report comes in of terrible atrocities and suffering. What do the young people feel, I ask myself almost every day.

In Stockholm, the UN conference is drawing ever closer. Posters are designed. On two or three afternoons after school and university, the young activists walk through the city and stick them to walls and bus-stops, bridge parapets and houses, in the most impossible places, sometimes under the noses of the police. And the police don't know whether they should intervene when they see the familiar Fridays For Future faces. "You're making sure you follow the rules, right?" one of the police officers puts it, after some hemming and hawing. They laugh and move on, heading for the universities. These poster actions are a way for them to release the tension of the war and the crises, I think to myself, as I try to stick a poster to the wall of the university and end up sticking my thumb to the wall instead. But there are only a few funny moments like this which make the heavy atmosphere all the more noticeable.

The UN conference in June runs its course without any results at all. For the meetings of the Treaty Group, the conference is relevant, even though many of them take place outside the conference site. New cities soon join the global social contract.

Fridays For Future combined with People For Future are actually able to mobilise people on Friday the 3rd of June. The march starts close to the grand building of the city library. The MAPA activists from Brazil, Argentina, Kenya, Mexico, the Philippines and all the countries which are already affected by the crisis walk at the front. Behind them walk the Stockholmers, thousands of them.

A democratic global convivialism would approach and enable security and peace in a new, transnational manner. How would it be established?

First, an "Article Zero" in the UN Charter ought to describe us as an interdependent population on an earth which we must look after together, with all the consequences this would have for a stewardship of the commons (Dixson-Declève et al. 2022: "Earth4All") and for the global rights of citizens. Security should be redefined in all documents and realised through peaceful approaches, guaranteed through a democratically reorganised Security Council.

The General Assembly can be expanded through a "second chamber": through a Global Assembly, a grassroots democratic reflection of populations, which would also integrate care of children, non-humans and future generations, as well as scientific insights into the crises; and so would add a substantial crisis task force dimension to formal democracy, which would protect people and not just states, and legally control the implementation of the Paris Agreement. This "second chamber" (citizens' assemblies guided by scientists, which would pay attention to planetary limits and the needs of all, as well as incorporating responsibility for future generations, children, and non-humans), can also become part of the structure of local and national parliaments. This would mean that an element of "shared humanity" would be integrated into existing structures of political decision making.

Secondly, cities, regions and countries can join together "from below" without waiting for the UN, and can establish a new international social contract, a treaty to dismantle fossil fuels (www.fossilfueltreaty.org), which is already in place in hundreds of locations. It firstly demands an immediate moratorium on all new fossil projects including their financing; secondly the downscaling of existing infrastructure, following the Paris Agreement, and thirdly, that this should take place on a fair basis, within and between countries, so that all people have fundamental security during this transformation, perhaps through universal unconditional basic services which would be financed

> by defining critical ecosystems globally as "commons" and ensuring that they can only be used if a fee is paid (Earth4All).
>
> But thirdly, and most importantly, pressure needs to be built up from below, from the streets, and grassroots movements for climate justice such as FFF, XR and PeopleFF must be strengthened so that they can use civil disobedience and other nonviolent means to enable the masses to stand up for a peaceful, globally democratic society and establish a crisis plan (see Appendix). Local and transnational democratisation can be regarded as the core of these demands – as well as being the key element of internal organisation in the movements themselves.

In March 2022, Antarctica is suddenly 40 degrees warmer than normal, completely beyond what was imaginable (Samenow/Patel 2022). The new IPCC report chronicles the global injustice dimension associated with the climate crisis (IPCC 2022). The Amazon is still being deforested at record speed (Spring/Kelly 2022). Europe and America are relying on new oil and gas infrastructure to compensate for the lack of gas from Russia. At the same time, millions of dollars in daily payments continue to be made to the regime in Russia for fossil fuels, while the country continues to wage war on Ukraine (AP 2022).

In the weeks of July, the whole of Europe suffers from enormous heat, exceeding 40 degrees. Many young people want to take part in the FFF meeting in Turin, but change their plans because forecasts predict 38 degrees.

In Europe, large rivers are almost drying up, including the Rhine and the Loire. The Yangtze in China is drying up (EUSI 2022).

Research shows how emissions and the climate crisis are making such disasters more frequent and more intense (Abnett 2023).

By the end of August 2022, a third of Pakistan is under water. Thousands of people are killed or injured, 33 million are affected (UNHCR 2022).

The forests in Portugal and California are on fire. Thousand-year-old trees are being burnt to ashes. New record temperatures are being measured in the UK and many places around the world, obliterating the old ones (Copernicus 2022). Soils are degraded; whole crops and the basis for the coming years are being damaged, especially in the Horn of Africa. Enormous hunger is spreading (WFP 2023).

Chapter 9: Education in Times of Crisis – Learning from Young People on the Way to "Centres of Sustainability"

January – August 2022: How to change schools and universities – (eco)philosophy for a sustainable democracy

At the university – a story and a fundamental challenge

Sometimes it is time to look back and look forward. What could and should happen in all educational spaces?

This time I do not go alone to Stockholm University. I present my teaching rooms proudly to some of the activists who are about to start studying.

All this is familiar: getting out of the underground and walking into the wind which tries to blow us down the endlessly long escalators. And then passing all the institutes and departments.

In these conversations with the young activists – globally, too – a new idea gradually emerges. The university could react to the crisis to an entirely different extent. A huge "lever" of change becomes visible: as in Gender Studies, whose intersectionality courses must be attended by most students in all subjects in Sweden, there could be sustainability studies and a centre for sustainability at every educational institution, every school and university everywhere in the world. This could quickly change all subjects, teaching methods and research. We could make that happen, I suggest.

So, some of the managers of the Department of Child and Youth Studies commission me to write a paper. It is to outline how democracy, sustainability, care, and education are connected, and how these connections could shape teacher training. Soon I will be switching to the Marc Bloch Centre in Berlin as a researcher linked to the Climate Change Centre, and I want to use the chance to put all these ideas together.

And so, at a meeting, I begin to tell the story of the last four years, the story of four or five young people who started to strike. How they built up the movement, together with their peers worldwide, who worry so much every day about the world and their future. How they became more and more familiar with these topics, from climate science to global theories of justice. How they themselves sought to work with hundreds of scientists and organised webinars – to save the forest, to influence agriculture, to reshape energy and education.

I continue the story, explaining how they introduce new young people to this knowledge; how they share their panic over the crises, as well as their grief. How they invite their friends to Stockholm, including Maria from Mexico, Eric from Kenya, and Yusuf from Balochistan, so that they can pass on insights into the crises from the perspective of those who are most affected. And how all of them work together with us researchers almost daily and strike every week.

In particular, I emphasise the fact that the original scene of the strike in Mynttorget, when the young people fought with their whole beings to stop things continuing as they were, gives rise to a task for all of us. Or, as the philosopher Levinas (1969) would probably say, a "demand" which comes from a human face: to care about giving them a future without this deep fear, to do everything for change, including as researchers – and that means reshaping the universities now.

It is wrong, I claim, just to continue as we have been, or only to study these young people from a sociological and empirical point of view. We must react to them and research together with them to find out how we can respond to the crises fairly. And from a global perspective on young people, this means that researchers in Child and Youth Studies and Democracy Studies ought to train their perceptions so that they see all people, including all children globally, as equal and free, with the same dignity. That isn't happening, even at our institutions, I think to myself. We are still clinging to a status quo which is already destroying the livelihoods and lives of hundreds of thousands through drought and floods.

And I draw attention to a scandal: all students at most universities can complete their education in all subjects (whether they study economics or law, and whether they want to be teachers or doctors) without receiving deeper knowledge about the crises and their causes, or learning the skills and values with which they could now immediately build a really sustainable society. In ten years, emissions have to be reduced almost completely, in all areas, at least in the Global North.

How can those responsible allow something like this? Why do the directors of the Stockholm universities boast about their work at a joint sustainability conference ("Sustainable Planet, Sustainable Health", 1st of June 2022) when they should actually be apologising to these young people for the fact that in the last decades since the climate summit in Rio almost nothing has changed? And the whole scientific community ought to be apologising along with them for not taking the crises seriously for so long.

How can schools and universities instead take their task as democratic institutions seriously (on this, see Raffoul 2023; Barry/McGeown 2023; Barrineau et al. 2021)? How can all disciplines redefine their content, their teaching, their ethics, their research, and their institutional framework so that all students and teachers are involved in a project which leads the way out of the crises?

And how can we reach a new understanding of the project of modern science, which played its own role, together with industrialisation and the establishment of a capitalist market economy, in creating a reductionist, mechanistic worldview and thus driving the climate crisis?

The idea – sustainability centres as the core of education

It is still late winter and cold – the ground is frozen – when I walk with the young people across the campus. We almost slip on the icy ground and the wind briefly makes it seem as if we could fly away.

I think about what I myself can contribute through my research from the last few years. The idea is simple: what if we had a centre for sustainability here, a prototype? But not one which only brings together knowledge or produces specialist knowledge on topics such as education for sustainability, as is the case at so many universities, or which sets goals for fewer flights and for renovating buildings on campus. (Though that is also necessary – immediately, in fact.) This would instead be a centre to research and present the sustainable core of education, so that all institutions can make use of it, at every school and university. In developing it, we could work together with researchers across the world. They can bring in indigenous traditions of knowledge, decolonial arguments for reshaping universities (Patel 2015), intersectional theories of justice, and interdisciplinary knowledge about the climate crisis.

We could research a new relationship to the world, with creative and practical means of teaching it and making it accessible to all disciplines. How to interact sustainably, democratically and in a caring way. Then I wouldn't have to walk past institutions with the young people which seemingly have nothing to do with each other, even though all of them ought to be connected by the crises: political science, geography, teacher training, and so on.

The rebellion of the scientists

Globally, too, a lot is going on. More and more researchers across the world are becoming dissatisfied in their offices and joining Scientists For Future and Extinction Rebellion, forming Scientist Rebellion and no longer just writing articles but going into the streets in acts of civil disobedience: in Zurich, Copenhagen, Berlin, and many more in so many cities across the world (Kalmus 2022). At last, I think to myself. At last, something fundamental can happen. However: something is still wrong.

I certainly agree with the arguments of Scientist Rebellion: it is no longer enough just to write articles and books. The situation is getting worse daily, emissions are rising, and global injustice is increasing. The earth is burning. Scientific climate advisory boards and task forces are – in contrast with the situation during the corona crisis – not being listened to. It is appropriate, then, at least for those whose living situation allows it, to go out into the streets, take action, and participate in civil disobedience, particularly as university researchers. And I think back to the last four years in Mynttorget, to our cooperation with researchers such as Isabelle, who works in Youth Studies, and the strikes and street blockades.

But I still keep thinking that this is not enough. It misses what might be the central point, because it implies, at least in some cases, that we, the researchers, are the "good" people. What is missing is a sense of what is wrong within educational spaces, academia and at schools and universities themselves, the extent to which they, in their current "logic", are part of the nonsustainable society: from the curricula which ignore the crises, to the teaching methods, which reproduce the upper middle class and are focused on learning texts by heart.

We must instead look to an activism which addresses both society and politicians with civil disobedience, I say to my colleagues, but which also addresses internal structures and aims to change the universities themselves.

The Friday strikes, or non-cooperation on Fridays, combine both these two directions. They are directed both outwardly at the powerful, disturbing the status quo, and inwardly too; a refusal to let things continue as they are among students and teachers. If thousands took part in this refusal to cooperate, something fundamental would change.

But what should we be fighting for in these internal protests? How should the transformation of the institutes look; what kind of centres should be created? What could – for example – be demanded at every school and university in a joint citizens' assembly, a democratically organised gathering of teachers and students?

While I open the doors to the "stables", the theatre, music, and dance space of the university, all of this is going through my mind. What would be the core of a centre for sustainability and thus the core of transformative and regenerative education (Van den Berg 2021)?

The core of a centre for sustainability – "regenerative metabolism"

Isn't there something simple which could be the priority, even if it's not easy to realise: understanding how we can democratically create a secure, fair, and flourishing space for all humans and for nature, in which all of us can have a good, dignified life? That would mean (as the previous chapters have outlined): understanding the domination of nature and other people; seeing through these mechanisms, structures, and relationships; dismantling them and replacing them with knowledge about how democratic and regenerative spaces are created, as a joint project; through mutual help and by being in contact with each other and with nature in such a way that we strengthen one another. In that sense, we could and should incorporate intersectional justice and sustainability into our thinking from the very beginning when developing such a centre: it would be a centre for sustainable democracy.

From this, however, a new basic attitude to each other would emerge among people in educational spaces. We could see each other as vulnerable social beings, as creative beings who live, body and soul, in problematic power relations with each other and with nature. However, we can also enter into a regenerative exchange. That is where all education could start; as presented in the previous chapters. Gradually, I make contact with others who are working on similar ideas: with the CEMUS institute in Uppsala, with the Stockholm Resilience Centre; with the teaching department of the Doughnut Economics

Lab; with researchers in Wuppertal, Lüneburg, and Hamburg; with the Wyss Centre in Bern, with the new institute at Columbia University in New York, but also with smaller institutes across the world, and with the researchers connected with them – and of course with For Future groups such as the Fridays, the Scientists, and the Teachers For Future, who organise lecture series and annual education conferences at universities in the German-speaking region. But most of them are focused on researching socio-ecological systems. Barely any see themselves as centres which can influence every course of study and the whole of school education.

Back to Glasgow – the "Faculty For A Future"

During these conversations with the activists, a memory comes back to me. Fog blows through the narrow streets of Glasgow. Only a few weeks earlier, we marched through the city with tens of thousands of children and adults during the COP meeting. After the march, I went in search of a small café. Drizzle had set in and transformed the streets into a dark, uncanny film set. I was to meet a certain Jordan Raine. We had contacted each other when exchanging texts about reshaping the university landscape. He said something about a faculty – a "faculty for a future".

In the café, I am sinking into an oversized armchair when the door opens and Jordan appears, full of energy, and takes a seat. How are these people aiming to deal with the problem: that we lack the knowledge we would need to perceive the crises as socio-ecological systemic crises; and at the same time, we need to learn the skills to build a sustainable society in ten years, from agriculture and sociology to the economy?

After his doctorate in bioacoustics, I learn, Jordan worked in London as an editor at *The Conversation*, perhaps the most important online platform for academic communication. He was increasingly frustrated by the inaction of schools and universities worldwide regarding the climate crisis, and collected researchers around him who felt the same. Thus, a small crew came together to take on the project of a "faculty for a future": Clara, Josephine, James, Wolfgang, and many others. They have formed a platform online (www.facultyforafuture.org) which everyone can take part in: a kind of grassroots movement for researchers and also students. The focus is on all the areas we keep coming back to: curricula, teaching methods to ensure that we are present with our hearts, minds, and hands; a new research ethics, and the institutional, theoretical and social framework for an education which itself becomes regenerative and transformative, making prosperous lives possible.

But how do we set out the criteria, we ask ourselves in the café, which is full of damp Glaswegians who have taken refuge from the rain. How can someone decide what really contributes to sustainable education and is not just "greenwash" drivel? Or pure ideology rather than free science? The network of the "faculty" has established six principles, with the most far-reaching being the decision to talk of a sustainability crisis and not just a climate crisis.

The principles are: recognising that we are facing multiple systemic, urgent and interdependent crises (ecological, social, economic and so on); that these crises are caused by systems which dominate humans and nature and exploit them; that severe damage has already been done; that communities who have contributed the least are typically the first and the worst affected, especially in the Global South; that we are all worse off without urgent action, risking bigger catastrophes that overwhelm our capacity to adapt; and that if we take on our shared but differing responsibilities to act, as a joint global task, then rapid transformation is possible.

And soon the Faculty collects all the existing, relevant MOOCs (massive open online course): public lectures accessible to everyone, incorporating sustainability into all disciplines, including in pharmacology, sociology and architecture. The question comes up: what would need to be taught in a "super MOOC"? By popular demand from students, Barcelona is currently introducing a compulsory "Studium Generale" focusing on sustainability, which tries to answer this question (Burgen 2022; see also Thunberg 2022, an anthology in which one hundred scientists present their knowledge in two or three pages, with the aim of communicating this kind of fundamental information).

The members of the Faculty also work with specialists worldwide on a freely available collection of resources on sustainable teaching, which can be found on the website: how to teach so that students are involved, body and soul, with hearts, hands and minds, as the research on sustainable development says (ZDI 2022), acting as equal partners on their path of learning? And while we drink one cup of tea after another in the winter air of Glasgow, we begin to talk about more existential topics, about the way in which universities keep so many of us in uncertain situations through temporary contracts. That connects everything, we say: sustainable democratisation must also apply to the institutions themselves.

A tour through the rooms of the prototype centre

The memory fades, and I find myself once again with the activists in the theatre rooms of the teacher training institute. We open costume cupboards and rummage through the fabric supplies, take the musical instruments off the walls and move partitions around to create a stage set. I imagine a compère, a guide, who could lead people through the university campus, or rather not through the campus but through the fictional sustainability centre which I am sketching out – like a bizarre version of Steve Jobs and all the other tech bosses who proudly present their new products and features. But this is not about the technical details of something which is basically the same across all products, but about what makes us humans unique and democratic.

I continue to play with the thought. "Roll up, roll up," he could call. "Come closer! Look around you! The presentation is about to begin!" And in my imagination, he begins to lead us into one research room after another, from the smallest neuropsychological room to the biggest room of global cooperation: the diverse abundance of the core, the substance of democracy.

Room 1: A different understanding of animals

"Welcome to the first room!" calls the compère. "Welcome!" In the paper for the Department of Child and Youth Studies, I argue that democracy, sustainability, and care belong together, from kindergarten on. Because some researchers say that already as small children, we are democratic animals – or we can be when social spaces allow it (Stern 1985). But first of all: what does it mean to be an animal? A pig. A moose. A rat. Something some people like some species of, while others are killed mechanically and bureaucratically by the million, and then eaten? What is our place within this horde, and therefore in nature itself?

We are all very similar, according to the researcher Jaak Panksepp (2004), the founder of "affective neuroscience". He says: there are a few basic aspects that characterise all mammals which are bigger than mice. In his research lab, he discovered something which he called the basic equipment of animal life: seven "affective systems of action". What we call "feelings" should, according to his idea, be seen as aspects of practical exchange with the environment, as relationship systems. They enable an exchange beyond the obvious "systems of needs" – for example, depending on the theory: the search for food, reproduction, or temperature regulation. There is the "seeking system", and the curious "discovery system", which is so obvious in dogs when they go around sniffing everything. Then the anger, desire, fear, and care systems. And finally, two which have for a long time not really entered the mainstream image of animals in the Global North: the play system and the bonding/grief system. Even rats play, says Panksepp. They giggle and bond with each other. They establish relationships and miss their parents when they are away, and complain when they feel lonely.

According to this theory, the question of whether animals have a consciousness is absurd. All these systems are nuances of a conscious process of relating to the world. When we rear animals and slaughter them or imprison them by the million, we are not only contributing to the climate crisis because of the methane which is released and the forests which are cut down (Foer 2019). We are treating living and feeling creatures, "earthlings", as the young people also call them, in a dominant way which causes terrible suffering. The various disciplines at universities, not only biology, physiology and neurology, ought to deal with this knowledge and prioritise its practical consequences for our relationship to nature, I propose.

On the contrary, many theories of consciousness miss the main point, according to this tradition, namely that consciousness is linked to us (and animals) as embodied beings in practical, existential exchange with the environment, as meaning-making and beings which connect with others. That is at least the theory of philosophers such as Hubert Dreyfus, Sean Kelly, and Evan Thompson (2010), who link their arguments to Merleau-Ponty's gestalt theory. From this perspective, it is absurd to think that non-biological entities such as computers, robots, or any form of artificial intelligence could have consciousness, or experience pain or happiness, compassion or attachment.

But everything goes wrong when we reduce people to this neurological, biological equipment. The next point is to ask what it is that defines us as potentially democratic creatures.

Room 2: The animal which can lose contact – the human spirit and imagination (the neuropsychological foundation of "being connected/democratic exchange")

I look around in the theatre spaces. So often, the students here, who are training to work in kindergartens or as teachers, have looked at different ways in which we can communicate with small children so that they feel safe and free – and are taken seriously as free and equal beings.

So that we can be seen as democratic animals in evolutionary history, we need the Pankseppian "infrastructure". This is what allows us to lead a life which is oriented towards relationships with the world. But that is not enough. Relationships can also be undemocratic. Rather than being on an equal footing, we can dominate others or be subjugated. What happens in a democratic encounter, when we meet each other as equal and free people; including in social movements or in educational spaces?

"Let us visit the world of David Bäckström in northern Sweden, and that of the American researcher Mary Immordino-Yang, and of Daniel Stern, who filmed the interactions of small children throughout his life," the compère could now call out. Let us look at their theories about what happens in a democratic encounter.

A small child drops a ball on the floor. That creates a particular rhythm. "Boing, boing!" Pause. "Boing!" An adult can repeat this, bouncing the ball and copying the rhythm. And the rhythm can be repeated again, but this time the last bounce can be delayed. And in a funny way, this can create playful contact. This often works even with children who are only a few months old and cannot yet speak.

Stern (1985) discovered various "channels" of such non-linguistic communication: rhythm is one of them, but there are also patterns of intensity, and shapes. We can throw the ball with the same force and intensity – or more softly, or harder. This "language" is located at a "transmodal" level, transcending the individual senses such as hearing, sight, and touch, connecting them together, according to Bäckström (2022), and making them translatable into each other. The result is a "gestalt": a meaningful "whole" which is more than the sum of its parts, manifesting itself in front of a background (Merleau-Ponty 1974). This is the fundamental way in which we perceive and experience as embodied beings, searching for and creating meaning in terms of meeting the world. If we hear that the sound of a train is getting quieter, we immediately connect this with a visible train which becomes smaller as it travels away from us. Patterns of intensity agree with each other – a "synaesthetic attunement" emerges. The getting smaller and the getting quieter match each other. And this, according to Bäckström, defines the core of the mysterious ability known as imagination. His research shows that most people, in common with other primates, have a well-developed infrastructure for this "synaesthesia". But in comparison with other primates, and also with dolphins, for example, it is often particularly pronounced in humans.

Bäckström (2022) attributes our ability to be conscious of ourselves, to be very aware of our contact with ourselves and with others, to this same structure. It is as if we can "meet" ourselves so well because our individual senses can meet each other well. And – I would say, going beyond Bäckström's analysis – our senses can also "disintegrate", becoming disconnected from each other. An example for this disintegration is the split between the visual form of letters, for example an "A", and its auditive qualities. Probably no one would say that the form is in "synesthetic attunement" with the sound.

What matters most here is not our ability to perceive patterns of rhythm, for example, but to integrate the experience of different modalities (seeing, listening etc.). In some sense, the perception of an integrated gestalt is much more than just noticing a pattern: it is "meeting" (as we meet someone when we explicitly say that we affirm the other as a whole person, beyond all good or bad actions).

In this perspective, losing contact or creating it with ourselves and others can be described in terms of becoming present as a whole person by integrating the modalities, the different parts of the brain, and the different embodied parts of the lived body (in detail: Fopp 2016). Someone says something which scares us (or we encounter violence), and we begin to look away, without quite realising it; our ears become closed. We become tense, diffuse – and lose contact, perhaps as a defence mechanism (Broberg et al. 2006). With the tradition which explores the emergence of "authoritarian characters" (Winnicott, Adorno) this ability to be integrated or not could be seen as relevant for the possibility of creating democratic encounters, or encounters shaped by domination. The main point here is not so much one of – individually different – infrastructure, but instead relates to the question of how we might create social spaces in which no one is forced to lose contact. In this sense, here we could already jump forward to room 8 and look at the societal transformations needed for everyone to live a life in dignity.

Room 3: The democratic animal – forming bonds (the socio-psychological foundation of "being connected")

At our institute, we also use the books of Dion Sommer (2012), who describes something similar. He says: all of this can happen through a "look" of recognition between parents and child, or between kindergarten teachers and children (Stern 1985). A "look" of affirmation (as a metaphor for an attitude and a way of behaving) opens a space in which people can feel free and secure. It can also consist in the sound of a voice. The point is not which senses it depends on, and in fact no one can be forced into eye contact. Such a "look" has the same structure as the affective, synaesthetic "attunement" which takes place when the ball is bounced or a spoon is hit, and it is basically democratic, one could say. It establishes a relationship on an equal footing, beyond domination. It expresses, as many philosophers since Hegel have said, freedom and love simultaneously: it sets free, but in a caring way. With my gaze, which aims to dismantle domination and accept the other, I see that you see me with the same gaze, and that we see each other like this; which is not only, to use the philosophical terminology (Schmid 2013; O'Madagain/Thomasello 2021), a shared but a reflexive intentionality, or what Kierkegaard and Hegel call "spirit" (Hegel 1986); it is therefore not only consciousness. Every theory of consciousness could start from this phenomenon of "full contact" or "spirit", to avoid the debate about its location "in" the mind or "outside" of it (Noe 2010, Chalmers 2010). For an idea, thought, or emotion to become conscious can in this approach be seen as a function of interfering with or repairing the social fabric of integrity (Fopp 2016; Thompson (2010) makes a similar argument).

The opposite would be an attitude towards the other which demands submission. Domination is often literally connected with making others small, pushing them off balance, and so on; or ignoring and neglecting them (this is why Marx uses "indifference" as the main concept in his analysis of the alienated capitalist society; see Lohmann 1991). This makes them withdraw. Sometimes, withdrawing and breaking off contact also has to do with one of the seven "systems" mentioned above, which according to Panksepp are found in all mammals bigger than mice.

One of those is bonding behaviour: children, for instance, do many things to ensure that their primary carers protect them and remain close to them (Bowlby 2010). And if these older humans turn away when intense feelings and needs are expressed, children will stop expressing them, stop showing them, perhaps consciously at first, but then increasingly unconsciously: their faces might show contentment (or fear) which is not really felt. They no longer cry when their parents go away. And with time, such "masks" solidify as patterns of muscle tension and habits which we no longer notice – and which mean that we are no longer completely aware of our own impulses, ideas, or feelings. In the long term, this can result in serious illnesses.

Subconsciously, we "prioritise" closeness and protection rather than real contact, because we have to withdraw, out of fear and distress. That is why I keep emphasising that adults in social movements should be aware of their position of power in relation to young people and should use it in such a way that the young people dare to say and express what is important to them. Education could be focused on this dimension of democratic care, I think to myself in the theatre rooms. This is also – I suspect – the basis for a sustainable approach to the world.

Room 4: Nonviolent communication

"You are now entering an intermediate space," the compère calls. When we look at ourselves as animals with our affective behavioural systems: fear, anger, desire, bonding, playing, and so on, what can we conclude from this? Now we can look at the ways in which we can behave democratically or not, including in chats or generally in educational spaces and social movements.

We can focus on what happens when we don't just follow our impulses to act, but act in such a way that we make contact – or not. We can be afraid, angry, and so on, and still flee, freeze, or make others flee, but instead seek contact even in these situations – if it is appropriate and possible. Sometimes it might be better to run away or to tense up (see Fopp 2016). If we are at a strike in the square in front of parliament or blocking a road, and we feel exposed knowing that passers-by are looking at us and might be doing so with hate, threatening us, then it may be that we freeze and that we actually want to flee. For a long time, hours or days after the protest action, we might remain in this state. But we do it because we believe that through the protest, in the long term, a more democratic society can emerge.

But let us look at the means with which – if circumstances allow – we can actively make contact. It is not something (like "resonance" in Rosa 2019) that simply kicks in based on the mechanisms of mirror neurons when we walk into the right surroundings. Naturally, it can also happen spontaneously, but whether we are in contact or not is also up to us and in the hands of others. Rather than taking revenge or answering domination with domination, it is possible to seek contact; this is what the theory of "nonviolent communication" proposes (Rosenberg 2015). In a conflict, it is then our task to steer the focus to concrete ways of behaving which create discomfort; to the emotions which this provokes from our own perspective; to our own fundamental needs which are not being met; and to concrete suggestions for changing behaviour. This can mean that people are affirmed even when their behaviour is being stopped and changed. This, too, seems to be central within intergenerational movements: affirming other activists as people, even when their behaviour is problematic.

Room 5: Alexander Technique and improvisation (the physiological foundation of "being connected")

"On we go!" calls the compère. "Now it's about putting all of this into practice creatively!" Frederick Matthias Alexander studied the process of making contact among children and developed his Alexander Technique (Alexander 2001). It has become part of the English health system, for example (NHS 2021). It can help to explain how patterns of muscle tension arise through physical or psychological pressure, and how subtly these can shape our perceptions. Alexander's remedy is connected with the imagination: by imagining the space around the neck, and then all the joints, and by remembering the orientation of the body (that the head points upwards; that the shoulders point outwards, and so on). For this, it is crucial to pause, to stop everyday habits and spend a short time finding a new direction with less tension.

Many people, through a life of psychological pressure or physical strain, have developed a false picture of the way their joints sit, especially the joint of the neck between the ears, and this means that they "use" their bodies "against" the body's own structure, so to speak.

We are often "anaesthetised" and don't even feel how tense we are, and this makes us suffer, especially in an education system or a job market which does not leave us any time to do so. At the university, I have so many misgivings about the courses I teach, which confront the students with more and more material and the pressure to get the right grades.

If we practise pausing and turning in a new direction, we become increasingly aware of the patterns and can let go of them. It is then no longer about doing something, such as establishing new patterns, but about letting go of something; finding our way into a given energy. An energy can develop in such a way that we don't have to "create" it, but rather "accept it": by accepting, for example, that the ground is supporting us. And when we have spaces in which we can allow this – depending on the social conditions – we can simultaneously listen to external stimuli and to impulses from within, without one of these having to dominate the other. Precisely this is what happens in theatre improvisation under good direction: for instance, with the help of ideas from Meisner (1987) or Johnstone (1987). Meisner creates his exercises from the point of view that good improvisation "begins from the other." Johnstone focuses on the way people "block" and "help" each other. In the theatre spaces, we explore a whole range of artworks, films (by Daldry, Scorsese, Anderson, Aschan, Bergman and so on which are about (not) meeting on "eye level" and equal footing; see Fopp 2021), music, and images, and see how they combine the visible aspects of the fabric of integrity (materiality, viscosity, texture, colours and shapes) with the invisible – but still very palpable – elements of the dominating social forces and worldviews.

Or the same idea expressed in terms of pain and suffering: so many people have no idea that the muscles in their face and neck (and whole body, because they are part of a "whole") are tense, often in a very subtle way and permanently, in characteristic patterns that have become habits over the course of their lives. This phenomenon challenges our traditional modern concepts of freedom, "well-being" and so on (see Pettit 2015), because one could argue that it is – at least in some circumstances – in itself a reaction to suffering; or a state of suffering itself, making contact to oneself and others diffuse. And creating this contact is in this view more a task of letting go of the tension, if that is possible; not learning new patterns or habits, as so many popular books recommend. Thus, being humane, connectedness, the space of integrity and the fabric or material of integrity are connected: in humane social spaces, people don't have to cut themselves off from the dimension of the person with its inviolable dignity, because they can look at and undo domination; they can fill, unharmed, their space of integrity. (The concept of „being connected" tries in this sense to avoid the dilemmas of the concept of authenticity which Charles

Taylor (2018) describes as a foundational idea of modernity from Rousseau and Heidegger to analytical moral philosophy.)

Room 6: Creating social spaces democratically (the social foundation of "being connected" and meeting on equal footing)

I think back to the previous years, and especially to how different approaches in the grassroots movements or among individual NGO workers have given the young people space to develop – or not.

"Welcome!" Once again, the compère leads us into a new room: this is where social spaces are organised and created, including at school, in the health system, and at work. How can we create spaces so that the democratic exchange and contact described here are actually possible?

"There are a few rules here. Listen up!" "Firstly: all ideas are allowed inside our heads; no censorship. It is not about obeying other people's expectations. Even ideas that seem completely unoriginal or politically incorrect are valuable. No question is too weird. Mistakes are allowed."

"Two: but not everything can be put into action. Respect other people's integrity. Be careful with each other, really listen; try to understand needs and articulate them."

"And three: everyone should ensure that everyone is okay. It is not enough to say that everyone is allowed to take part. That is true, a liberal inclusivity, but it's too little, because it does not affect the dominant power structures." Because this has been shown in many teaching sessions at Stockholm University: if we all look after everyone's wellbeing, the atmosphere in the room often changes instantly. It is no longer about being better, cleverer, more original; and precisely through this, something valuable is created.

Many researchers agree (Samuelsson 2017): it is important to establish such rules explicitly for a space – and of course to react if they are broken, so that everyone can rely on them.

But rules are not enough. It is also about the attitudes of those who have leadership roles (which in democratic spaces might be all of us) and who establish these rules. One: "Try to see everyone; and as equals, with no one being preferred. Give everyone the necessary space and time to express themselves in a fair way, appropriately to their needs." Two: "Rotate cooperation in smaller groups so that no informal centres of power are formed." Three: "Establish the difference between people and actions. Criticise and praise actions, not people; make sure people are affirmed, even when their behaviour needs to be stopped."

Here, too, improvisations can help, including status exercises (Johnstone 1987). Through such improvisations, we can test these principles playfully. Someone takes on the role of a teacher and talks to a headteacher, with both of them taking on the same status initially – and then the teacher rises up and treats the headteacher condescendingly; until the two of them switch round and the headteacher's status increases. This can take place in a friendly way, so that the person with the higher status tries to lift the other to the same level, or in an unfriendly way, through domination, by keeping the other person down (Johnstone 1987; see also the descriptions in the first chapters of this book). It is always astonishing to see the energy released by this. The aim is to see through domination – and to learn to establish relationships which are at least on an equal footing, or are even beyond the struggle for status; this makes them humane (see Nussbaum 1992) and allows people to fully connect to each other.

What can happen in improvisations, in these playful encounters, if we succeed in listening to impulses from others as well as our own ideas, is the integration of these interdependent systems: the subconscious self and the conscious; the deeper, older parts of the brain (which Panksepp describes) and the cortex; and within the cortex the areas linked to different senses; and between the attuned senses and higher level reasoning (Fopp 2016). This is matched by the integration and cooperation of body parts, expressing the transmodal integration of the "connecting" self, one could say; a self which is to be treated as being there even if all of these integrating processes cannot occur. This is one meaning of treating someone with dignity (Bieri 2016).

Precisely this attitude and these rules are the basis for the "democratic leadership" of groups, which relies on training in collective decision making. It would be so important to try out all the democratic processes from early on which the activists have encountered over the last four years, all the collective decision processes. Because the consensus-orientated decision-making which the young activists have practised so often at meetings of Extinction Rebellion and Fridays For Future requires everyone to be cooperating affirmatively. Collective autonomy and self-limitation (Heidenreich 2023) becomes possible without this having to be experienced as a loss of freedom, quite the opposite. All of us can participate, bringing in our experiences and needs, and trust that we will be taken seriously.

Room 7: Sustainable exchange in all sectors and areas of work (the interactive foundation of "being connected")

"On we go!" calls the compère. "Let's get to work!" Now, with this compass (of being in full contact), we can explore all the work that is necessary to build a sustainable society. Because what is missing is the understanding that we, as democratic animals, are embedded in the nature and culture which surrounds us: the rivers, the forests and the cycles of growth and decay, of planting and harvesting, the building of houses and cities, repair and care. "In short," says the compère, "welcome to the space which is here to explore democratic-regenerative exchange with nature and the culture arising from that!"

I think back to all the struggles which the young people have become involved in over the last years: the fight for sustainable forestry and agriculture based on conservation; and the fight for a sustainable energy system and a sustainable finance sector. So we have to go out into the world, into the space where we now have to explore what is needed so that we can extend and preserve the democratic energy which comes from making contact with each other. This can also happen with the help of experts, not only from the Global North, but also indigenous experts who work daily on creating a regenerative relationship to the world.

That means building furniture, apartments, houses, and cities with materials, ideas and concepts that correspond to this democratic energy: from the way in which workplaces are organised, to the idea of really sustainable, sufficient and efficient cities and houses. When we are in contact, everything can be measured by a new standard. And the same goes for the broad areas of planting, caring for nature, fields, and forests, but also for producing food, cooking it, and producing clothes (see Pelluchon 2019 for a detailed theory of such a practice in relation to agriculture, cooking and eating, and treating animals). The third realm is that of caring for animals and humans and looking after their health. The fourth is that of child-rearing and childcare, looking after children and older people. The fifth is our treatment of economic means. The sixth is that of developing technology and engineering, through which our relationship to nature and to other humans can be explored and developed (Vetter 2023), including research into solar, wind and water power, rather than nature being treated as a burnable material. Then there is the realm of political participation and leading political processes, and the realm of philosophy, which develops new concepts of what all this means for our understanding of nature as (non-)property (von Redecker 2021). And so on.

When these realms are shaped in a way that is guided by democratic contact and exchange, and when the latter is also at the centre of education, then not only do we combine theory and practice – minds, hearts and hands (ZDI 2022), but we really learn to deal with life in concrete ways and build our lives individually and together. With our students, we also incorporate the development of mathematical and linguistic skills into their fundamentally productive exchange with the world; into cooking and playing, making clothes, and looking after children.

If we know how to do this, we can relate it to what is now happening and what has happened historically: how political processes currently don't happen in this way; how we are not brought up on these principles, how the production of food currently looks, and so on. But learning all this – that is, the current, often non-sustainable content of our education – only makes sense if we know and learn how sustainability works. Otherwise, we can neither make sense of any of this, nor criticise it.

What could also be redefined in this way is the relationship between education and work. Work is now not entirely separate from the realms of living together, education, playing, regeneration, creativity, and self-expression. It is part of a movement for democratisation: "creating the resources together" which all of us need (see von Redecker 2021 for similar arguments). It is then no longer about submitting to non-democratic command structures in workplaces, nor about dominating nature through processes of exploitation, nor about polluting or burning, but about building up regenerative energy; it is not about appropriating or taking control of nature, and also not about self-actualisation or expressing ourselves through products, but about weaving and repairing the shared fabric of integrity.

Back at the Glasgow café – the problem with "education for sustainable development"

Back to Glasgow. The evening in the café grows ever longer. The big educational concepts of the last twenty years begin to swirl around the two floors of the café.

What was it that ended up being so strange about one of the biggest educational projects, we ask ourselves. It is called "education for sustainable development," and it is still one of the projects showcased by the UN, as well as by many countries and their education systems. Somehow this project has failed, we think to ourselves. And still we have great sympathy for those who are working on it, with whom we ourselves are in contact (Sterling/Huckle 1996), as well as with the cooperation of all universities worldwide and with UNESCO as the organ which coordinates this global initiative for sustainability education. Many of them would like to work together on a "Faculty For A Future." Because they themselves draw sobering conclusions at the UN global conference of the universities in May (UNESCO 2022): the climate crisis is continuing unabated. Educational institutions are changing too slowly.

Is this failure (if one wants to use this description) due to the fact that the project seems to have fallen apart into two sections, even though the interesting part is actually the connection between the two, on which our lives depend? The two extremes are: the practical heart-hand-mind understanding of ecosystems, meaning the understanding of how we can act sustainably in our everyday lives, which is important knowledge but only plays a role as a niche area that exists alongside all the other disciplines, such as history, philosophy, economics, and so on. And then the project also has an abstract section: understanding global sustainability goals such as the SDG goals and the UN Agenda 2030. This is also important, but as well as being generally abstract it is a bleak kind of knowledge which we can barely absorb. Poverty and hunger still shape the lives of large parts of the world population; about a third of people have no secure access to clean water (WHO 2019). Many girls and women still have no access to education. Understanding all of that, we say to each other in Glasgow, ought to be accompanied in quite concrete terms with an understanding of our own place and our own role in world history – in such a way that we can really change things. (One solution to bridge this gap is the model by De Haan (2010) to develop twelve basic sustainability competences.)

For that, the concept of "development" is not helpful within the project of "education for sustainable development": the whole thing is conceptualised as a program of hundreds of small steps which citizens can learn and put into practice in the existing late capitalist market context – while, as the terms imply, everything basically remains the same. But that is an illusion, and the clever people and texts within "education for sustainable development" are aware of that. It is not about "development", small steps of improvement, as if we just needed to recycle a bit more rubbish, or plant a few more trees, or build a few more wells and solar panels. That is not a scientifically sound description of the problem. The problem is systemic, as UN organs such as the UNFCCC, the WHO and so on have said when analysing the climate crisis and the biodiversity crisis. And it is not a "development" which would be sustainable, but a systemically different way of living together. That is why this is about transformative and regenerative education, not about education for sustainable development.

Room 8: Democratic system change (the political-economic foundation of "being connected") – the Diabolo model

This raises the question for the next and final room: which underlying political and economic conditions, as well as cultural conditions, would need to prevail so that this kind of democratic exchange and humane contact would even be possible in the other rooms?

"Welcome!" calls the compère one final time and leads us to the "laboratories" of political science, economy, sociology, philosophy, and the humanities (as described in previous chapters). There, theories come up which aim to think about society in a new way, so that it is oriented towards the needs of all people within planetary limits, not only towards economic growth as traditionally understood (see e.g. Raworth 2018; Göpel 2016; Hickel 2021). But an important role is also played by the research which approaches this transformation on the basis of intersectional justice (class, gender, ethnicity, etc.), which aims to replace structures of domination; as well as theories of global democracy and ethics.

All these approaches fundamentally aim to create a framework which could become a new system, compatible with what has been described in the other rooms: non-dominant exchange and cooperation to provide most important resources to everyone in a way that is sustainable.

At the same time, we now have a compass for these new ideas and for the systemic transformation that would be required: they have to correspond to the enabling of democratic meeting, contact and being connected in a humane way to oneself, each other and nature. That is the "dough" of the "doughnut" (Raworth 2018), meaning the core of a secure social space. The definition of education in economy, law, or sociology can be newly investigated from that point. "The point is to keep walking between the rooms and find out how they rely on each other and require each other!" The compère hurries from one room to the next: "All of them build on each other. Together, they form the structure of what defines substantial democracy!"

The image of a diabolo presents itself – the toy which looks like two bowls joined together which can be catapulted into the air on a string and caught again. On one side, there are the conditions which are necessary for a globally, nationally, and locally sustainable life in the crisis: the diagram of the doughnut, the idea of a CO2 budget; the democratising systemic transformation of the economy; the political action plans in relation to regenerative exchange in all sectors. This means combining research in climate and biodiversity studies with studies of socio-economic system change and global justice.

The other bowl is formed by knowledge about caring encounters on an equal footing, humane energy and being in contact (what I have called spaces of integrity and the fabric of integrity in previous chapters), from neuropsychology and drama education to the educational organisation of social, intersectional democratic spaces. In the centre of substantial democracy, they are connected: through the bodily and practical understanding of relationships of domination and the structures they create; and through the understanding of how we replace them through democratic relationships and through inclusive, democratic decision processes.

Connecting all the rooms through education – the science of democracy

Such an approach to education at school and at university would bring together the research of climate scientists with that of social psychology; it would allow the free exploration of the link between disciplines. It would not only be transdisciplinary but would aim for something like a shared core of everything at a "subdisciplinary" level: a sustainable "being towards the world" (to use Merleau-Ponty's concept inspired by gestalt theory).

As I walk across the campus with the activists, the lack of such a centre is so obvious. The institutions each stand by themselves, without seeming to be related to each other at all, or to the democratic task in the sustainability crisis. Being in the world in a regenerative way, and the theoretical and practical-creative knowledge discussed here, often have no role to play. They literally have almost no space at schools or universities. They still – in contrast with the approaches of drama education – reproduce Descartes and Kant's separation of body and mind and the Aristotelian distinction between practical/technical, theoretical and ethical knowledge (Gustavsson 2017). But the aesthetic, the theoretical and the ethical cannot be separated, nor can the content of what we learn from the teaching methods.

That is why we do not just need institutes for social and ecological system theories, which is what many geography faculties are gradually becoming, but also a new "science of democracy" which would hold the university together (see the similar approach by McGeown/Barry 2023; Raffoul 2023).

Theories of sustainable democracy

Developing theories for a centre of sustainability could also go in the same direction. The three perhaps most important theories which have set out to understand sustainability and democracy during these years and which dominate the humanities and social studies – system theory, actor-network theory, and posthumanism – do not seem to do justice to the core of democratic spaces which has just been discussed. Many of the most important books on the climate crisis operate with system theory (including the research of Göpel, Raworth, Rockström, Hickel, and Monbiot). In them, among all the parameters and variables, the experiences and structures of domination or of meeting on an equal footing hardly play a role. In the case of posthumanist or network approaches (Braidotti 2017; Latour 2017), in which the boundaries between humans and nature are rightly questioned, the perspective on experiences of our own shared humanity is problematically neglected – "being humane" defined as a characteristic of spaces and practices as well as approaches and structures which make it possible to dismantle domination and participate in affirmative contact; transforming and sublating the three concepts of "being human": the ethical of being compassionate; the anthropological of being different from other animals; and the moral of being full of weakness and failure (Fopp 2016).

Finally, theories of democracy, which can be seen clearly in Heidenreich's 2023 book on democracy and sustainability following Habermas and Rawls, often focus one-sidedly on formal aspects of human rights and collective decision processes without paying attention to the substance of democratic encounters – and with that the key question of how the substance and form of democracy can be newly connected, nationally and globally (for a similar critique, see Young 2000). Postdemocratic theories, meanwhile (on this, see Marchart 2018) emphasise often their disagreement with established formal democracy, without explicitly discussing its substance. The theoretical task would instead have the goal (as described in earlier chapters) of pursuing these substantially democratic spaces, open to new theories and methods (see the research in previous chapters on eco-feminist and eco-socialist as well as degrowth approaches), so that it becomes clear what the consequences would be for a deepening of democracy.

Extra room – the foundation of it all: from moral and political theory to (eco)philosophy

If we follow this approach, we need philosophy to act as a foundation for moral and political thinking and acting, as well as a foundation for all sciences, but we also need it to be inspired by the sciences.

But what is philosophy? Evidently, it involves working with concepts and reasons, arguments; exploring what "justice", "suffering" and "happiness" are, and what a "prosperous society" is or could be. But with the approach sketched here we can get a new view of what philosophy is, can and should be.

One main task is indeed to work with concepts which match reality better, helping us to understand ourselves and our world, especially by making dimensions visible which are otherwise overlooked; for example, making everyone's dignity visible, and what I called the common fabric of integrity, this strange material which links us to nature, but also transcends us as unique souls, as persons beyond actions which may be good or bad.

In this view, this enterprise is already something different from the simplifying definition of analytical philosophy as "analysing meanings", namely: working with concepts, their premises and implications (see Dummett 1996, Glock 2008 and Brandom 2022 for the difference between working on concepts and analysing meanings).

But what I would like to stress and what seems often to be overlooked in universities: philosophy is (as all sciences could and should be) creative work with exploring new concepts, not only dealing with traditional ones and existing theories. In this sense, it is about letting oneself be inspired by reality, so that we are compelled to describe it in a new way: defining "being humane" for example, in relation to three concepts of "being human", transforming and integrating them: not only human in an ethical sense of being compassionate and trying to care for the ones who suffer; not only human as "a-moral" in the sense of the spontaneous following of all impulses; not only human as an anthropological feature of humanity as a species which can reason; but humane in the sense of integrating all these three but transcending them: understanding impulses and structures of domination, not denying them, but going beyond them into creating loving relations to persons beyond good and bad deeds; making visible the space and fabric of integrity, and so on. This conceptual work does not take place in a vacuum, but reacts partly to existing, important theories, for example in moral and political philosophy: thinking about how to frame our compass for actions and political structures; and about why we should act ethically in the first place.

But by focussing on creating contact with the other and the idea of being humane, this approach to philosophy encompasses our practices as social, embodied, creative human beings, and is therefore not only concerned with theory and concepts. But before exploring this thread, a short exploration of how this approach reacts to existing theories and tries to "solve" their challenges.

In terms of moral and political philosophy and their relation to each other (in this section I am referring rather to the analytical tradition, while the whole book is inspired by the continental approach and tries to see bridges between the two): many classic problems can be reframed or even solved by the introduction of concepts like the idea of "being humane" as seeing through domination and creating social spaces in which no one has to disconnect and everyone can fill the space of integrity, repairing and "weaving" a common fabric of integrity.

With such an approach, the idea (which is only sketched here) is to avoid the opposition of utilitarianism (minimising suffering and maximising happiness for everyone) and Kant's deontology or other approaches focusing on inner motivations and human rights (as guaranteeing integrity and dignity) – which all remain in the frame of what I called "formal" democracy.

A similar analysis can be applied to classic texts about justice (Rawls, Nagel, Sandel, Nussbaum, Sen etc.), either in terms of distribution of resources, in terms of vices and virtues, or in terms of guaranteeing freedom: we can see how the intersubjective and structural work on repairing the space and fabric of integrity and dismantling domination in some cases "unites" these approaches but also goes beyond them in terms of substantial democracy, focusing on the quality of relations and the routes of the ethical dilemmas and political challenges (and not only on the distribution of resources, the application of rights, the procedures of decision making etc.) – similar to the approaches of ecofeminism/postgrowth-socialism/postcolonialism etc. (discussed throughout the previous chapters: from the ideas of Angela Davis, Iris Young, Judith Butler and Nancy Fraser to Eva von Redecker etc.).

For example, we can now focus not only on suffering and happiness but on (de-)connectendess and domination (which seems to be morally and politically problematic even if no one suffers; see Pettit 2015). We can work on repairing the fabric of integrity which opens up space for a new (non-libertarian or liberal) definition of freedom. Using this terminology, intergenerational justice for example is not primarily about „future people" (Parfit 1986), but about our common fabric of integrity which links all dimensions of time and space.

Or we can develop a better compass of "doing well" which is a intersubjective and social category – instead of "happiness" or "well-being" (see the chapter on economics, and Fopp 2015). In this way, one can establish an objective foundation of moral and political reasoning action without neglecting the dimension of subjective experience. And finally: the idea of justice, even in the sense of non- domination, cannot be seen as the only fundamental compass – it has to be complemented by the idea of being humane (or "love" as Martha Nussbaum 1992 defines it), which could mean: going beyond relations and structures of domination, affirming everyone as a person beyond their actions, and making a life in dignity possible, including providing resources, for all.

This critical "expansion" of the philosophical work (that it is not only about making concepts explicit but inventing new ones which correspond better to our lives and the challenges we face), can – as mentioned above – be complemented by a second one. From this perspective, philosophy is not only conceptual work. To be able to work with concepts, to present arguments, justifications and giving reasons, we must be connected with others, ourselves and the realm of ideas, with the fabric of integrity: and this is a social, embodied, creative practice which takes place in social spaces affected by problematic power relations. This work of connecting ourselves to ourselves and the environment is not an arbitrary step, but a part of the fundamental work of making visible how the world can be and "is" when we are not forced to disconnect. Pelluchon (2020) uses Lévinas to make the argument that the foundation of ethics are not rules or ideologies, but the perception of the other as other, as "Antlitz" which demands of us to care. In this sense, philosophy becomes a practice of making a phenomenon visible, as Simon Critchley argues in his book about the relation between analytical and continental philosophy (2001), reframing Merleau-Ponty´s phenomenological "gestalt theory" approach and demanding the integration of "wisdom" and "knowledge".

Or in terms of Hegel (1986): we try to get into and operate in the realm of "Vernunft", in which we are linked to a common spirit, not only in that of the argumentative "Verstand", presenting reasons and arguments. Philosophy can itself be or become a substantial democratic activity, rather than only analysing what democracy could be. (This is reflected in the idea of "being humane": describing social spaces and relations as well as attitudes and structures; see Fopp 2016.)

This view has a double political consequence: philosophy (and its institutions at universities and schools) needs to be a very specific activity which is carried out with a practical knowledge (or ethic-aesthetical wisdom) of how to create the (social) circumstances which allow connectedness to the realm of ideas, the others and nature. This should be a premise, a prerequisite and an ingredient of philosophy and all science. And, second, this implies that philosophical and academic institutions – if they want to live up to the scientific methods and have a reliable compass – need to become outward-facing: working on transforming society (and educational institutions) so that these spaces and activities become possible in society; in which we can connect to each other and to the realm of ideas.

In this sense, we can say with Feuerbach (and Marx) that philosophy (and academia) shouldn't only give an interpretation of the world but change it. But there is a second step: highlighting that this transformation is also done by creating concepts and practices which help us to make everyone's dignity (or the "demand by the face") visible; to open us up to the shared fabric of integrity; and to lead to a change in how we see each other, nature, and the rules which should govern our shared lives on this living planet.

Back to the question of what philosophy and academia is. In this sense, philosophy is needed as a foundation for all sciences; including neurophysiology, literature, history, education, economics etc. Hegel (1986) called it the work of the "absolute spirit", which consist of art, religion and – philosophy. We can describe his main idea by going back to our rooms 1 to 8, or the life of the "subjective" and "objective" spirit, as Hegel calls them (see Taylor 2015 and Theunissen 1980 and 1984 for a similar analysis of Hegel's basic idea): our relation to the environment which enables us to go beyond domination ("Herrschaft") and establish a relation of freedom and care (subjective spirit); and the ethical and political relations and structures which are informed by this movement and make it possible (objective).

I want to translate this idea into the insight that ethics, moral and political theory – with their idea of the good and justice – are not enough: in order to create even a basic foundation for ethical and political theory and action, we need the idea of love and the fabric of integrity which art, religion and philosophy could develop (and should be measured against; for this aspect of religion, see e.g. the work of Martin Buber, and that of Dorothee Sölle and Catherine Keller).

Notions of what is good, of what is morally right, and of justice can produce results which lead to the opposite: the "tyranny of the good", as Corine

Pelluchon calls it, totalitarianism, violence and so on, if pursued only for themselves. Nussbaum (1992) shows in her analysis of Charles Dickens' *David Copperfield* how ethics can become deadly. Some notions of ethics are inconsistent, Hegel might say, and lead to and require a foundation: the humane spirit, which means freedom, mercy and love, acknowledging impulses of domination, playing with them and going beyond them; allowing us to create affirmative connections with others and with ourselves, as persons with inviolable dignity beyond the realm of actions.

However, against the tradition of idealism, all this work is not carried out by an anonymous reason or spirit itself, but is a task for all of us: it is not something which makes its own way through history, leading by itself to a better world, but is dependent on us human beings (in a similar way, Menke 2023 reinterprets the concept of spirit and the task of the humanities).

In this sense, philosophy (or for Hegel: "the absolute spirit") is not just about looking back at all that happens in rooms 1 to 8, from the subjective to the intersubjective, social and political; it is not only a reflection on all the other topics, but a work in itself: making – by developing adequate concepts – the fabric of integrity visible and experienceable for every one of us, exploring ways to describe what is already there, and how we can repair and renew it; as persons with equal dignity (which is more than relating to psychological states of "intuitions" and "moral sentiments"). Recognition in social relations and the material provision of resources are interdependent, to solve an opposition between two strands of Critical Theory (see Honneth and Fraser 2004), if our approach is rooted in a phenomenology of being humane.

And accordingly, "doing" ethics, politics, or moral and political philosophy is ultimately not something elitist; it is not just something for upper middle-class seminars with their analytical terminology; it is not only what you hear in (often privately owned) prestigious halls of philosophy etc. It is about connecting to and articulating the idea of being humane, which is already there, a humane energy for which we must rather remove the aspects of domination which stand between us and it (and thereby let it be a guiding "value" which produces duties and obligations). In this sense, this approach is a non-relativist and objective one in its approach to moral and political values and principles. We don't have to choose a morality, ethics, or political philosophy; and we don't have to find reasons to engage in this project in the first place, because not engaging in it, not making the fabric of integrity visible is often itself already an effect of deconnectedness (to react to a discussion about the objectivity of norms and values; and the question of why we should create and listen to ethics in the first place).

Moving through the drama rooms at Stockholm University, we try to connect to these dimensions. Art and aesthetics become important even in university spaces (as part of "Hegel's "absolute spirit") if this is true: that all sciences and all education, not only social sciences and the humanities but even law, theology, psychology, literature, economics, etc must be rooted in this embodied social practice of creating humane spaces in which democratic meeting becomes possible. Otherwise there is no normative foundation for the scientific enterprise.

We move through the drama rooms, watch films, improvise and explore ways of meeting and doing art. It is possible to transform the analysis of the works and the history of art, from Bergman, Daldry and Aschan to Paul Thomas Anderson's films: it is possible to make the fabric of integrity visible, everyone's dignity beyond the realm of actions: to show how the main characters in the movies are driven and disconnected from themselves by "forces" (of the diagonal versus the vertical; the roundish versus the edgy; and so on) which can be analysed in terms of politics (patriarchal or late capitalist structures etc.), ethics (being selfish etc), psychology ("the "doubtful" versus "the determined"), gestalt theory (these "forces" being deviations from the gestalt structure), and so on; and by doing so, making visible what it means to meet in a non-dominant way, literally "at eye level", which is humane, affirming, unconditional and universal. In a paradoxical shift (of "mercy", one could say), the person and soul becomes visible, to which we can connect even if it is struggling with life and not connected to itself – because nature, the environment, the fabric of the world in its transmodal texture (beyond colours, forms, and so on), takes upon itself the "forces" of disconnection (see Fopp 2019; also for the argument that we cannot describe this process in terms of metaphors, forms, signs or structures as aesthetics often does). But art, as Hegel would also say, in this classical "Western" form of commodified works of art, can also be criticised (and is criticised by contemporary art itself) if it leaves the political and economic spaces untouched which need to be changed and transformed so that we human beings can live our lives without being structurally excluded from humane relations in the first place.

Five perspectives on ecophilosophical thinking

We could – in a very simplifying and incomplete way – identify four perspectives (or even „epochs" and paradigms) of ecophilosophical thinking; at least in the academic world of the Global North. They partially overlap and sometimes compete with each other. I tried to sketch above a fifth alternative, using traditions which have been overlooked by these approaches. All of them react critically to the "mainstream" tradition of sustainability research which is linked to ideas such as neoclassical market economics of monetary values of ecosystems, nature as capital, green growth, the reduction of humans beings to consumers, and so on (see Neumayer 2010; Grunwald/Kopfmüller 2022).

For the first one, influenced by the „deep ecology" movement, represented by Arne Naess (see for texts: Birnbacher 1997), the idea is to explore a new way of understanding our place in nature. This is done in something which Birnbacher (1997) describes as a mixture of ethics and ontological-metaphysical enterprise, often referring to philosophical approaches by Spinoza and Whitehead as well as poetical works by Thoreau and the Romantics. Some of the discussed problems are: is there an intrinsic value of nature and species; and how can we develop a lifestyle which acknowledges this presence of an own value (and sometimes: of panpsychism) of nature (see Birnbacher 1997 and 2022 regarding the philosophical incoherencies and problems of this tradition).

During the second era, the methodical and thematic interest shifts to ethics, more precise to applied ethics (first to environmental ethics and sustainable thinking, very much inspired by Rachel Carsons „Silent Spring"; and even more specific to climate ethics, see for example Moellendorf 2014), as well as to metaethics and questions of the justification of our norms and values.

This „subjective" shift („how should we act in an ethical appropriate and reasonable way" confronted with the destruction of nature) is then in a third epoch expanded and criticised by the posthumanist tradition (Braidotti 2017), going back to more metaphysical-ontological questions about the problems of separating humans from the rest of nature. A fourth tradition, inspired by ecofeminist/socialist (D'eaubonne 2022; Shiva 2020), postgrowth and decolonial ideas, focuses on the intersectional destructive effect of the economic and ideological system as a whole; and emphasises – sometimes against the posthumanist "association" of humans and non-humans – the specific human responsibility for changing these norms and structures (see for this argument: Hornborg 2017, Malm 2017).

Finally, the approach presented here as an ecophilosophy of democracy starts with (post-)phenomenological theories by Levinas, Merleau-Ponty, Pelluchon (2019), the indigenous thinking by Kimmerer (2015), and the traditions introduced above: especially the experience of another individual as ethical imperative; but expands it with the help of empirical sciences (psychology, neurophysiology, sociology, education, politics and economics etc.) to an understanding of the creation of non-dominant relationships and structures of „being connected". In that sense, the ethics of „being humane" can be linked to an ontology of our sustainable „being towards the world" as embodied, vulnerable beings which can lose or create affirmative meetings and connectedness, to others and the realm of ideas. Maybe we can talk about a fifth epoch which focuses on these questions of global democratic convivialism and develops arguments about justification of (ethical and political) norms from the philosophical interpretation of experiences as well as the scientific explorations, including the tradition of indigenous knowledge. This tradition questions even the difference between theory and practice by pointing to the challenge of making the dignity of everyone visible and to change our perception as one of the philosophical tasks.

Ecophilosophical debates and an adapted theory of science

In Mynttorget, debates often arise on the following question: what is our place in nature? The question is taken up by many in their talks on the climate crisis. It also shapes discussions in connection with the struggle for ecocide legislation, which would expand the offence of crimes against humanity to environmental crimes against whole ecosystems. In that context, some argue for a form of anti-speciesist thinking (for a discussion of this, see Gunnarsson/ Pedersen 2016), claiming that humans should not be in the centre of our worldview. They argue that we ought to be seen as one of the animals which form a complex whole together with all the others, and not as the exceptional animal which can place itself above all the rest.

But although such arguments seem very plausible, they can also be problematic, depending on the way the theory is developed. The approach described above – building democratic spaces – is in this sense also meant as a proposal for this field of eco-philosophy. It tries to retain the intuitive impulse to locate ourselves in nature, but avoid a reifying view. Because that is what seems to afflict many of these approaches. They rightly locate us humans in socio-ecological systems, but take on a problematically objectifying approach towards us, as well as to all other animals and nature; they look at the whole system from the outside. It seems more appropriate to me to start with our democratic or dominating relationship with our environment. From this we can derive a yardstick of values that goes beyond an abstract adding up of suffering. And from this perspective philosophy and science would not be an objectifying project, but would be connected with insight and interest, as Habermas has argued in the context of Critical Theory (Habermas 1987); but not only in the sense of emancipation, of shaking off dominating structures, but also in the complementary sense of building what we need for everyone to live a dignified life.

This does not mean that the standard of objectivity or truth must be given up. Quite the opposite: it moves to the centre when it is sensibly defined as approaching reality with integrity, self critically and with transparent methods, in an open, fair exchange with fields of research, without favouring certain positions one-sidedly. But it will no longer be acceptable to ignore the key research suggesting what it would mean to live a dignified life as free and equal people within planetary limits.

The path to this flourishing life is then not just part of theory but can only be understood in practice. Many researchers point this out, including in art-based research (Leavy 2009), which critically extends the concept of Critical Theory: away from the interest behind theory (emancipation) and towards education as a democratising, global creative practice of transformation and regeneration (Raffoul 2023).

From a decolonial and indigenous perspective (Patel 2015; Kimmerer in Thunberg 2022), researchers also alert us to the fact that the fundamental framework of science must be redefined. The existing distinctions made by the theory of science, which are drummed into first-year students across the world, including here on the campus of Stockholm University, thus turn out to be problematic. Epistemology (how does individual knowledge come about), value theory (which norms are we working with) and ontology (what are the characteristics of the world) can no longer be abstractly separated from each other. How we see and perceive the world; what causes suffering or enables contact; and how the environment and our own bodies are composed: all of this belongs together in a permanent exchange which can cause suffering or repair the shared fabric of integrity (see Merleau-Ponty 1974). The point of departure is that we are already deeply entangled in dominant relationships – as has long been emphasised by education theorists from Paolo Freire, Ellen Key, and Augusto Boal to Gert Biesta, as well as a large part of the feminist theory of science and "critical university studies".

Transformative and regenerative education

But such democratising education can and does provoke resistance (see Van den Berg 2021) – at the level of teachers and students, and from institutions. That seems to be an important element of education, and this would apply to a sustainability centre too: dealing with this resistance, leading transformative processes and thus building democracy.

Here it is important to learn from those who have long been leading and reflecting such transformative processes, for instance in relation to racism (e.g. Ogette 2018). If, for example, a person realises that they are taking on a higher status than others and subtly keeping others down, this realisation is often a shock. We have to learn to deal with this, and have the courage to look at all of this together and dismantle it, and to try out the opposite: daring to interact in caring ways, to improvise together, and to help each other with resources in difficult situations.

If we succeed in building this project – not moralistically, by simply condemning domination and demanding empathy, but in such a way that we set off on a path together on which everyone makes mistakes and is allowed to do so, this can become transformative, changing the whole person as well as the group which is learning.

And so, transformative education can become regenerative. The transformation applies emotionally and cognitively to our values, our worldview, and to a change in attitude to learners and teachers. The whole of the educational process and the institutions can become regenerative. In this approach, the aim of education can be defined as building up democracy and regeneration, and thus as weaving the shared fabric of integrity. Science then has the function – among other functions, at least – of pursuing this regenerative coexistence in democracy, researching it, and enabling it in concrete terms. The ethics behind the whole project, which all researchers should already subscribe to now, could thus be defined as follows (Raffoul et al. 2021): It is not just about not causing harm, but also about repairing the harm that has been done, and making a dignified life possible for everyone.

Science and activism

While I present these thoughts about reshaping teacher training and creating a new pedagogy of sustainability to the management team, the young activists are preparing for the next global strike, and scientists across the world are beginning to take to the streets and to get involved in civil disobedience. The question is: how can we make all these ideas a reality?

At this time, a debate is developing on social media and in politics over the extent to which scientists can be committed activists. Because everywhere in the world, more and more teachers are saying that they can no longer just watch while forests burn, emissions continue to rise, and science is not taken seriously by those in power. We join these actions in Stockholm, and in April we block the centre of the city with several hundred people (Platten 2022); many more teachers should be involved, we think to ourselves. Doing nothing is activism for the status quo.

And still, there is a lack of change at the universities themselves. The image which emerges is one of middle-class professors blocking the way of workers. And it is professors and the upper middle class who produce the most emissions. Among all the possible forms, we in Mynttorget still see the "strikes" as the most appropriate one, at least the non-cooperation strikes on Fridays, as a demand to end "business as usual" – making a statement to the institutions that we can no longer teach or learn in the same way. At some point, whole universities or schools would have to strike, to insist that this cannot go on. Even after four years of school strikes, still almost nothing is happening, even at the universities with the most progressive sustainability departments (see HU 2023). Neither the curricula are changing, nor the things which could be changed so easily, such as the enormous number of academic flights.

But the dichotomy of the existing discussion also ignores a fundamental dimension. There is not just research and teaching on the one hand, and street activism on the other. This notion of inside and outside the universities is abstract and wrong: if we look at education itself, at teaching and research, and if we change the "social logic" of the spaces of education, we are already working on societal transformation. And by the same token, this transformative education is only possible if universities are changed, in terms of their structure, their management and the societal and political conditions underpinning them.

I come back to where I started: "We all ensure that everyone is okay and develop an inclusive attitude. All ideas are allowed, no censorship, but not all actions. Have the courage to try something new. Help each other." Something in the room changes, something that does not happen when they see each other as lone fighters, these students, who are afraid of not understanding something and of getting bad grades. That is what they are made into. So this kind of teaching is resistance against the logic of educational spaces, but also already a change, a transformation of society.

The dual way into the future

In my conversations with the institution's management, I keep returning to the "original scene": the time in 2018 when the five young people sat on the ground on Fridays. How this action gives us a task. How this task consists of working together with them to guarantee a secure future for everyone worldwide. And how this requires changes to schools and universities.

And so a double perspective opens. On the one hand, all of us can continue to build sustainability centres in every school and university (and a prototype for these), as well as virtual variants such as the Faculty For A Future: platforms which place the focus on the crises, their urgency and interdependence as social and environmental crises, by changing what is learned, how it is taught, and the institutions where this takes place.

On the other hand, it is about expanding activism. Otherwise, neither universities nor society will change. All those who have the resources can take part in the movements of pupils, students, and teachers, who strike and rebel through civil disobedience; ideally as entire institutions. Two dimensions can be combined: one which is shaped by communication (Scientists For Future) and blockades (Scientist Rebellion), which address society; but also a new, key dimension which incorporates and addresses educational institutions through non-cooperation such as the Friday strikes. These two strands can be connected through citizens' assemblies to reshape institutions: combining the Faculty For A Future with Scientists For Future and Scientist Rebellion.

If only all the tens of thousands of scientists would begin to go out on Fridays and stand in front of their institutes, I think to myself, and join together to form assemblies with the students who want to take their education into their own hands and implement sustainability. Then the educational institutions could make it clear to politicians that life at school and university will not continue in the same way and that they must respond to research; to "listen to the science". That is what all states, we all have promised to do by accepting the Paris Agreement with its focus on stopping the emissions in a fair way.

And so we meet: some of us from Scientists For Future and Scientists Rebellion, some Faculty people and even delegates from the youth movement. "We have to make clear that the universities are ours," a professor says. "We should say: 'We take them over. We are the universities. And we won't allow them to be places where the fossil society reproduces itself,'" says the youth delegate. We need disruptive direct action, combined with assemblies; we all agree. We start to create a basic document, basically a flyer, for all schools and universities, saying: as an emergency reaction to the interdependent sustainability crises, we are taking over; we, the students and teachers, the workers at educational institutions. This is about democratisation: away from the tendency to commodify everything. Universities should be public places serving the local and global community. This is what sustainability is about. And we start to create a "toolkit" on how everyone can lead democratic assemblies; how to react to resistance; and how to go deeper than talking about UN goals for 2030. Teachers and students at every school and every university could stand up like this and fight until the institutions are changed; until there are centres for sustainability and the idea of sustainability and democratisation is in the core of curricula, teaching methods and research.

Occupying schools and universities – what are the demands?

While we scientists take to the streets in this spring of 2022, pupils across Europe begin a wave of occupations at many schools and universities (SZ 2022). They go into the biggest auditoriums or sit in front of the doors of the head-teachers and rectors and refuse to continue cooperating. The "wave", as they call themselves, begins in Portugal and the United Kingdom; then Austria follows, as do Switzerland and Germany. A few groups gather under the label "End Fossil", while others operate more independently, although they are connected through the chats which soon include more than a thousand participants; many from Fridays For Future are involved.

Every group comes up with its own demands, partly in cooperation with scientists. Some are concerned with the content of teaching: the crises should be part of the introductory courses, and the perspective should be decolonised. Others care more about practical questions: no flights, vegan food. A third group is concerned with criticising the fossil industry and its connections with institutions and their funding; a fourth is focused on educational justice and the inclusion of the whole population.

Young activists address academia, tutors, and teachers directly

Winter is over. The puddles have thawed. It has become warmer, and we are once again walking through the campus to the theatre rooms of my institute. With some of the activists, I am making a film for an interdisciplinary conference in the Aula Magna (on the topic of "Young People, Power, Societal Change", Fopp 2022/3). They are now addressing the university directly, and the whole research community, and calling on them to strike. They say that we are in such an existential crisis that we all have to work together. And we are the ones who are so privileged that we can choose whether we get involved or not. And they point out that for many millions of people, the crisis is already an unavoidable part of their everyday lives. For them, too, we have to stand up.

We begin to speak about the first days, four years ago, when they suddenly left their schools and sat in the squares in front of the parliaments and town halls. They say that when children are sitting on the ground and fighting for their lives, that obviously has a strong effect. Just sitting there, silently, is a strong act in itself: doing nothing. Not cooperating, not acting like we could buy our way our way out of the crises. Stopping, pausing, opening up for new directions.

Schools and universities – strike, occupation, transformation: Five possible demands and perspectives for assemblies

1. Introduce an obligatory "studium generale" for all, focusing on the sustainability crises
A. Discuss the origins and interdependences of the crises (climate, biodiversity, social justice, …);
B. Teach the skills, knowledge, values and attitudes needed to create a sustainable society within very few years (and about the urgency: tipping points, CO2 budget, etc.);

C. Focus on climate justice and the knowledge in social sciences about the intersectional analysis of injustice and democracy beyond structures of domination (class, gender, ethnicity etc.: anti-racist, anti-colonial etc.).

2. Transform all curricula in all disciplines
When implementing A, B, C (perspective 1), adjust it to every discipline (economics, history, education, geography etc.). Linking content to UN Agenda 2030 goals is not enough.

3. Focus on and teach transformative, healthy, and regenerative teaching methods and didactics
A. We are embodied, social, imaginative, creative, vulnerable creatures living in problematic power relations to each other and nature, even within educational spaces – this should be the starting point for didactics and teaching methods;
B. Incorporate knowledge about (mental) health into curricula and teaching methods and focus on a "sustainable relation to each other and nature" as the core of education: care, sustainability and democracy are linked;
C. Focus on the practical, hand-heart-head sustainable "metabolism" with nature: teaching in real life contexts (in the areas of building; agriculture/forestry; arts; care) the basics about creating, repairing and sustaining social, ecological and economic sustainable work;
D. Focus on democratic leadership within school/university classes and spaces, creating democratic relations beyond domination and being able to handle existential and emotional transformative processes related to the crises in a regenerative way.

4. Specify Research Ethics – with sustainability criteria
A. Specify the three obligatory ethical principles which are guiding all research today, so that the educational sector contributes to a sustainable democratic world. "Not doing harm" or "not harming integrity and well-being" is not enough;
B. Create clear funding flows into sustainable transformative research and stop all research funding by fossil society (fossil industry, banks etc.).

5. Create Institutional Change – towards sustainability
5.1 Within schools/universities:
A. Install a creative transdisciplinary sustainability centre which can help with all these aspects (curricula, teaching, institutional change, research) in every university/school;
B. Give students influence and co-leadership over their sustainability education;
C. Stop emissions caused by universities/schools (transport/flying, buildings, food etc);
D. Stop all institutional funding by fossil society (fossil industry, banks, etc);
E. Democratise the workplace: 1. Move away from the (new public management) logic behind the organisation towards democratic decision-making by all involved; 2. Open up educational institutions with an intersectional approach (to BIPOC, working class people etc.);
5.2 This implies but also presupposes a political and economic system change within society – for (educational) democracy:
A. Away from a political and economic framework of competitive, exponential growth towards a post-growth, regenerative, cooperative, sustainable one;
B. Focusing on educational justice (class, ethnicity, gender etc.) which is linked to the democratisation of the society in general.

Chapter 10: Global democracy, the Elections, and the Future
September 2022 – June 2023: On forming a global, democratic, sustainable postgrowth society

The plan

It is spring. 2023. A Monday evening, around five o'clock.

The young activists are sitting around a huge table. But they are not alone. It is not only the young people who are now meeting, but also representatives of Extinction Rebellion; churches and other faith communities; the Mothers Rebellion; all the For Future Groups including People For Future; the people from A22 are also here, and so are Amnesty, environmental organisations, and many small and bigger antiracist, feminist and social organisations. Every second Monday evening, they meet to prepare an action week in September. Internationally, too, groups are working on a similar plan.

"What is the ground we're standing on?" someone asks. "We need a name." After a thorough discussion, the name is decided. The project will be called the "Week of Action for Social and Climate Justice", and – as in September 2019 – it will be framed by big Friday strikes at the beginning and end of the week. A sense of confidence takes hold. The procedures of consensus democracy are clear. After first exchanging information, their second step is to gather everyone's ideas. Thirdly, they collect arguments and listen to everyone's needs; they then make a proposal for a consensus in terms of values and actions for the week. They do a "temperature check" to reflect opinions in the room, before allowing everyone to express their smaller and more serious worries, and then they make changes to reflect them. They confirm whether there are any vetoes and whether a text needs to be rewritten, and check how many people are not entirely content but can still support the action or text – and so, after four or five meetings, they have a clear statement and a definition of the forms of action. All of them should be able to contribute to the week in their way, as long as peaceful and democratic means are used; that is the final text.

Images of the last few months go through my head and for the first time in a long while, tension and grief dissipate and are replaced with a kind of excitement about the next months. The elections in Sweden half a year ago were very depressing, and the apocalyptic "battle of Lützerath" (Schüpf 2023; BR 2023; Maier 2023) only took place a few weeks ago. I ask myself: What can we learn from these months for the time that's now approaching? How can we organise the solidarity which is so urgently needed?

The evening of the elections – a world falls apart

Because six months earlier, on Sunday the 11th of September, 2022, at eleven in the evening, something happens which shocks the whole movement in Sweden (and in Europe); it is symptomatic of the elections in other countries in the world, from Italy to the Philippines and Turkey.

It is exactly four years since the beginning of the strike and the founding of Fridays For Future. A new parliament is being elected. There has been a strange election campaign. It put the environmental movements and the justice movements in a seemingly impossible situation. They emphasise again and again that no party is offering policies that implement the Paris Agreement, including the left-wing and green parties. They have been striking in front of the left-wing and green parliament for four years, after all. And still: a win for the right-wing conservative parties would be even worse – especially for those who are already marginalised and for a planet worth living on. What should people do when what they want to fight for is not represented? A new party has formed, the "climate alliance", and doesn't even gain one percent of the vote.

The first numbers begin to come in at around eight. It looks good for the red-green block. The greens ought to make gains, it seems. No more talk of them losing their seats in parliament, as it looked for a long time. The Social Democrats are stronger than anyone has ever imagined. They are expected to get around 30 percent of the vote. But then it happens. The numbers – and with them the power balance – begin to collapse. The right-wing "moderates" suddenly look as if they are going to get 19 percent rather than 17. This changes the whole structure. The centre-right block, together with the Sweden Democrats

(SD) get the majority, though only just. This becomes clearer and clearer as the night goes on.

The SD party, which was founded by Nazis (Ekström et al. 2020), has become the second biggest party in the country, because of the votes of the male population. It gains power and now belongs to the "base of the government". A Sweden Democrat candidate calls out something like "Sieg Heil" to the television cameras (SVT 2022). More and more speeches and proposals pile up during these days, about how some people who come from other countries and look different will not get permanent residence anymore; that if they "behave badly", they will immediately be deported (DN 14.10.22); and that the climate crisis does not exist (SVT 19.10.22). One of the first measures of the new government is to close down the Department for the Environment (EnvironmentAnalyst 2.11.22).

Many of the adults try to comfort the younger people. And each other. What is the best way to do that? One of the central questions of the last four years. Listen to them. Make it possible to understand the situation, and offer protection, security. Security. That is the word they're twisting round, the ones who are now coming to power, I think, with enormous grief. They talk about nothing but security; and what they mean by it is SUV machines, nuclear power plants, razing forests, segregating cities, and so on. And through that, they make so many things insecure and fuel the crises. They probably don't see the security which really holds: that of caring encounters on an equal footing. Because care is needed. So many of the fears are justified, after all: living costs are rising, enormous social inequality is spreading (Cervenka 2022), as well as violence in the society (Ahlander/Johnson 2022); there is increasing segregation between city and country and between the centres of the cities and the outskirts. In the countryside, especially, climate policies meet with scepticism; everything just becomes more expensive, and no one really offers policies to integrate this into real help. Someone would have had to present a plan for the next ten years to guarantee basic security to everyone, I say to myself.

Formal and substantial democracy

It all started so well. Only two days earlier, the Fridays activists marched through the centre of Stockholm with the adults of People For Future, thousands altogether, ending up back at Mynttorget. But the most crucial debate seemed to be missing. What was clear to many people two years ago during the pandemic, namely the fact that work and life should be the core, supported, healed, educated and repaired by care, so that everyone has enough resources, has disappeared as a basic idea. The idea of deepening democracy has also disappeared: ideas about redistributing power, breaking down the segregation of cities, redistributing the enormous wealth, and radically reducing emissions as well as protecting the soil and the forests; as the project of building a new, flourishing, inclusive democracy.

And when the media report on the strike in the evening, new reports of drought and floods appear on the same day – in Pakistan and China, for instance (CarbonBrief 2022). Many people are losing their homes forever. Terrible droughts are spreading across Italy and Spain, and huge rivers are running dry in the worst drought in 500 years (EUSI 2022). In France, too, they are suddenly gone: the Loire has disappeared. In China, there is a drought followed by gigantic floods. And coal production continues. Liz Truss' new UK government announces that fracking is allowed, and that oil and gas will both be extracted from under the North Sea.

Four years have passed since the youth began to strike. Most citizens have accepted that humans are having a devastating effect on the environment and the climate. And still, the extent of the crises and the effects of tipping points still don't seem to have got through to people; or else, they are in denial.

Not only authoritarian regimes but also formal democracy – when it is not tied to substantial democracy – can lead to election results which destroy everyone's living conditions and with them the possibility of democracy itself. As societies, we seem not to have understood that. To that extent, elections are only a small part of democratic activity. We need to root the formal in the substantial aspects of democracy.

After the evening of the elections, it is the overt racism that scares people the most. On Tuesday morning, the idea of a spontaneous demonstration comes up. It is to be organised together with all groups across society which see a connection between anti-racism and climate justice, in solidarity. Who will contact whom? When and where? And so, the embryo of the broad movement emerges, which then organises the strike week in September. A few days later, the demonstration takes place. The fear of attacks by right-wing groups hangs in the air. The official organisers are: FFF Stockholm, People For Future Sweden, Black Lives Matter Sweden, Extinction Rebellion Sweden, "No one is illegal," Greenpeace, Amnesty, the Kurdish Women's Council, Friends of the Earth, Doctors Against Racism, Queers Against Fascism, and many more. In the speeches, there is a whole series of descriptions of everyday racism and the underlying structures. The common fabric of integrity, the substantial aspect of democracy, is damaged, everyone's equal dignity. How to repair it?

#BeyondGrowth in the EU parliament – a new vision

How can we change the way we live together nationally, across Europe, and globally, so that real substantial democracy can be established?

While some set off on the journey to Lützerath and others work on bringing together all the climate and justice movements in Stockholm, still others are working on political and economic solutions during this spring 2023, and on changing the structures which create injustice in the first place.

It is still early morning at the EU parliament, one day in May. Adelaïde has the floor. In front of her sit hundreds of young climate activists and researchers. Already in autumn 2018, together with Anuna and many other Belgians, she reacted to the Mynttorget group and helped to establish the school strike movement. Now she is making a speech, and after her one speaker after the next comes to the podium. Together, they outline the concrete elements of a new way of living together, a new view of what economics and politics are. #BeyondGrowth is the name of the three-day conference (https://www.beyond-growth-2023.eu). The president of the commission, von der Leyen, says the opening words and then hands over to those who are demanding a way of living which would be very differently structured from the one her commission is holding on to.

Many follow these speeches online on the live stream across the world during the three days. And slowly, the feeling takes hold that something is shifting, here and now. A new future is becoming visible to those who have gathered. The 300 or so professors and young people paint an astonishingly consistent picture. Kate Raworth says – firstly – that we can align our society with a "doughnut" goal, rather than with abstract GDP growth, which means focusing on a "secure and fair space" to guarantee that all needs are met with enough means for a dignified life, without breaking the planet's limits. Jason Hickel adds – secondly – that for this, basic services can be established collectively and provided to everyone unconditionally (not only regarding health and education, but also transport, energy, and housing). Thirdly, the debts of the Global South should be cancelled, and reparations should be paid. Raj Patel calls – fourthly – for this to be combined with a transformation of the economy in the Global North, meaning that people and nature in MAPA regions would no longer be exploited. Julia Steinberger then describes how resources worldwide are enough for everyone, if we pay attention to the logic of sufficiency. She also tweets that she feels like Pippin in *The Lord of the Rings*, achieving historic deeds beside Gandalf, Aragorn, Gimli, and Frodo – though such comparisons make it particularly noticeable that so many female professors and young people are setting the tone here. For a few hours, it feels as if a convivial, globally just postgrowth society might be a possible future.

What would have been unthinkable four years ago is now right in front of us: a vision, and not a niche phenomenon but a potential new mainstream. What has long been discussed at universities and among youth activists, is now entering legislative spaces: queer feminist texts in the tradition of Silvia Federici, inspired by the early texts of intersectional feminism by writers such as Angela Davis, and by decolonial theories and the ecological theories of Vananda Shiva.

And still, during these hours it also becomes so clear that barely any party is standing up for this vision at all, including the left-wing and green parties who are hosting the event; they could change. And soon, we also hear from people such as Emma River-Roberts (2023), who point out that the perspective of the workers' movement is still mainly lacking, and is at most mentioned generally, for instance in the context of trade unions. If this vision of society comes across as being focused on limiting the use of energy and the throughput of material, then hardly anything will be gained; that is the argument. That is why the voices seem so important which emphasise all aspects of democratisation (including Bell 2020): from organising workplaces, to property relations. And that everyone, ultimately also on a global level, has access to the basic services which are necessary for a secure existence, particularly workers in the fossil sector.

Many of those who then take the floor (including Johan Rockström) also explain what the alternative would be – what it would mean to continue as we are. The climate crisis is currently growing at a brutal rate. Southeast Asia is breaking all records. And the global meteorological organisation WMO suddenly shocks everyone by saying that there is a 32 percent chance that the next five years will, on average, already exceed the 1.5 degree rise in temperatures (WMO 17.5.2023).

On the way to global democracy

These ideas are also discussed by the activists who have gathered in Stockholm when they found the new united movement for the September week and – even globally – for the next years.

It is no longer only about movements protesting for general demands. And it is not about parties or organisations standing up for a concrete program. It is about combining both to make a movement with a concrete global "system change program" for intersectional justice. This raises two questions: how this new movement could be organised, and which forms of action are appropriate – so that the "BeyondGrowth" society can truly be realised with the "doughnut" goal.

People are afraid that the next four years in the media and in politics – globally, too – will be shaped by a false, contrived conflict between two worldviews: between the perspective shaped by authoritarian nationalism, a male, patriarchal, and conservative insistence on the superiority of one's own culture and nation, and the neoliberal perspective which is focused on growth and global markets. But this depiction of the problem is wrong, according to many people. Neoliberalism has abandoned people to a competitive fight against each other within a capitalist market over the last thirty years, public organisations have retreated from the countryside, the cities have been segregated, schools and the health system have been weakened, and all of this has contributed to the shift to the right, according to a large part of the research (Brown 2019). Right-wing populism has grown rapidly, including in countries such as Germany and the USA.

So we need a different description of the actual conflict, some say. With the insights of the #BeyondGrowth meeting, this could be described in the next months and years as a conflict between those who legitimise domination (over nature and other humans), whether authoritarian or neoliberal, both relying on fossil fuels – and those who want democratic, humane sustainability.

And again, as in the first days of the war, which are now already almost a year ago, I think of this idea: that we can only stop global heating and the extinction of species if we realise this idea of one shared humanity (see the chapter about the war and the new global order).

The activists are in contact with the Fossil Treaty group, which is driving fair global transformation in all areas beyond national borders; with the "commons" group around the Earth4All project, which stipulates that the most important ecosystems belong to, or must be looked after, by all people transnationally (Dixson-Declève et al. 2022); with the Doughnut Economics Lab (with its focus on everyone's needs within planetary limits) and the Faculty For a Future; and with the organisers of the "Global People's Assembly," as well as with those who are working to transform UN structures on a radically democratic basis.

These initiatives are open for all people and organisations to join; they complement each other, and all aim to democratise the global transformation. Those who are interested in stopping the fossil fuel industry can join the fossil Treaty Initiative. Those who want to change universities and schools as students or teachers can join the Faculty for a Future or Scientists for Future and Scientists Rebellion. Those who want to create sustainable societies locally and globally can join the Doughnut Economics Lab. And of course, there are justice movements everywhere which focus on global transformation, like Climate Justice Now, and international labour movements; and finally, democratic grassroots movements such as Extinction Rebellion, the For Future groups and for the younger ones Youth For Climate and Fridays For Future. Without these movements, it is not very likely that fundamental aspects of our societies will change, if we look back at history.

We can change. Societies can change; they have changed before. But we have to dare to raise our voices, to organise and unite – against what amounts to a crime against humanity but is presented as "normal" by so many people; the parts of society which follow a logic of domination.

The workers' movement seems to be divided on this, between the official direction of the trade unions, which globally has a social democratic orientation, and the groups in the tradition of an international struggle for a just, postcapitalist societal structure, in global solidarity. Somewhere in between, the newer movement for a global "Green New Deal" is located, which brings together figures such as Bernie Sanders, Alexandria Ocasio-Cortez, Caroline Lucas, and Asad Rehman, and continues the work of groups in the Global South and indigenous populations, who have already been fighting for so long for a change in thinking (Taylor 2021).

What shows itself in all these social initiatives and movements is a spirit of global democracy: to make a common fabric of integrity visible, everyone's dignity. In these movements, many are already trying to implement this approach. We could call it a kind of global "performative" approach. They already want to see themselves and treat each other as global citizens, so that they can measure legislation against this and demonstrate through civil disobedience that the current situation harms our common humanity.

The (not really) final strike

And on Friday the 21st of April 2023, they gather from all corners of the city once again, as they do every week, to strike in Mynttorget. This time, there are around a hundred young people and adults. The atmosphere is similar to spring 2019. They have spent more than 250 Fridays together since then, almost five years.

For the first time, as a definite sign of spring, the ice cream stall is once again in the square, in all its glory, just as it was in spring 2019. It is warm, a wonderful spring day, and everyone sits on the wall in front of the canal which separates Mynttorget from the parliament and brings the water from the Mälaren lake to the sea. And the first images of the strike appear on my social media, as they do every Friday. From Japan, Brazil, Finland, while the climate scientists on television predict the warmest summer since records began.

During these days, it is warmer than it has ever been in human history, for a whole week (WMO 31.7.2023). One day after another. The world is burning this summer. Italy reaches 48 degrees – a first in Europe. More than a hundred people die in Hawaiian wildfires. The sea is warmer than ever before; it has warmed up much more quickly than expected; with implications for the Gulf Stream and the future weather patterns around the globe (Readfearn 2023). Soon, a large part of the world that has been a habitable "niche" for humans could be uninhabitable (Lenton et al. 2023). And the profits of the oil companies are at record levels again (Ambrose 2023).

And then it comes, the school strike on the 9th of June, which is technically the last one for some of the young people who started the strike movement. They will be finishing high school before the summer holidays, and thus their school days will be over. From now on, for some of the young activists, it would have to be called a "uni strike" or Friday strike. They have held out for five whole years. Every Friday, they have left school to address the public, and with them so many of the global group.

On those cold November mornings five years ago, I said to myself that I would mobilise adults so that the children would not be left alone on the hard ground. That I would not leave them alone. And so it happened, one Friday after another – and through that, the chance opened up to cooperate on founding Scientists For Future and to help form Extinction Rebellion, to work together with the best scientists and contribute to democracy research – and to build an intergenerational global movement with the young people, fighting for their lives and for those of their peers. "Until politicians adhere to the Paris Agreement" – that was the official formulation at the start of the school strike. That has not happened – yet.

What they have achieved in the last five years is incredible. I am often not sure they really understand how they have changed all our lives and what they have done for all "earthlings". What a difference they have already made through their bravery and courage. Through their friendships and their imagination. I'm not sure they understand how much work they have put into this fight for their peers, and how grateful so many of us are; and how concerned we are for their health. And how many of us wish that all the elders would take responsibility. We should say: we understand; we see you; we get the problem – we are taking action.

In the last three years of the strike, half of the entire quantity of CO_2 has been emitted which could still be emitted within a 1.5-degree budget, which is the limit mentioned in the Paris Agreement (Forster et al. 2023). In three years, it will probably all be gone, this budget, forever. Tipping points of the earth system are being reached. And as if nothing were going on, politicians are still just talking about "net zero emissions in 2045 or 2050." As if the earth would not be unbearably warm by then, with further hundreds of species going extinct; and fellow humans suffering around the world. Ultimately, this is a denial of reality, an activism for the inhumane status quo.

The big picture – into the future

New York suddenly goes dark because of forest fires in Canada. The image goes around the world: UN Secretary General Guterres stands at the window and looks out at the filthy yellowish air (IPCC 2023).

The 1.5-degree boundary could already be crossed once for the first time this year or next year (WMO 17.5.2023); soon it would also be exceeded as an annual average. The countries of the world, including all democracies, are not adhering to the Paris Agreement they ratified.

Emissions are still rising worldwide. They ought to stop. The news arrives that the Arctic ice will disappear completely during the summers, regardless of how humanity responds (Carrington 2023). And this will continue to fuel the self-perpetuating processes.

Why are politicians and politics still not reacting and changing, some people ask in the newspapers. There is anger everywhere about the situation – and love, seeing all the creatures suffering, and all the people who are struggling to stop it and create better structures.

It seems that we are still trying just to understand and to establish what substantial democracy would be – in our societies and in activism, locally and globally. The idea only spreads slowly: these are structures of domination which shape society to a large extent, structure huge parts of politics and the economy, lead to crises, and even control forms of activism. They can be replaced by another world view, new knowledge, structures and practices.

The goal is to create structures and circumstances in movements and in society, schools, universities and workplaces so that no one is forced to give up the connectedness to oneself, others and the idea of being humane. This means especially that adults must create a society which enables young people and children to live in circumstances which don't force them to lose contact with themselves and with humanity, through fear, anxiety, lack of resources, and the sense that the future is insecure and existentially threatening.

There is work for everyone there; a whole range of different kinds of work; from organizing strikes to helping with the fundamental transformational initiatives.

But this demands new knowledge, new practices of what I called here "substantial democracy" on a global scale: how to create those social spaces and restructure the economic basis (see the chapter on Davos and economics); and redefining our relationships as a sustainable "being towards the world" (see the chapter on the "many fights"; and the chapter on education for the description of a prototype centre and the diabolo-model).

A "one people, one planet" movement for all of us could be organised around these new centres for knowledge and practices. Every school and university could develop one, every village and city; not as elitist locations for very few, but as people's movements for the broad public.

These centres can bring together the initiatives which have been presented throughout this book. They can foster a bottom-up movement for a fundamental transformation of our local and global society, ending domination (class, gender, ethnicity etc.) and replacing it with substantial democracy and care: the Doughnut Economics Lab (new postgrowth, regenerative economics, forestry, agriculture etc.); the Fossil Treaty Initiative (from fossil to renewable society); the faculty for a future (education); the concept of bioregions; the Earth4All creation of commons; and so on. In terms of governmentality: they are all open to everyone, all individuals on this living planet, as well as to organisations and movements. They act on the level of citizen movements but work with the creation of new binding rules.

But this is not enough; the movements which act as handbrakes are still needed. Real change comes from open, democratic people's movements – not to stand in the way as privileged, is central; not creating pressure to fit in with middle-class NGO culture and communication skills or drop out with shame (see Bourdieu 2010).

A school strike by children is still one of the strongest political forces we can imagine. And we need a complementary movement of older activists to disrupt the status quo for example at workplaces, through non-cooperation on Fridays or other similar peaceful means. New centres could provide the safety of mutual aid which is necessary to organise strikes and non-cooperation in workplaces, to stop the machinery of the fossil society.

We could unite this function of movements as emergency brakes with the task of building a society which makes everyone's dignity visible and experienceable; fighting for the new political framework which guarantees unconditionally and without structures of domination the resources to live a life in dignity (see appendix for the three pillars and the two principles).

In the same week that the fires are raging in Canada and darkening the UN headquarters, I take part in a "stress research day" at Stockholm University, together with Isabelle. Loukina, from Switzerland, will be visiting in a few weeks; the two of them have now known each other for more than four years and have organised so many trips and strikes together. At the university we talk about the experiences of young people worldwide and about the responsibility that falls on all of us. About anger and fear. About exhaustion and grief, the enormous grief. The emissions must stop. The forests must be protected. The fossil fuels must remain in the ground. And this has to happen fairly. And quickly.

We know that this is possible. There are enough resources for everyone here on this living planet. What is needed is that we, the people, stand up for a global, humane democracy; not only as professionals in civil society, creating "narratives" and "campaigns", but as human beings organizing ourselves to fight for new democratic rules and structures, in a transparent way which is open to everyone. Everyone is needed.

In a report on the 17th of May, 2023, the World Meteorological Organisation states that one of the next five years is very likely to see temperatures more than 1.5 degrees higher than preindustrial levels. There is even a 32 percent chance that the average will be above 1.5 degrees in the next five years. This is drastically worse than previous prognoses (WMO 2023).

The world's seas are also suddenly heating up enormously. This, too, is a development which was not predicted (Readfearn 2023). The razing of the Amazon rainforest is continuing unabated (Grattan 2023).

In the first quarter of 2023, the oil companies BP and Shell make unimaginable profits of around 15 billion dollars (Ambrose 2023). Similar profits are made by the financial sector investing in the fossil industry (Bukold 2023).

"The richest 1 percent grabbed nearly two-thirds of all new wealth worth $42 trillion created since 2020, almost twice as much money as the bottom 99 percent of the world's population", says Oxfam (2023), while showing that the richest 10 percent account for about half of all carbon emissions in the world (Oxfam 2015).

And carbon emissions are at an all-time high in June 2023 (Harvey 2023).

In the journal *Nature*, an article points out that current policies will lead to global warming creating lethal living conditions for billions of people, during the lifetimes of the children who are now protesting. A third of the global population live in regions which will then no longer fit into the "niche" of what is bearable for humans (Lenton et al. 2023).

> # Part Three: Facing the Future Together – A Conversation with Isabelle Axelsson and Loukina Tille

Contents

Chapter 1: What the Climate Strike Movement is About

Chapter 2: On the Relationship Between Young People and Adults
How We Can All Write History

Chapter 1: What the Climate Strike Movement is About

The compass

What is it all about? That is the question I ask Loukina Tille and Isabelle Axelsson in summer 2020, when we look back at the first two years. The two of them are among the young people who have made the global movement possible. By now, they have begun their degrees in climate studies at the ETH in Zurich and human geography at Stockholm University. They have been involved in the movement since December 2018; they joined when they were still at school, got in touch with each other already very early on in the global chats, and met for the first time in Strasbourg in March 2019, together with sixty other young Europeans, shortly before the first global strike. Later, they took part in the World Economic Forum together in Davos. And it was partly thanks to them that four hundred young activists gathered to develop the movement further at the Smile meeting in August 2019.

If somebody – without knowing about the new climate movements – wanted to build a movement to change the world and unite millions of people in the fight against the ecological and climate crisis, which principles would they decide on? What should be at the core, guaranteeing that hundreds of thousands of people want to join, and ensuring that it is also quite clear why and how our societies have to change? What kind of movement have the young people actually created?

The Fridays For Future document which comes closest to being an explanation of their principles is the Declaration of Lausanne. However, as it was written only by mainly white young Europeans from 28 countries, it does not claim to speak for the whole movement; Isabelle and Loukina also emphasise that. But it is a good starting point for a better understanding of the basic issues.

During these weeks in summer 2020, I think about the fact that Greta and her fellow activists could have sat down in front of their school; they could have adopted a political program, and they could have represented an NGO or founded a new one. But they didn't do any of those things, and instead concerned themselves with the core of democracy. They call attention to the publicly funded research institutions of democracy, the universities with their academic researchers, and from the beginning they have been thinking globally, drawing everyone in with the hashtag and the English name. And the basis they start from is the document, the Paris Agreement, which all countries have in principle already signed and recognised. What they are demanding from the world is nothing radical, but is only what was already promised to them and not delivered.

In Lausanne, they summarise this as follows:

1. Keep the global temperature rise below 1.5 degrees compared to pre-industrial levels!
2. Ensure climate justice and equity!
3. Listen to the best united science currently available!

The demand to "follow the science"

The fact that the biggest environmental movement in history has agreed on these exact principles is not something to be taken for granted. What does "listen to the science" actually mean, I ask Loukina and Isabelle. And what is "climate justice"?

"We begin from the facts about the ecological crisis and the climate crisis; that's our starting point. We are not primarily concerned with opinions or convictions, like other movements or parties. We start with the crises, their scientific background, and the knowledge that they are bad for our future and are already doing damage now. Humans and animals are suffering and will continue to suffer." These facts about the world cannot be explained away. That is why the young activists start from that point. "For the people who call that into question or deny it," says Loukina, "we can explain that the context behind the crisis has been demonstrated scientifically again and again. This is about the laws of nature. And those are different from subjective beliefs. There is an essential truth beyond our wishes." Of course, people can be wrong, even scientists, I think to myself, but with our attempts at explanation we can get better at approaching the world. And with "follow the science", the movement means above all the fundamental research, not every opinion expressed by every scientist. That is why the special report IPCC-1.5 is so important for these activists. It was published by the UN right at the beginning of the strike as the politically legitimised result of hundreds of research groups, and it shows why it is so important not to surpass 1.5 degrees of global warming, not least because otherwise potentially disastrous tipping points could be reached. The activists are referring to this consensus in the scientific community when they talk about following the science.

"Nature is something that lies beyond us human beings and beyond our projections. We can't suddenly say that there's no more gravity, or that higher levels of CO2 don't cause global warming. It's really about molecules and their behaviour. And they always behave the same way, again and again," Loukina continues. "But in Europe and in the world at the beginning of the 21st century, we have rather a self-centred attitude, which means that people are quite confused when they're confronted with something that is beyond them and doesn't depend on their opinions. And doesn't fit into the algorithms that are tailored to them by social media. They often push the facts away and ignore them." But is it necessary for all people to understand the fundamental science behind the climate crisis before they can react to it? Or is the point that they should at least recognise the facts? "I don't even understand all the scientific processes, and I'm studying them at university," says Isabelle. "But I understand the correlation between our emissions and global warming, which then in turn leads to possible disasters. Not many people will actually understand the chemistry and physics behind all of that, but we should recognise the consequences of our actions. In that sense, everyone has to understand what we cause when we (as humanity) produce emissions, on an individual and a systemic level." "Yes, exactly, and it's also the task of scientists to communicate that," adds Loukina.

But again, I insist: isn't it astonishing that the whole movement agrees on this motto, "Listen to the science," and directs people to the universities? "It's also a reaction to demands which are too specific," explains Isabelle. "The point is, it's not about our opinions or about party politics. It's about understanding the crisis as a whole and realising that we are changing the 'earth system', and that we have to stop doing that quickly. It also gives us legitimacy. We're not claiming that we have all the answers. We're pointing to the research. And we're saying to politicians: hey, you have to respond to this knowledge which is recognised at the universities as institutions that are accepted as publicly funded parts of the democratic system, and you have to translate it into real life, through the appropriate measures." "And that would mean," Loukina continues the thought, "that politicians accept that this kind of research exists, and that we can't just claim that something else is true and refuse to take responsibility for the state of the world. We have to push back against this kind of relativism, the way people don't actually argue or attempt to get a better understanding of the world, but instead immediately claim that everyone has a right to their own version of the world. 'Listen to the science' is a way of countering that." It starts a conversation, I think to myself, it calls on people to work together to get a clearer perspective and look at the world more closely – and to build the democratic process of decision making on this exchange.

The demand for "climate justice"

And how does the demand for climate justice relate to all that? How does it come into play? After all, we can understand that the young activists want to insist that governments follow the Paris Agreement and the science such as the IPCC-SR-1.5 report, which says exactly what it means to keep global warming below two or one and a half degrees. That we basically cannot produce any more emissions. But we can imagine climate and environmental movements – such as Extinction Rebellion, for instance – demanding this without going into the dimension of justice. "That has to do with the way the world works," says Isabelle. "That all aspects are connected. If we only focused on one political measure or demand, we'd immediately see that this is not enough to deal with the climate crisis. There are so many aspects contributing to the crises that you soon see that it's a whole system; these things are connected." Isabelle and Loukina point to our societies, how we eat, how we dress, how we transport things and how all of that is connected; but also how our behaviour in the Global North is connected with the situation in the Global South. "For example, the overconsumption of clothes and habit of entrusting big companies to 'recycle' them in countries such as Sweden in reality contributes to landfill and pollution in countries like Ghana, negatively affecting the life quality of the locals."

"We have to organise the transformation in a fair way. Otherwise, it will hit the people who already have the least say the hardest," says Isabelle. "They are the people who have already been treated unfairly for a long time, and who already suffer from the situation." "So many things are really connected with everything else in this global economy, in the way we've built it up since the industrial revolution and colonisation. A change that would be as drastic as the one we need now could also have very unfair consequences. Listening to the science and organising the transformation in a fair way means that we shouldn't just pay attention to a few countries or a few people, but to everyone, and organise it fairly and safely for everyone. Maybe that's also why Fridays For Future became a global movement so quickly, because this perspective was there from the beginning. To not only listen to the western science, but also to the people disproportionately affected by the climate crisis." "In Switzerland, we discussed the demand for climate justice for a long time," Loukina explains. "After all, at first it just looks like two concepts glued together. But it makes sense. We have to get from A to B quickly. The way we choose to go from A to B is in essence an ethical choice. We have to make sure that we keep everyone in view."

But what is justice? I think about the facts, such as the fact that each person should only be able to emit about 1.5 tons of CO_2 per year if everyone in the world is allowed the same amount. This would mean that the richer Europeans would have to reduce their emissions by 90 percent within ten years. Instead, they want to use offsetting to buy their way out by paying poorer countries and maintain their standard of living. Is that fair? The most affected MAPA and BIPOC people are often emitting the least (see the chapters about the economy; and about intersectionality).

"The way you define justice is based on your values," says Loukina. "Yes, and different people have different values," adds Isabelle, "depending on their political philosophy." Does that mean that there is no shared understanding of climate justice? There is something like a fundamental understanding of the concept, both of them agree. That richer countries have to pay more than poorer countries, through something like "fair shares," and that they have to reduce their emissions more quickly (http://civilsocietyreview.org/report2018). And that there is something like a historical burden when it comes to emissions, but also the exploitation of the regions known as the Global South by the Global North in fossil societies for the last few centuries. And that the individual steps of changing society, which we all have to take together, must be arranged to protect people with fewer resources within each country, as well as people in the Global South. That also means financial reparations and help with the damage that has already occurred, and which continues to increase. Together with the other Mynttorget activists, Isabelle has already been in contact with Isak Stoddard, who has worked out a detailed model to calculate all these individual aspects of justice (Anderson et al. 2020).

"Obviously, different people have different concepts of justice," the two of them continue. "Some of them might emphasise that the state should guarantee everyone's freedom. Others would say, no, it's mainly about ironing out the unequal starting conditions and reacting to structural discrimination when it comes to gender, class, and ethnicity." "That is why it's important that Fridays For Future addresses all these dimensions. Young people with different value systems should be able to join. But it is difficult to understand why some people should have fundamentally better conditions than others; for instance, when it comes to overall CO2 emissions or ecological footprint. That's a question of dignity and decency."

"Within the movement, we also have to check the arguments again and again and challenge each other," says Loukina. "What does your freedom mean if other people's freedom is limited by your behaviour? Not all concepts of equality and fairness are equally good. We can make progress there. Seeing freedom as being based only on conscious decisions, for example, is too simplistic; what if there are options you know nothing about? We need philosophers who define concepts such as justice and freedom in a new way for our time, with the arguments that are relevant to the global climate crisis." "Yes, and it's about deciding criteria based on what we want to achieve. We can measure justice by that." "No one should suffer. Everyone should have a decent life. No one has the right to exploit others. Why should some people have it better than others, or live at other people's expense by using up a bigger share of the emissions pie?"

The connection between the three demands and the question of system change

So, all three demands are connected, I think to myself as I look back at our conversation. After all, the Paris Agreement says that the actions of individual states should be measured against criteria and principles of fairness and justice. All countries have committed to that, without taking it seriously so far, given the enormous emissions which continue to be accepted and hidden behind the concept of "net zero goals 2050".

"Follow the Paris Agreement and the science which explains what it means to keep global warming well under 2 and if possible 1.5 degrees, and put the principle of justice into action." What makes the construction of FFF's demands special is that each individual element looks innocent. Democratically legitimised politics (the Paris Agreement); science (IPCC-SR-1.5) and ethics (justice). But the combination of all three is so explosive! If we want to transform politics fairly in such a way that global warming and the ecological crisis are kept in check, we must reduce emissions in richer countries as fast as possible, while distributing power and guaranteeing everyone a decent life. That is tantamount to system change, a push for democratisation which the thousands of young activists are getting behind: attention must be paid to the needs of all people, power must be shared and influence on legislation has to be broadened through participation by larger parts of the population.

In these summer weeks, reports come in from the WMO, the UN's global organisation of meteorologists (WMO, 8.7.2020), saying that the temperature between June 2019 and June 2020 was already 1.39 degrees above the temperature in pre-industrial times, and that there is a 20 percent possibility that in the next five years at least one year will be more than 1.5 degrees warmer. The Paris Agreement can still be followed, but the pressure to stop emissions immediately is enormous.

"In Switzerland, we also discussed and introduced other basic demands," says Loukina. "System change is necessary if we want to transition to a just and sustainable global society. So, we also mentioned that system change might be needed. At the same time, we tried to enable as many people as possible to take part in the movement and still establish an internal compass. When it comes to defining and establishing that compass, we can keep developing better arguments." "Talking about systemic change or transformation is also sensible because it means recognising that social and political change has to be holistic," says Isabelle. "It's not enough to come along with a demand about farming." "But most people would probably still vote for parties that don't believe in that kind of system change or in making sure the principles of justice are put into action," says Loukina. "We have to understand their way of thinking and arguing, so that we can get them involved." How could system change come about on a just basis, the two of them wonder, and think back to what they experienced in Davos. We have to redefine economics and change our concept of a good life for everyone. Could that happen at a meeting of the most powerful people, like the World Economic Forum? Could those people design and realise a fairer and more sustainable system? "They have it much too good, they're stuffing themselves with chocolate in the Swiss mountains and they're never brave enough to aim for real change."

How will change come about? If it doesn't come from cliques of powerful people, where will it come from –governments? "Yes, this change could come from governments," says Isabelle. "But only if populations, all of us, raise our voices."

Grassroots movements, court cases and the new spirit of the time

That is why grassroots movements are so important, I think to myself.

"They are put under pressure not to cling onto the status quo. It's much too easy just to carry on in the same way. We have to make them admit that this is an emergency. Bringing in laws through normal processes takes about five years. By then, we ought to have stopped emissions to an enormous degree. We can't go by the usual rules. We need a climate law immediately to really bring change. How is that supposed to work?" they ask. "I come back to the idea of citizens' assemblies, or something similar," says Loukina. "The way we tried it out in Davos. The public have to speak out and come together, show that we can come up with different rules. We can decide all together how we want to live and that we want to reshape the framework of our everyday lives, and actually do so."

The courts and the justice system could play a role, I add, through their understanding of justice, particularly intergenerational justice, and the young people's right to a secure future. "On the other hand, they are limited by the existing laws, a court just gave the green light for the Preemraff oil refinery in Sweden to be expanded, which will raise the country's CO_2 emissions massively," says Isabelle. "In Switzerland, a court sided with the activists when they played tennis at CreditSuisse to draw attention to the financing of the fossil industry," Loukina counters. In America, a federal court has just ruled that the Dakota Access pipeline must stop operating, in a reaction to years of activism. And another US court has just declared that half of Oklahoma belongs to indigenous people, including the city of Tulsa with half a million inhabitants, in a landmark ruling (Healy/Liptak 2020). And the court cases brought by Greta and fourteen other children against Macron and Merkel are still underway. Denmark has established a new climate law, including mechanisms to keep it on course for the next ten years, regardless of changes in government. In Ireland, a court has ordered the government to develop a clearer and more effective climate policy – as in the Netherlands a few years earlier (Göbel 2019). "Ecocide" may maybe soon be recognised as a crime by the International Criminal Court.

With the climate strike and their daily campaigning, maybe the young activists are changing the zeitgeist, changing what is seen as reasonable, I think to myself. To the point that carrying on in the same way can no longer be justified ethically, not by any government or any court. Like the women's movement a few decades ago. Really changing power relations. That is what they are achieving with their courage, these two, and with them millions of other children and young people, before our very eyes: worldwide, together with Vanessa and Hilda in Uganda, Howey Ou in China, Mitzi in the Philippines and Arshak in Moscow, with whom they are in constant contact, they are shifting all our standards.

All of them are prepared to go to the public with their strike and be met with punishments and sanctions. They are making a statement that it is no longer okay to destroy the rainforests in order to produce meat and palm oil; to build coal power stations; to run banks that maintain this system, so that power is concentrated with a few people and businesses. The people behind the fossil society, who are often rich white men, suddenly seem like grotesque creatures holding fast to the old ways. Maybe the young activists are mainly working on that: on making sure that one day, digging up coal, oil and gas and burning them will be seen as a criminal act, as ethically completely wrong; as will holding onto business models that concentrate power and do not serve everyone. A few brave people say: No. From now on, that's absurd. We are equal, free, one human race on one planet. It can't go on like this.

In these summer months in 2020, the permafrost in Siberia melts rapidly. Suddenly, it is 38 degrees in the north of Russia, and for weeks there is unbearable heat that has never been recorded before. Scientists talk about their nightmares caused by the release of methane from the thawing ground. And the destruction of the rainforest continues 50 percent more quickly in Brazil under the Bolsonaro regime in comparison with the previous year, while the villages of the indigenous population around Manaus are hit hard by the corona crisis (Phillips/Maisonnave 2020). At the same time, researchers point out that the likelihood of reaching tipping points such as the thawing of the permafrost is already increased by surpassing 1.5 degrees of global warming, and not only when two degrees are reached (Lenton et al. 2019).

New studies show that the glaciers in the Alps are melting by about one percent per year (Sommer et al. 2020). By the end of the century, they will almost all have disappeared forever, and with them part of our drinking water.

The German government decides in June to keep coal power stations up and running until 2038, for almost another twenty years, and promises to pay operators such as RWE billions in compensation. The new coal power station "Datteln4" is added to the grid.

Kevin Anderson's (2020) study is published in June and shows that emissions would have to decrease by more than 12 percent every year from now on in European countries, for reasons of fairness. In its new CO2 law, the Swiss parliament instead aims for a decrease of about 25 percent within Swiss borders over the next ten years; this is less than a third of what the Paris Agreement would require.

The climate crisis increases the frequency and intensity of natural disasters worldwide. In Uganda, Hilda gets in touch when Victoria Lake floods. In China, flooding is so bad that millions of people lose their homes. Howey Ou, who is one of the bravest of the global band of rebels, writes a despairing tweet in the middle of July in the midst of the floods and raises the alarm. On the same day, Arshak is arrested in Moscow for participating in the climate strikes, and taken away by the police.

The Keeling curve in August shows that CO2 concentration in the air has reached 417 ppm, the highest level since about 3 million years ago, when sea levels were several metres higher. Despite this, during the corona crisis, the richest countries invest much more in the fossil economy than they do in renewable energy (Simon 2020).

Chapter 2: On the Relationship Between Young People and Adults
How We Can All Write History

A wakeup call – how do we get through to the adults?

What is needed now? A wakeup call for everyone, those in power and the wider public? "An alarm clock can just be switched off," say Loukina and Isabelle, "set on indefinite snooze." "We need something else."

That makes sense to me: the danger is that parliaments and governments – and with them a large part of the public – will simply continue to delay everything. That they will say: "There is a problem, and we're taking it seriously. We have goals (such as 'net zero and climate neutrality by 2050') and we will give the markets incentives to change things as necessary. We are investing more in green projects." At the same time, they refuse to base their plans on concrete emissions budgets. This means that young people cannot know whether emissions will really go down. It is totally unclear what their future will hold. Until now, this hesitant political strategy has not achieved much, particularly on a global level. The CO2 concentration in the atmosphere continues to climb, and accordingly, so do global temperatures – with all the devastating consequences, including drought, floods, and ruined harvests, in the warmest decade since records began.

It's "wilful procrastination," as Isabelle says. "We have to open everyone's eyes and keep them open, not just wake them up," Loukina adds. "Their noses, their ears, their sense of taste – so that they can't start dozing again." What would be needed so that all of us could once again sleep peacefully and breathe deeply? She recalls a quote from Joanna Macy (2014): "Of all the dangers we face, from climate chaos to nuclear war, none is so great as the deadening of our response." We need to make sure that a gigantic social project has been set up which will really take the crisis seriously and stop global warming and the loss of species. We all ought to be sitting down together, admitting that there is a crisis, and defining and implementing a new framework. That would be possible within a few weeks or months if we were allowing ourselves to feel and connect to the crises in the first place.

Publicity

"But how do we change public opinion? There are so few public spaces," says Loukina. "We need citizens' assemblies in which everyone can take part. People don't want the foundations of their own lives to be destroyed." As can be seen from opinion polls in April and June 2019, at the time of the first global strike and the XR blockade of London, public opinion is shifting. Suddenly, more than half of British people believed that politics should react quite differently to the ecological and climate crisis, including by declaring a state of emergency (Barasi 2019). A "disruption" led to a change in public opinion.

"We children and young people can make a unique contribution to solving the crises," say Isabelle and Loukina. "We're still prepared to think unconventionally, or rather, to think sensibly. We are not caught in the mechanisms that explain away the dangers, developed by those powerful people at the Davos WEF – and this is what makes our contribution necessary." Isabelle and Loukina think back to the first strike, to the moment when they made that step. "I shouldn't do it, I shouldn't do it, I thought, but it is so important," says Isabelle. "And then it almost becomes a routine, and when there was one time that I didn't go on strike, it felt so wrong." "We always went on strike from 10 am," says Loukina. "So we really went out of the school, which is a big step because the teachers are right there. You take a megaphone, and you call: leave the school, go on strike! We knew we wanted the right thing, that we didn't want to avoid going to lessons but to draw attention to the crisis. We explained to all classes in the school why it was important, and then we called out: It's time to strike!"

The role of the adults

What is the role of the adults, of the broader population? "You could use the image of a hand with fingers. Fridays For Future is one finger, the one that was initiated by us young people and led by us. Adults can help by being supportive but not suddenly taking on leading roles," the two of them say.

"Their task is mainly to believe in us, to trust us," Loukina explains. "We have our unconventional ideas, we do things that might be wrong, but we need this freedom. Adults can get fully involved in the parents' groups or the scientists' groups." "But that doesn't mean that we don't need everyone. We need them as parts of a comprehensive climate justice movement, a hand with all its fingers," Isabelle continues. "Unfortunately, it often seems like only young people and grandparents are really involved, but not the middle generation, which is in power. That's why we tried to make a general strike happen, to bring all working people into the movement," says Loukina. "We need everyone. Many people may have briefly thought about the climate, but everything happens so fast, and people forget again what's important. That was the good thing about the Week For Future in September, when all of us protested together in the streets."

But isn't there still a problem that it's the adults who have caused this dire situation, and that they are now the ones with power, casting a shadow over the children's future? "That dynamic does exist, and sometimes we're incredibly furious, including when we're standing in front of politicians. But that does change with time. Sometimes we no longer expect anything from them, because they have such a narrow view of their own options." "It's as if they can't grasp the crisis as a crisis. They're trapped in their way of thinking." "But this definitely can't be about the younger generation rejecting and despising the older one. We've never said that, we've just said that we won't forgive them and we won't accept their actions, unless they start looking for a solution with us." "We need the power and strength of the adults on our side. It also isn't about mistakes by individuals; it's a systemic problem. And some of them have been fighting for decades in all parts of the world for their survival and for the protection of nature."

"But we children still relate to the world in a different way. We're more free, more independent, and often more open. We're not as corrupt, and as a group we don't care that much what adults think. Adults should listen to us much more," Isabelle continues. "Especially if we want to change all the underlying conditions, we have to bring in young people much more," says Loukina. "Basically, politics and the economy and even the whole of society should be based on principles which young people fundamentally understand, otherwise something is wrong," she adds. "That is also a kind of power struggle, because it's a question of worldview."

The best thing that can happen at these moments, in my experience, when young people and adults work together, is that we manage to see what is most important, to see the whole picture, and to see what would be needed in order to change all the conditions, so that it would work out well for everyone. We should include the young people much more, I think to myself, and show them real respect; we should lower the voting age and give them a voice in all institutions.

"When we spent a few weeks working at all levels, locally in Lausanne, nationally and globally, we were able to ask ourselves this question: what does the global justice perspective mean here in this city or in the region? And vice versa," says Loukina. This, I think to myself, is the perspective we need now, as we develop new rules for our shared life. We shouldn't just be continuing as usual with a few corrections, but should have the courage to set out the necessary framework, to outline what is "really needed." That is similar to what Scientists For Future are doing with Maja Göpel and Kate Raworth, who has been testing out her doughnut model in many cities together with the people who live there (Boffey 2020).

If we now sketch out the common principles and agree on them democratically, locally, and globally, we need this holistic compass. A compass for children and adults everywhere, pointing towards what we would need for a dignified life, for all of us. We have to hold onto this global and local holistic view and make it accessible to everyone in such a way that a continuous movement emerges which everyone can get behind. Then young people could relax, knowing that the adults are taking care of the world and trying to create as much security as can ever be possible in interaction with nature.

Conclusions for all of us – into the future

How can adults just carry on as usual and not try to ensure that this fear disappears now, I ask myself. We could sit down as an international community, together, and change the underlying conditions so that productive energy would emerge; that's the utopian ideal. That is how we humans work, my students at the university and probably most other people: as soon as we have a shared framework, principles and rules, and we can rely on people following them, we have energy – for instance, that is how it was with the first global strike, when we really only knew the date and the basic demands.

"Change doesn't just happen when we get new information," Loukina adds, "or when we're called on to behave differently, but when expectations change in terms of what's seen as normal. We can shift the expectations of the whole population, including the people in power." Often, these young people are presented in the media as a moralising or puritanical group. But they are not accusing other people; their target is the framework or the mindset (Göpel 2016); they want to shift normality so that we can learn to live in a sustainable way that works in the long term.

Those are the two alternatives, I think to myself. Either we go on as we have been for the last thirty years. With diffuse goals for "climate neutrality" in 2050, annual COP meetings where barely anything happens, market incentives for consumer-focused change, slightly different flows of investment, hopes for new negative emissions technologies, everything only coordinated loosely on a global scale, and fossil fuels being treated as goods to be managed nationally – and the likelihood that emissions will only fall slightly, if at all, and global warming will continue.

"There always seems to be a higher authority," says Loukina. "The people in power give you the feeling that they can't actually do anything. So that you ask yourself the whole time: but then where is the power that could change something? They often act as if they aren't responsible at all. Barely any of them say: yes, that's right, we'll bring in new legislation now. I might not be elected again, but who cares, I'll do my best to act on the science and try to mitigate the crisis." "And many of them are so convinced that they're doing enough, or they claim that this isn't their area, but instead has to be solved at a global or local level," Isabelle adds.

The alternative: together, as a global population, living together in such a complex biosphere, we establish a new set of rules to respond to the crisis as a crisis. We decide, as we did with a contract for nuclear disarmament, that the coal, gas, and oil have to stay in the ground, because the GAP report says that the existing infrastructure will produce so many emissions in the next ten years that it will be impossible to uphold the Paris Agreement, because the earth will become more than 0.5 degrees warmer (UNEP Production Gap Report 2019). And we agree on how this can happen fairly. We decide on emissions budgets at a global, national, and local level (and perhaps an individual level), so that we can be sure that emissions in richer countries will be reduced every year and will soon be stopped. We decide together that the rainforest in Brazil, Congo and Indonesia will not be razed any longer to meet western demand for meat or palm oil, but on the contrary, that forests will be protected and expanded, that powerful countries like Germany and China cannot rely on coal power any more, and that no new investment by Swiss banks can go into the fossil society, because that is not compatible with the emissions budgets. We understand ourselves as being an integral part of nature. We reposition ourselves in a non-extractive and non-abusive relation to our surroundings and other beings.

In this way, we will create trust in each other. And a gigantic shared project for the whole of society. We will give influence to those who have knowhow when it comes to sustainable agriculture, town planning and so on. But we will not start with these sectors, but with the basic underlying framework (see Appendix). It is this crisis management which we now have to achieve together, fairly, through discussion. And at the same time, we can formulate it in such a way that we can stand up for it, young and old, and work every day to realise it; through disruptions, but also by building a broad popular movement on the other, and through jurisdiction and education.

This, I think to myself, is the wonderful thing about FFF and S4F, and this is what makes the story of the last two years so important. They have not only shifted public discourse through their strike and made millions of people aware of the ecological and climate crisis. They have also worked on this framework the whole time, by uniting with us, the scientists. That was our shared project, which actually started already after a few days in September 2018, when we were inspired by Greta's idea and sent the first #ScientistsForFuture email to various professors in different disciplines in different cities and said: we need to support these young people, and we need to show that what they want is not impossible.

This whole time, we have been using public funds to work on ideas for the sustainable transformation of society. We are able, at least at some institutions, to bring together climate science and ethics and develop precisely the framework needed as a result of research, globally and locally. And we are able to think systematically, as the young people are asking – when it comes to the financial sector and the economy in general, for example. How exactly the overall picture of this transformation of society will look is something we might not yet be able to know. There are countless scenarios. But the framework can be sketched out. "And to think of the framework, we need to face three well-known counterproductive feelings: judgement, cynicism, and fear. From the moment that we manage to hold back, even for some seconds, our inner voices of fear, cynicism, and judgement, we will be able to let the sketch of the framework emerge in us. But that requires a lot of courage, compassion, and curiosity (Scharmer 2009). That is part of our democratic duty and our central task, all of us who work, teach, and learn at universities. We need to let the future emerge," says Loukina.

In the weeks after two years of joint activism, there is a sketch of such an alternative, when Greta, together with scientists and other FFF activists, sends a letter to the EU, and twenty FFF activists from the Global South send a similar letter to the G20 (as described in the chapter on the corona crisis and in the Appendix). So many discussions were needed before we reached this framework, both among the young activists and among the scientists. Precisely that is the story of FFF and S4F, as well as the story of the discussions between Isabelle and Loukina. They have had so many conversations, already at the beginning in Strasbourg and then in Lausanne and in Davos. How exactly should we stand up for these principles? Should we negotiate climate action plans with scientists, or only rough outlines? What should these look like; what does climate justice mean for these plans? In spite of differences of opinion, they have not lost themselves in the details, but have kept their shared project in view. As Reto Knutti says (Ryser 2019), the big environmental problems (such as the damage to the ozone layer) have never been solved through incentives. We all need to be given a jolt which brings us onto a different level by changing the underlying conditions as a response to the crises. As soon as this has happened, we can start on the real work. We need to change our lives. And this process must centre those, and be led by those, who are most affected by injustice and crises.

Taking the crisis seriously as a crisis, as the corona crisis was taken seriously: that means changing the fundamental rules and adjusting them to reality. And it means agreeing on the principles and establishing them democratically, locally, nationally, and globally. By admitting that we are vulnerable and totally dependent on each other, across national borders. And now we need to bring ourselves onto this new level of living together. Two years ago, there was a sign and a sheet of A4 paper which Greta took with her. Then five young women had the courage to join her, week after week, even when no one took any notice of them and nothing happened, in the cold and rain. And now a global movement exists, with tens of thousands of young activists and scientists who have set up a program we can build on, and which can be joined by all people who are active in other movements, whether for justice or for the climate, such as XR or the NGOs, to form a huge shared popular movement.

On a Friday, at some point, we can say to each other: we agreed together to protect the rainforests, to stop emissions, to keep fuel in the ground and to look after everyone's basic needs worldwide on a just basis. Painting this picture of a shared new crisis agreement might be an important part of what Loukina and Isabelle describe as "opening people's eyes": not only waking them up and making them aware of the crisis; not only protesting, but making it clear to them that we will only be quiet when we, all of us, have changed the rules. And not in any old way, but in the way that we all need, as interdependent creatures on a living planet. The idea for the mechanisms needed already exists (see Appendix).

Then we can wake up in the morning and know, or at least hope: oh, the oil will stay in the ground in Venezuela and Saudi Arabia, and the coal will stay in the ground in China and Germany. The forests will stay standing. People are eating plant-based diets. Energy is being produced more decentrally and sustainably. And above all: no one can dominate others or exploit them. Along with the oil drilling towers, we will dismantle the injustice which oppresses many people (especially children, women, and BIPOC), and which has oppressed them for so long, through this fossil society which benefits just a few people. On the contrary, we will make sure all people have what they need to live, as equals. We have the tools and the ideas to do it. We have a democratic framework; and we are millions of people, young and old. We will not give up until this has been realised. And along the way, we will go on inventing new ways and means to realise this together, with disruptive, nonviolent direct action at the centres of power, so that those in power can no longer keep going as usual, and with broad educational efforts. Everyone is welcome, everyone is needed.

Appendix and Summary

What a Global United Climate Movement Could Fight for
The Basic Principles for Social and Political Change

What would a global, united climate movement be about? Which new rules do we need as one people on one planet?

The following outline was inspired by the cooperation across generations between the young activists of FFF and all the other climate and justice movements at the COP meeting in Madrid. It is based on the conviction that we need a broad, global, democratic popular movement which everyone can join. In this sense, the ideas go beyond those of FFF and XR, but adopt key points from them. The following proposal shares the same basis as most of these movements: what follows here spells out the Paris Agreement, to which almost all countries and parliaments have already committed themselves.

The world is burning. What we need is a global movement, political in the broadest sense. We cannot continue to watch while the Amazon rainforest is razed to produce meat for the Global North; as coal power stations are kept running in Germany and at the same time tens of thousands of jobs disappear in the solar and wind energy sector; as billions of animals are killed in Australian bushfires; as millions of people in Bangladesh, China, and Mozambique lose their homes to floods, and as Swiss banks continue to make money by investing in the fossil industry.

We need a collective global reaction, as one people on one planet, a cooperation between all movements, big and small. They can present a joint alternative to the approaches which are currently doing such damage to the world, to children and young people's futures.

What could be the core of such a social and political change, the key ideas that all people could stand behind? Ideas which would shape the foundations of our shared life sustainably, within ten or at most fifteen years, worldwide?

All movements, governments, and organisations can be measured against this framework, and they can be called upon every day to realise it – while we can work every day to contribute to it.

Overview

We should build on three pillars and two principles:

(1) Fair global, national, and local greenhouse gas emissions budgets with action plans and legally anchored regulations which immediately bring in dramatic reductions to emissions in all sectors, rather than abstract goals such as "climate-neutral, net zero 2050", which mean nothing in absolute numbers; so that worldwide in a few years almost no emissions are produced, because this is the only way to stay within the 1.5 and "well below 2"-degree limit. This includes fundamental protection for forests and oceans, as well as shifting to regenerative agriculture and focusing on secure plant-based nutrition rather than on animal "products".

(2) Local and global organisation, for example with the help of a global treaty ("Fossil Fuel Non-Proliferation Treaty"), to keep fossil fuels (oil, coal, gas) in the ground; and at the same time an immediate stop to the financing and building of all fossil infrastructure, while continuously downscaling and dismantling the existing infrastructure by around 7–10 percent each year on average (different countries have greater or smaller obligations), as a reaction to the "UNEP Production Gap Report 2019" (on this, see https://www.fossilfueltreaty.org and Newell et al. 2018).

(3) Resources should be provided to meet everyone's basic needs, in the form of global basic services and a basic income, transforming ecosystems into commons. A rapid, publicly and democratically funded transformation of the energy system towards 100 percent renewables, inspired by global, national, and local scenarios from researchers such as the Stanford Group (Jacobson et al. 2019) and the group around Teske et al. (2019), which calculate such an immediate shift for 150 countries and describe it in detail. This funding plan should be combined with aid for populations who have already suffered damage and are expected to do so in the future, and who are already most affected by the climate crisis.

The following two principles of substantial democracy should apply throughout this transformation:

(A) social justice (global, national, and local, to dismantle inequalities and injustice; Global North and South; BIPOC leadership) and equity or fairness (as stipulated in the Paris Agreement and the Convention on Climate Change) should guide every step,
(B) leading to and based on the grassroots democratic principle of non-domination from an intersectional perspective (eco-feminist, anti-racist etc.), in relation to all areas of society (gender, class, ethnicity, etc.), with a focus on the democratisation of politics and the economy, and on creating humane social spaces, in which everyone can live a life in dignity, repairing the damaged common "fabric of integrity".

These three pillars and two principles could produce the framework we all need so urgently, which would take the ecological and climate crisis seriously as a crisis. What does this mean in detail?

The three pillars

(1) Zero emissions budgets according to IPCC-SSR15-Scenario1 (rather than setting goals for particular years such as "climate neutrality in 2050"), including the corresponding action plans for all sectors

Every political measure should focus on the overall quantity of greenhouse gases being emitted and on reducing them drastically every year from now on, if global heating is to be reduced. And if we are to stay within the 1.5-degree threshold worldwide, only around 300 Gt CO_2 can still be emitted (from the beginning of 2021; around 45–50 Gt are emitted annually (Anderson et al. 2019). In a few years, all of this will have been used up and we will be drawing even closer to dangerous irreversible tipping points (the Arctic ice; the Amazon rainforest; permafrost; etc.); we have probably already passed some of them, with disastrous consequences (Rockström et al. 2023). Adhering to this tiny global budget (without relying on unrealistic reductions by negative emission technologies, which we do not have), means that richer states must reduce their emissions by more than ten percent per year and reach almost zero emissions towards the end of the decade.

The budgets can be broken down to the global, national, and local level, and can become relevant as a political framework for all legislation.

This requires systemic transformation across all sectors of society within about ten years, a system change, by making standards stricter every year by several percent, for example. According to the UN research reports, this means an immediate shift to renewable energy, including massive reductions in energy usage; it means switching to mainly vegetarian (or vegan) diets and establishing regenerative, careful agriculture; it means shifting to public transport and to fossil-free modes of transport in general; it means sustainable construction with less cement; and it means approaching the political economy, the finance sector and the monetary system in such a way that the economy does not force us to exploit nature and other people (concrete measures are sketched out here under the heading "Possible policies"; see also Hickel 2021 and Raworth 2018).

By 2030, at least 30 percent of the world's land and sea must be designated as commons through UN rulings and protected (Rockström 2020); forests must truly be preserved, "rewilded", and expanded. Burning "biomass" cannot be classed as CO_2 neutral.

States, communities, and institutions must immediately develop such action plans for all areas and calculate budgets for emissions within each country, as well as for emissions from consumption. (Even the countries which are supposedly the most progressive, such as Switzerland, Germany, and Sweden, still do not obey emissions budgets. Plans are not transparent in terms of real emissions; emissions are calculated in such a way as to allow loopholes, often meaning that only half of the real figures are included; Thunberg 2022.)

The details of what is to be achieved are a matter of democratic discussion, following the two basic principles of fairness and non-domination (see "Possible policies"). However, this means democratically reshaping the public sphere and the definition of participation, to ensure that media and education do not primarily serve private interests, and to prevent the lobbying that damages democracy. Because richer nations have already emitted a disproportionate share of greenhouse gasses, they must – through a joint fund – contribute their fair share to ensuring that poorer countries can establish sustainable energy systems. This fair share can be calculated state by state (Anderson et al. 2020).

(2) Stopping the building, financing, and running of new fossil infrastructure, and dismantling existing structures on a globally just basis; perhaps through a global treaty analogous to the nuclear Non-Proliferation Treaty

According to the UNEP Production Gap Report, the fossil infrastructure (oil, coal, gas) we are currently building or planning to build will already make it impossible for us to adhere to the 1.5-degree limit. In the next ten years, up to double the quantity of fossil fuels permitted within the 1.5-degree limit will be extracted and prepared for burning.

That is why we need immediate political action, which could take the form of a global treaty, so that we as global citizens can stop this process in a socially just way, leave fossil fuels in the ground and mark them as "toxic". This could be modelled on the Non-Proliferation Treaty on the production and spread of nuclear weapons. We can and should define the extraction and burning of fossil fuels (as well as the profits made from them) as toxic and potentially analogous to pressing the red button. There is already a scientifically supported proposal suggesting how this global treaty could look (Newell et al. 2019). States must implement these demands as quickly as possible and initiate a process introducing the treaty (see https://www.fossilfueltreaty.org). It would be even better, even if it sounds utopian, would it be to organise a global conference – representing the global population – which decides how much oil, gas and coal can be taken out of the ground, and where exactly, in a fair way; so that we can guarantee that most of it stays in the ground.

The first step in this direction is to put an immediate legal stop to the financing of the fossil industry (by large banks, pension funds, universities, etc.) and to ban the construction of new fossil infrastructure across the world.

(3) Building a renewable, global-local energy system (inspired by the Stanford Group around Jacobson (2019) and Teske et al. 2019), and creating joint global sustainability legislation which – as an expression of global democracy – provides the most important existential resources to everyone unconditionally ("global basic services/income")

But it is not enough to stop emissions and keep fossil fuels in the ground. This can be complemented by working together to build security and to provide the most important resources for a dignified life to everyone unconditionally.

A public fund could be created, through which an immediate transformation of the global energy system can be realised (through fair shares and the idea of charging a fee for the use of ecosystems as "commons"; Dixson-Declève et al. 2022, as an example). There are already detailed calculations and plans, fully costed in terms of public funds, for the construction of a global-local renewable energy system, planned out for 150 countries, combined with calculations to improve public health (Jacobson et al. 2019; Teske et al. 2019), without relying on non-existent technologies or nuclear power. It is essential to focus on saving energy, above all among the richest ten percent of the population. This funding must be connected with aid to help countries and communities which are already affected by loss and damage caused by climate change and which will be increasingly severely affected. On this, the Least Developed Countries have themselves made concrete proposals (www.reeei.org).

So that this drastic reshaping of societies can succeed worldwide, we need existential security for everyone, particularly the workers in fossil sectors, which could be provided through unconditional basic services and perhaps a basic income (see Schmelzer et al. 2022). This can also be seen as a step towards the project of a global democracy, which treats all world citizens equally (on this and debt relief, see e.g. Fraser and Lessenich in Ketterer/Becker 2019).

The two principles

Throughout this fundamental transformation of our societies, we can be guided by two principles which both stem from the same idea. This idea is that we are all dependent on each other because we live together on a living planet, and the task is to care for this home and for each other (which also ought to be stipulated in a new version or "article zero" of the UN Charter).

(A) The principle of justice and fairness

The Paris Agreement already demands that all governments adhere to a principle of "equity" throughout their legislation. "Social justice" is also mentioned in the context of shifting to a sustainable society. All states have committed to this. This principle has a historical dimension (based on previous emissions and on colonial profits), a global dimension (of the Global North and South) and a national dimension (of unequally distributed resources).

Globally, this principle implies that richer states must contribute their "fair share" (see civilsocietyreview.org) and help poorer states both with the transition to renewable energy and with current and future loss and damage caused by the climate crisis; as well as listening to the knowledge and leadership of indigenous people. At a national and local level, for example, there must be efforts to counteract unjust distribution and unequal access to participation and the resulting inequalities in relation to CO_2 emissions.

This principle of equity or justice also means that richer countries have smaller emissions budgets than poorer ones; emissions must be stopped in these richer countries through new regulations as well as lifestyle changes, and not "bought" somewhere else. This means that geoengineering, carbon offsetting and carbon trading are not fair options. This is about real changes, not about the outsourcing of reductions. In particular, non-existent technologies for negative emissions should not be planned into calculations by any government, as is currently the case everywhere. This has already had dramatic consequences for the public understanding of the necessary emissions reductions.

The standard against which this should be measured is that of intergenerational justice (as a principle of freedom set down by the German Constitutional Court). Children are rightly demanding that the generation of those in power should finally take on responsibility.

This general principle of justice can also be understood in such a way that the MAPA communities which are especially affected by the crises should be protected the most. This goes both for indigenous populations and for BIPOC people worldwide, as well as for the poorer sectors of society and for those who live with disabilities.

(B) The humane principle of non-domination and democratisation

As we transform our societies, we can implement the fundamental principle of democratisation, which demands that we dismantle existing relationships and structural relations which rely on domination, and prevent similar ones from developing; both in relation to nature (ecocide; soil leaching; livestock farming; forest clearcutting) and in relation to other human beings. All humans should be able to meet each other on an equal footing. Research in the humanities and social sciences has provided intersectional analyses of power structures for this transformation (in relation to gender, sexual orientation and ethnicity, etc., and in terms of overcoming class structures); and corresponding proposals for how humane, democratic structures can be realised so that no one is forced to give up "connectedness" to oneself, others and nature.

All the proposed solutions for the global, national, and local transformation required should satisfy these demands. And transnational solutions (such as the problematic Green Climate Fund) can be organised in such a way that particular segments of society or individuals do not profit from transformations at the expense of others; and that structural power relations are instead dismantled. The structure of the economy and the finance sector must become sustainable and be adapted to this principle of non-domination (see the chapter on the economy). This idea thus leads to a demand for the substantial democratisation of politics and the economy, and a deepening of democracy which expands democratic participation far beyond participation in elections and involvement in political parties. This means that education also has to be organised correspondingly, and young people and science have to be given a clearer place in the collective political decision-making process across society (perhaps in the form of a climate task force).

Through this, we can turn our attention to the shared fabric of integrity, which connects all of us with each other and with nature – and which was and is damaged by colonial history, indifference and oppression.

The goal

Often, the goal is formulated as follows: this is about a dignified democratic shared life for everyone on a habitable planet. Or, in slightly more detail: it is about really working together to meet everyone's basic needs (from nutrition, equality and housing to education and political involvement) in a way that does not exceed planetary limits (climate, biodiversity, pollution, nitrates...; see Raworth 2018; Göpel 2016; Hickel 2020). This fundamental direction – along with the humane democratisation principle outlined above, which abolishes problematic power relations – could define the direction of all policies in global, national, and local documents, and replace existing political and economic goals (especially the goal of exponential GDP growth).

That is the framework which could guide everything.

Possible policies enabling the transition to a sustainable global society

The question of how exactly this framework (the three pillars and the two principles of justice and non-domination) should then be developed in reality – this is the democratic debate which we should now be conducting in all possible arenas (in citizens' assemblies, parliaments, media and the public sphere, and within the grassroots movements).

One proposal among others, which is emerging as common ground in the climate movement (see Hickel 2020, Petifor 2019, Göpel 2016, Hornborg 2017 and Raworth 2018, as well as Hällström 2021) would implement the three pillars (global, national, local, and perhaps individual emissions budgets; a global fossil fuel treaty and a global renewable energy system) and two principles as follows:

- It would include global (!) unconditional basic services to ensure survival beyond hunger and poverty, and possibly a basic income (through a fee for the use of ecosystems as "commons", see Dixson-Declève et al. 2022); this would make it possible to value care work more highly, balance out power relations (gender, ethnicity etc.) and really meet the needs of all people; it would partly be tied to location so that fewer goods would have to be transported (Hornborg 2017).
- This would be combined with the introduction of "positive" money, meaning that the arbitrary power of banks and the dynamic of debt and interest would be dissolved; central banks would be run democratically and would invest in sustainable transformation. Corporations would be transformed over time into substantial democratic organisations (cooperations; various forms of shared ownership; charitable foundations; state property, and so on), which would produce goods sustainably for a circular economy, particularly so that the 100 corporations which produce most emissions worldwide can be replaced with sustainable organisations, and sustainable products are valued economically; leading to immediate democratic control over the fossil fuel sector and its phase-out.
- At the same time, a green public fund (also created from fees paid for using the "commons") would help to develop a renewable energy system which would give people decentralised power over energy provision.
- In all sectors (building, transport, clothing, nutrition, and so on), sustainable models would replace current approaches (see e.g. the future visions

created by Scientists For Future, which sketch out climate action plans for all sectors). This goes particularly for agriculture, which UN reports have said must immediately move away from animal to plant "products" and look after the soil in the process; as well as for regenerative forestry policies, which must completely shift away from clearcutting and monocultures and no longer regard the burning of biomass as carbon neutral.
- Annual regulations can use increasing standards to reduce emissions in all areas (depending on a country's wealth; Anderson 2019). As outlined above, it would be necessary to use fair shares to guarantee poorer countries the possibility of developing their infrastructure to enable a dignified life for everyone.

Regarding international and transnational cooperation (see the chapter on a new global order), we could establish a form of convivialism which would conceptualise and facilitate security and peace in a new way. An "article zero" could be added to the UN Charter, describing us as an interdependent population living on an earth which we must all look after together, with all the consequences of that for stewardship of the "commons" (Dixson-Declève et al. 2022: "Earth4AAll") and for global citizenship rights. Security should be redefined in existing documents and realised through peaceful means, and the Security Council should be reshaped on a democratic basis. The General Assembly could be extended with a "second chamber", a Global People's Assembly which would be a grassroots democratic reflection of populations and would pay particular attention to children, non-humans and future generations, as well as incorporating scientific knowledge about the crises, and thus expanding formal democracy with a substantial crisis task force, which would protect people and not only states, and therefore oversee the implementation of the Paris Agreement. This structure of a "second chamber" (citizens' assemblies led by scientists who would pay attention to planetary limits and to the needs of all, as well as to responsibility for future generations, children, and non-humans) could also be adopted by local and national parliaments (see Pelluchon 2019).

Regarding the health and education sectors (see the chapter on education), the result of this is that in terms of curricula, teaching methods and research, we must emphasise and investigate what has been defined here as spaces of integrity, "being in contact", and relating sustainably to the world. This involves seeing humans as embodied social beings with imaginative abilities, who live in problematic power relations with each other and with nature. Those power relations can be replaced with democratic relationships: so that education itself becomes transformative and regenerative, centring care and the competences and knowledge (which may be non-academic or indigenous) that enable an understanding of the crises and the development of a sustainable society for all students and teachers. Educational institutions (and health institutions) can thus become places where substantial democracy is practised and lived; they can change their institutional logic of competition, becoming cooperative places which contribute to the societal transformation towards a global, sustainable convivialism (McGewon/Barry 2023; Raffoul 2023). One possibility to achieve this could be by establishing a "centre for sustainability" at every educational institution, which explores this new sustainable way of being in the world and makes it accessible for curricula, teaching methods, research, and institutional change.

This means that as we shape our values and culture, a relationship with nature emerges which defines freedom in a new way (see the chapter on democracy): no longer following John Locke and defining integrity as the guarantee for abstract freedom on the basis of property rights, which in turn means the power to dispose of property, even in relationships of domination and abuse. Instead, freedom can now be understood as the enabling of integrity, the collective repairing and opening up of democratic relationships on an equal footing, as a new, structurally relational concept of democracy, which is complementary to the concept of human rights, and is focused on the democratic abolition of domination in relationships (gender, ethnicity, class, etc.), and on the caring, humane contact which becomes possible through that.

Such measures would correspond to the Paris Agreement. But there are certainly also other scenarios which would match the criteria of the framework. The central step is the anchoring of the framework in legislation and the creation of democratic forums which can discuss how it should look in detail.

The movement

The populations of this living planet could stand together for these changes as a global democratic political movement which not only addresses all governments directly, but also represents the prototype for a global democracy. The demands of the three pillars could be applied to all levels, from the local to the global. They provide a standard against which we can measure the political rules of governments, parties, institutions, movements, and the relevant actors in society, and correspondingly allow us to protest against the status quo.

As a framework – better than a list of specific policies, or conversely, a list of general demands – they set out what can be done immediately (for instance, stopping the building and financing of all new fossil infrastructure) and at the same time offer a long-term compass.

However, implementing them requires quite ordinary people "like you and me" to begin to rebel, and not let themselves be misled by talk of "climate neutral net zero emissions 2050" – which is far too late for richer countries. The governments of states such as Switzerland, Sweden, and Germany are miles away from the transformation sketched out here. Let alone the regimes in China, Russia, and Saudi Arabia. If we continue as the states with their NDCs (national reduction plans) are proposing, the young people who are now striking will live in a world which is two or three degrees warmer when they are older. A world which is hell for billions of people and would remain so.

We need an uprising of the hesitant and the fearful. We need new collective political rules, and for that we need a united movement. Worldwide.

The methods

In order to make this new framework (three pillars, two principles) a reality, to take the crisis seriously and really change politics worldwide, a spectrum of methods can be used.

Three processes can be coordinated. One of them is the development of the "emergency brake" function which is exemplified so clearly and so well by FFF and XR. We cannot allow the fossil machinery of society simply to continue as it is. In order to bring about real change, school strikes are one excellent method; blocking the main squares in different countries is another. Civil disobedience is legitimate; it exists as a democratic tool (Chenoweth/Stephan 2012). Historically, the best effects have been achieved by massive, non-violent blockades and protests in the centres of power, lasting for days.

It is crucial for the movements themselves to practise and organise substantial democracy in an inclusive way. That also applies to the relationship between the generations: the welfare of children must have priority; we must pay attention to unequal power relations. (Meanwhile, small organisations run "top-down" with enormous financial means can damage and divide not only society, but also the movements. The aim can never be to delegate activism only to the professionals.)

The second task is to draw attention to what is needed now in practical terms. Climate strikers and Scientists For Future are working on "visions of the future" which can be implemented immediately and would enable emission-free, fair societies by the beginning of the 2030s; these are shaped by a systemic thinking which does not simply list one hundred proposals as a catalogue of demands.

However, not only researchers, but all people as experts in their areas of work (Workers, Architects, Teachers, Economists, Designers, Doctors, and Nurses – united as People For Future) can immediately set out to contribute concrete ideas and actions to this transformation, and to report on their work and share it with others as part of the movement.

So that this can happen, we need a continually united movement which everyone can join. That is the third strand: this includes spreading knowledge about this shared framework (with the three pillars and two principles), so that all people can work for it in their specific local and global contexts. One implementation could be to create centres in every school and university, in every village and town, where all dimensions of substantial democracy are explored, and which can also serve as meeting places for movements.

And this is perhaps what is now needed most: the awareness that this is a crisis and that we need crisis regulations, meaning a new framework and action plans. This could be the way for us to ensure together, globally, that the rainforest is protected, that oil fields are not built in the Arctic, that existing coal power stations stop running and are no longer financed, and that in the same process a humane, sustainable, just society develops, in which all people can lead a dignified life, because we cooperate to provide and share resources to meet basic needs and repair the damage of the past (regarding gender, ethnicity, etc.; racism, colonialism). That is why those who are most affected by the ecological and climate crises, often people in the Global South and BIPOC people, should have the leading roles in this movement, while the biggest changes must affect the richest in the Global North.

That is what we can stand up for now, peacefully, together.

Epilogue

Many of the activists, including the younger ones, the older ones, and the scientists, are people who might prefer to sit under a tree at the edge of a field, chewing a blade of grass, looking at the world and dreaming about getting caught up in an adventure.

Many of them would like nothing more than to return to that place, to those summer evenings in August many years ago, when the heat was not yet so brutal. One day, maybe, all the "earthlings" who protest will set off in summer together and sit under a tree. And look back at the days when they were caught up in an almost absurd story, together with all of us. But the suffering cannot be reversed, and nor can the destruction caused by the crises. The summers will not suddenly become colder. If a radical collective transformation does successfully take place, only the worst can be prevented.

By then, we might succeed in changing politics fundamentally and stopping the corporations in the world which are responsible, through the trade in fossil fuels, biomass, and the corresponding financial instruments, for most of the emissions. All of us can then look at the development of a renewable infrastructure which connects the world, at the almost complete shift in nutrition from slaughtering animals to plant-based food; to a different school and university which include everyone in exploring what it means to be in the world sustainably; to cooperation on providing unconditional basic services and resources to all people; and to a fairer world for all of us creatures on this planet, "world citizens", beyond oppression and domination in relation to gender, ethnicity, class etc.; as humans who have dared to look at the damage caused by colonial history and who have tried to react to it by enabling justice and a humane convivialism (see appendix).

All we need to do is pause what we are doing, take a deep breath, and get organised. We can all become part of the movement for a sustainable, global democratic society.

Many people would like the world to change; they would like security, and enough to eat for everyone. Many have had enough of the terrible fires in the rainforests, of the droughts and floods. And they want to contribute – even on a long-term basis – to stopping all of this.

Fridays For Future (and Extinction Rebellion) has introduced the idea of a global united grassroots movement, with the help of scientific research and universities. Now what is needed is to strengthen this idea. Not by trying to intervene and tell the young people what to do. Older activists can take on their own responsibilities, as teachers, carers, parents, grandparents, or simply as people.

For that, we need a different form of organisation than simply taking to the streets now and again: a real popular movement which not only fights for but also embodies a substantial, humane democracy, with corresponding explorative centres. In most countries, politics has not changed. Emissions are rising worldwide year by year – except for the drop due to the pandemic. And fossil infrastructure continues to be expanded in many places.

How do we stop this process and deepen democracy? How do we approach that? Not merely by coming up with party policies; those just end up disappearing in the mass of proposals. But at the same time, abstract demands are too unclear; hardly anything changes. Seeing the ecological, social and climate crises as interdependent crises means drawing attention to a new framework with action plans and standing up for it, in an uprising of all of us, until it has been realised (see the appendix for the three pillars and two principles). During the last five years, young people, movements and scientists have worked on it, building on decades of research and the long history of the struggle against the ecological and climate crisis (a struggle which has often taken place in the Global South and been led by BIPOC communities).

What does it mean to say that we will stand up for this until the framework is realised? How? By continually working together and seeing ourselves as one humanity, part of the same "fabric of integrity", the "people... for future"; a movement which everyone can join, without membership, without giving up the identity of other movements or organisations. The core: we stand up for the three basic ideas and two principles for global, sustainable democracy. Every day, together. Towards governments, institutions, universities. Until we have made them a reality. With Friday strikes and civil disobedience in the centres of power. And with education, knowledge transfer, and continuous organisation.

All those of us (who live in privileged circumstances with the required resources) can take on three roles. We can be the emergency brake which strikes, signalling non-cooperation with the fossil society with all its structures of domination, and literally refuses to allow the machinery of the "Great Acceleration" to go on running, creating pollution, burning down the forests, and insisting on oil, coal and gas, producing so much pain and suffering.

As Workers, Artists, Developers, Economists, Teachers, and Parents For Future, as experts in our areas, we can build sustainability and not only dream about it; we can make it a reality and report back on this, collecting knowledge about the dimensions of substantial democracy.

And finally, everyone can contribute to this: we can continually make the new framework clear which we now need as a crisis reaction (see appendix). We are here and can help each other, building up centres and spaces for education and support. We can draw attention to crucial scientific findings. We know what it means to meet everyone's needs, within planetary limits. And we will make sure that happens, as an intergenerational, intersectional and global movement. Everyone is needed, everyone is welcome.

Illustrations

Many thanks to Jana Eriksson, who took most of the photographs in this book. As well as being a photographer, she is an important part of the climate movement in Sweden and Stockholm.

Other photos were taken by David Fopp (p. 32, 34, 37, 71, 79, 90, 94, 148, 151, 159, 173, 203, 213, 214, 256, 267, 274, 303, 314, 356, 364, 400), Isabelle Axelsson and Loukina Tille (p. 198, 220), Tonny Nowshin (p. 224), Carl-Johan Utsi (p. 221), and William Persson (p. 150).

Bibliography

Abate, Randall & Kronk, Elizabeth Ann (2013): *Climate Change and Indigenous Peoples*. Cheltenham: Edward Elgar.

Abnett, Kate (17.8.2023): "How climate change drives heatwaves and wildfires in Europe", Reuters.

Abram, David (1997): *The Spell of the Sensuous: Perception and Language in a More-Than-Human World*. New York: Knopf Doubleday.

Adloff, Frank & Leggewie, Claus (2014): *Das konvivialistische Manifest*. Bielefeld: transcript.

Adorno, Theodor W. (1995): *Studien zum autoritären Charakter*. Frankfurt: Suhrkamp.

Ahlander, Johan & Johnson, Simon (12.8.2022): "Analysis: Gang crime looms over election in Sweden as shootings spread", Reuters.

Alexander, Frederick Matthias (2001): *The Use of the Self*. London: Orion publishing.

Ambrose, Jillian (4.5.2023): "Shell accused of 'profiteering bonanza' after record first-quarter profits of $9.6bn", The Guardian.

Anderson, Kevin & Broderick, John F. & Stoddard, Isak (28.5.2020): "A factor of two: how the mitigation plans of 'climate progressive' nations fall far short of Paris-compliant pathways", in: Climate Policy.

Anderson, Kevin (24.1.2019): "Climate's holy trinity", Oxford Lecture. https://www.youtube.com/watch?v=7BZFvc-ZOa8, 24. January 2019.

Anderson, Kevin & Jewell, Jessica (2019): "Debating the bedrock of climate-change mitigation scenarios", Nature, Vol 573, 19 Sept 2019, S. 348.

AP (4.5.2022): "What's the impact if Europe cuts off Russian oil?", Associated Press.

Bäckström, David (2022): *Fantasi*. Stockholm: Fri tanke.

Barasi, Leo (10.5.2019): "Polls reveal surge in concern in UK about climate change", https://www.carbonbrief.org.

Barrineau, Sanna et al. (2021): "What could sustainable academic cultures be?" CEFO publication series number 3, Uppsala.
Bell, Karen (2020): *Working-Class environmentalism*. London: Palgrave Mamillan.
Bellamy Foster, John (2010): *The Ecological Rift: Capitalisms War on the Earth*. New York: NYU Press.
Bidadanure, Juliana Uhuru (2021): *Justice Across Ages: Treating Young and Old as Equals*. Oxford: Oxford University Press.
Bidadanure, Juliana Uhuru (2019): "The Political Theory of Universal Basic Income", Annual Review of Political Science, Vol. 22, 481–501.
Bieri, Peter (2016): *Human dignity*. Cambridge: Polity Press.
Biermann, Frank et al. (2022): "Solar geoengineering: The case for an international non-use agreement", WIREs Clim Change, 13:e754.
Birnbacher, Dieter (2022): *Klimaethik*. Stuttgart: Reclam.
Birnbacher, Dieter (1997): *Ökophilosophie*. Stuttgart: Reclam.
Biro, Andrew (2011): *Critical Ecologies: The Frankfurt School and Contemporary Environmental Crises*. Toronto: University of Toronto Press.
Boffey, Daniel (25.6.2021): "Environmental claims of new EU farm subsidy policy are questioned", The Guardian.
Boffey, Daniel (26.5.2021): "Court orders Royal Dutch Shell to cut carbon emissions by 45 % by 2030", The Guardian.
Boffey, Daniel (8.4.2020): "Amsterdam to embrace 'doughnut' model to mend post-coronavirus economy", The Guardian.
Bourdieu, Pierre (2010): *Distinction*. Abingdon: Routledge.
Bowlby, John (2010): *Bindung als sichere Basis*. München: Reinhardt.
BR (16.1.2023): "So reagiert das Netz auf den Mönch aus Lützerath", Bayrischer Rundfunk.
Bradbrook, Gail (18.9.2018): "Heading for extinction and what to do about it". https://www.youtube.com/watch?v=b2VkC4SnwYo.
Braidotti, Rosi (2017): *Posthuman Knowledge*. Cambridge: Polity Press.
Brand Ulrich & Wissen, Markus (25.1.2023): "Lützerath erschüttert Gewissheiten", Frankfurter Rundschau.
Brandom, Robert (2022): „What is philosophy?", https://www.youtube.com/watch?v=LedSYJHE8Zo.
Braune, Andreas (2017): *Ziviler Ungehorsam*. Stuttgart: Reclam.
Broberg, Anders; Granqvist, Pehr; Ivarsson, Tord; Risholm Mothander, Pia (2006): *Anknytningsteori*. Stockholm: Natur&Kultur.

Brown, Wendy (2019): *In the ruins of neoliberalism: the rise of antidemocratic politics in the West.* New York: Columbia University Press.

Brundtland, Gro H. (1987): "Our Common Future: Report of the World Commission on Environment and Development." Geneva, UN-Dokument A/42/427.

Bruno, Linnea & Becevic, Zulmir (2020): *Barn och unga i utsatta livssituationer.* Stockholm: Liber.

Bukold, Steffen (2023): "The Dirty Dozen", https://greenpeace.at/uploads/202 3/08/report-the-dirty-dozen-climate-greenwashing-of-12-european-oil-companies.pdf.

Burgen, Stephen (12.11.2022): "Barcelona students to take mandatory climate crisis module from 2024", The Guardian.

Burkhart, Corinna & Schmelzer, Matthias & Treu, Nina (2020): *Degrowth in Movements(s).* Alresford: Zero books.

Butler, Judith (2006): *Gender Trouble: Feminism and the Subversion of Identity.* London: Routledge.

Calliess, Christian (2023): "Der Klimabeschluss des Bundesverfassungsgerichts (BVerfG): Aufwertung des Staatsziels des Art. 20a GG und Intertemporale Freiheitssicherung als neues Grundrecht auf Klimaschutz?" Berliner Online-Beiträge zum Europarecht, Nr. 146, 2023.

Carbin, Maria & Edenheim, Sara (2013): "The intersectional turn in feminist theory: A dream of a common language?" European Journal of Women's Studies, 20(3), S. 233–248.

Carrington, Damien (6.6.2023): "Too late now to save Arctic summer ice, climate scientists find", The Guardian.

Carrington, Damian (19.6.2019): "Himalayan glacier melting doubled since 2000", The Guardian.

Carrington, Damien (21.5.2018): "Humans just 0.01 % of all life but have destroyed 83 % of wild mammals – study", The Guardian.

Celikates, Robin (2016): "Democratizing Civil Disobedience", Philosophy & Social Criticism, 42 (2016), 10, 982–994.

Cervenka, Andreas (2022): *Girig-Sverige.* Stockholm: Natur&Kultur.

Chalmers, David (2010): *Character of Consciousness.* Oxford: Oxford University Press.

Chappell, Zsuzsanna (2012): *Deliberative Democracy: A Critical Introduction.* Houndmills, Basingstoke, Hampshire, New York: Palgrave Macmillan.

Chemnitz, Christine (2019): *The agricultural atlas*. Heinrich Böll Foundation, Berlin, Germany; Friends of the Earth Europe, Brussels, Belgium; BirdLife Europe & Central Asia, Brussels, Belgium.

Chenoweth, Erica (7.2.2012): "Why civil resistance works". https://www.youtube.com/watch?v=EHkzgDOMtYs.

Chenoweth, Erica & Stephan, Maria (2012): *Why Civil Resistance Works: The Strategic Logic of Nonviolent Conflict*. New York: Columbia University Press.

Chestney, Nina (26.10.2021): "U.N. warns world set for 2.7C rise on today's emissions pledges", Reuters.

Chrisafis, Angelique (10.1.2020): "Citizens' assembly ready to help Macron set French climate policies", The Guardian.

CIEL (2021): "Leading Climate Scientists, Climate Activists, Indigenous Peoples, and Youth to Speak on the Risks of Solar Geoengineering". https://www.ciel.org/news/leading-climate-scientists-climate-activists-indigenous-peoples-and-youth-to-speak-on-the-risks-of-solar-geoengineering.

Civil Society Equity Review Group (2018): http://civilsocietyreview.org/files/COP24_CSO_Equity_Review_Report.pdf.

Civillini, Matteo (16.2.2023): "World Bank chief to step down early after climate controversy", Climate Home News.

Club of Rome (2023): "Reform of the COP process – a manifesto for moving from negotiations to delivery". https://www.clubofrome.org/cop-reform.

Cohen, Louis et al. (2017): *Research Methods in Education*. London: Routledge.

Collins, Patricia Hill (2019): *Intersectionality as Critical Social Theory*. Durham: Duke University Press.

Copernicus (2022): "Extreme heat, widespread drought typify European climate in 2022", https://climate.copernicus.eu/extreme-heat-widespread-drought-typify-european-climate-2022.

Cormier, Zoe (27.8.2019): "Why the Arctic is smouldering", BBC.

CorporateEurope (12.10.2021): "Leak: industrial farm lobbies' coordinated attack on Farm to Fork targets".

Costanza, Robert & Graumlich, Lisa & Steffens, Will (2009): *Sustainability or Collapse?* Cambridge: MIT Press.

Critchley, Simon (2001): *Continental philosophy*. Oxford: Oxford University Press.

Dahl, Robert A. (2015): *On democracy*. New Haven: Yale University Press.

Davis, Angela (1983): *Woman, Race & Class*. New York: Vintage books.

D'eaubonne, Francoise (2022): *Feminism or Death: How the Women's Movement Can Save the Planet*. London: Verso books.

De Haan, Gerhard (2010): "The development of ESD-related competencies in supportive institutional frameworks", International Review of Education 56(2), 315–328.

Dixson-Declève, Sandrine; Gaffney, Owen; Ghosh, Jayati; Randers, Jorgen; Rockström, Johan; Stoknes, Per Espen (2022): *Earth for All: A Survival Guide for Humanity*. Gabriola: New Society Publishers.

DN (14.10.22): "Här är det som sticker ut mest i Tidöavtalet", Dagens Nyheter.

Dummett, Michael (1996): *Origins of Analytical Philosophy*. Cambridge: Harvard University Press.

Eddy, Melissa (29.4.2021): "German High Court Hands Youth a Victory in Climate Change Fight", New York Times.

Ekardt, Felix (2020): *Sustainability*. Berlin: Springer.

Ekström, Mats & Patrona, Marianna & Thornborrow, Joanna (2020): "The normalization of the populist radical right in news interviews: a study of journalistic reporting on the Swedish democrats", Social Semiotics, Volume 30, Issue 4, 466–484.

Emission Gap Report (2019): https://www.unenvironment.org/resources/emissions-gap-report-2019.

Emmett, Robert & Nye, David (2017): *The Environmental Humanities*. Cambridge: MIT Press.

Engler, Mark & Engler, Paul (2017): *This is an uprising*. Avalon Publishing Group. New York: Avalon Publishing.

EnvironmentAnalyst (2.11.2022): "Sweden axes environment ministry", https://environment-analyst.com/global/108743/sweden-axes-environment-ministry.

Eumetsat (1.6.2019): "Maximum temperature records were broken in many parts of Europe in June and July 2019, due to a series of heatwaves".

EUSI (31.8.2022): "Rivers and Lakes are Drying as Europe Faces Worst Drought in 500 Years", https://www.euspaceimaging.com/rivers-and-lakes-are-drying-as-europe-faces-worst-drought-in-500-years.

Evroux, Clément & Spinaci, Stefano & Widuto, Agnieszka (2023): "From growth to 'beyond growth': Concepts and challenges". EPRS, European Parliamentary Research Service. PE 747.107, May 2023.

Extinction Rebellion (2023): "Self-Organizing Systems". https://extinctionrebellion.uk/act-now/resources/sos.

Extinction Rebellion (2019): *This is not a drill*. London: Penguin.

Felber, Christian (2018): *Gemeinwohl-Ökonomie*. München: Piper.
Foer, Jonathan Safran (2019): *Wir sind das Klima!* Köln: Kiepenheuer&Witsch.
Fopp, David (2022/1): "Climate, health, activism – the societal role of the university/researchers in times of an ecological and climate crisis". Stockholm: DiVA.org:su-201175.
Fopp, David (2022/2): "Greta Thunberg at Stockholm University, two conversations. Conversation: The crisis, democracy, and politics", https://vimeo.com/manage/videos/679533736.
Fopp, David & Axelsson, Isabelle & Tille, Loukina (2021): *Gemeinsam für die Zukunft. Fridays For Future und Scientists For Future*. Bielefeld: transcript.
Fopp, David (22.9.21): "How to meet (and get out of) the crises by focusing on intersectionality", https://medium.com/@davidfopp/how-to-meet-and-get-out-of-the-crises-by-understanding-intersectionality-16-steps-b588952b23eb.
Fopp, David (2020/1): *Changing the social imaginary*. Stockholm: Edition TP1.
Fopp, David (10.1.2020/2): "The Core of a United Global Movement Reacting to the Ecological and Climate Crisis", in: Resilience.
Fopp, David (2020/3): "Ledarskap". https://video.su.se/media/Ledarskap%2C+David+Fopp+%28med+text%29/0_kdkvwgqu.
Fopp, David (2019): "Ästhetik im Zeitalter der Globalisierung", Zeitschrift für Ästhetik und Kunstwissenschaft 2019/2, 146–158.
Fopp, David (19.1.2019): "Så visar historien att tonåringar kan förändra världen", Dagens Nyheter.
Fopp, David (2016): *Menschlichkeit als ästhetische, pädagogische und politische Idee*. Bielefeld: transcipt.
Fopp, David (2015): „Menschliche Energie", Internationale Zeitschrift für Philosophie und Psychosomatik, IZPP 2015/2, p. 1–15.
Fraser, Nancy (2022): *Cannibal capitalism*. New York: Verso books.
Fridays For Future Sverige (2022): "Fridays For Future Sveriges Upprop". https://fridaysforfuture.se/wp-content/uploads/2022/05/Fridays-For-Future-Sveriges-upprop.pdf.
Gardiner, Stephen M.; Caney, Simon; Jamieson, Dale; Shue, Henry (2010): *Climate Ethics: Essential Readings*. Oxford: Oxford University Press.
Garric, Audrey (9.11.2022): "At COP27, the battle for a fossil fuel non-proliferation treaty is underway", LeMonde.
Garza, Alicia (2020): *The purpose of Power. How to build movements for the 21 century*. New York: Doubleday.
Gates, Bill (2021): *How to avoid a climate disaster*. London: Penguin.

George, Susan (2010): *Whose crisis, whose future?* Cambridge: Polity Press.
Ghosh, Amitav (19.10.2022): "The Colonial Roots of Present Crises", Green European Journal.
Glock, Hans-Johann (2008): *What is Analytical Philosophy?* Cambridge: Cambridge University Press.
Giddens, Anthony & Sutton, Philip (2021): *Sociology*. Cambridge: Polity Press.
Göbel, Alexander (1.12.2019): "Klimaklage in den Niederlande. Ein historisches Urteil", Deutschlandfunk.
Goering, Laurie (31.3.2021): "Sweden rejects pioneering test of solar geoengineering tech", Reuters.
Gomez, Antionette M. & Shafiei, Fatemeh & Johnson, Glenn S. (2011); "Black Women's Involvement in the Environmental Justice Movement: An Analysis of Three Communities in Atlanta, Georgia", Race, Gender & Class, 18(1/2), 189–214.
Gonzalez, Jenipher Camino (5.5.2020): "Greta Thunberg asks UN to back lawsuit against Germany and others", DW.
Göpel, Maja (2022): *Wir können auch anders*. Berlin: Ullstein.
Göpel, Maja (2020): *Unsere Welt neu denken*. Berlin: Ullstein.
Göpel, Maja (2016): *The great mindshift*. Berlin: Springer.
Gore, Timothy (2015): *Extreme carbon inequality*. Oxford: Oxfam International.
Götze, Susanne & Joeres, Annika (2020): *Die Klimaschmutzlobby*. München: Piper.
Gough, Ian (2017): *Heat, Greed and Human Need*. Northampton: Edward Elgar publishing.
Graeber, David (2014): *The Democracy Project*. London: Penguin.
Grattan, Steven (7.4.2023): "Deforestation in Brazil's Amazon rises in March", Reuters.
Graza, Alicia (2020): *The purpose of Power. How to build movements for the 21 century*. New York: Doubleday.
Gredebäck, Gustaf et al. (2015): "The neuropsychology of infants' pro-social preferences", Developmental Cognitive Neuroscience, 12, 106–113.
Greenfield, Patrick (9.3.2023): "First Cop15, now the high seas treaty: there is hope for the planet's future", The Guardian.
Greenwell, Marianne (2019): *Fridays for Future and Children's Rights*. Frankfurt: Debus Pädagogik Verlag.
Grunwald, Armin & Kopfmüller, Jürgen (2022): *Nachhaltigkeit*. Frankfurt am Main: Campus.

Gunnarsson, Karin Dinker & Pedersen, Helena (2016): "Critical Animal Pedagogies: Re-learning Our Relations with Animal Others". In: Helen E. Lees; Nel Noddings (eds.).: *The Palgrave International Handbook of Alternative Education*. London: Palgrave Macmillan, 415–430.

Gustavsson, Bernt (2017): *Bildningens dynamik*. Göteborg: Korpen.

Habermas, Jürgen (1987): *Knowledge and Human Interests*. Boston: Polity Press.

Hagedorn, Gregor et al. (2019): "Concerns of young protesters are justified", Science, 12. April 2019, Vol. 364, Issue 6436, 139–140.

Hallam, Roger (2019): *Common Sense for the 21st Century*. White River Junction: Chelsea Green Publishing.

Hallam, Roger (1.5.2019): "Now we know: conventional campaigning won't prevent our extinction", The Guardian.

Hällström, Niclas (2021): "Economic Diversification and Just Transition to 100 % Renewable Energy". https://static1.squarespace.com/static/5dd3c c5b7fd99372fbb04561/t/6183c26f34f097407f20271c/1636024949787/FFNPT +Pillar+III+workshops+report+%281%29.pdf.

Hardt, Judith Nora; Harrington, Cameron; von Lucke, Franziskus; Estève, Adrien; Simpson, Nicholas (2023): "Climate Security in the Anthropocene: Exploring the Approaches of United Nations Security Council Member-States", Cham: Springer.

Hardt, Judith Nora & Viehoff, Alina (2020): "A Climate for Change in the UN Security Council?" IFSH Research Report 7/2020.

Hardt, Judith Nora (2018): *Environmental Security in the Anthropocene. Assessing Theory and Practice*. London: Routledge.

Hart, Roger (1992): "Children's participation. From tokenism to citi-zenship". https://www.unicef-irc.org/publications/pdf/childrens_participati on.pdf, March 1992.

Harvey, Fiona (20.3.2022): "Heatwaves at both of Earth's poles alarm climate scientists", The Guardian.

Haseman, Brad & O'Toole, John (2017): *Dramawise reimagined*. Sidney: Currency Press.

Hattenstone, Simon (25.9.2021): "The transformation of Greta Thunberg", The Guardian.

Hecking, Claus & Schönberger, Charlotte & Sokolowski, Ilsa (2019): *Unsere Zukunft ist jetzt!* Hamburg: Oetinger.

Hegel, Georg Wilhelm Friedrich (1986): *Enzyklopädie der philosophischen Wissenschaften*. Frankfurt: Suhrkamp.

Herrera, Andrea et al. (20.9.2021): "Sverige måste göra mer – vi kan inte vänta 25 år på att lösa klimatkrisen", ETC.

Hickel, Jason; O'Neill, Daniel W.; Fanning, Andrew L.; Zoomkawala, Huzaifa (2022): "National responsibility for ecological breakdown: a fair-shares assessment of resource use, 1970–2017", The Lancet, Planetary Health. Volume 6, ISSUE 4, e342-e349, April 2022.

Hickel, Jason (2020): *Less is more*. London: Random house.

Hickel, Jason (2018): *The Divide. A Brief Guide to Global Inequality and its Solutions*. London: Windmill books.

Hickel, Jason & Kallis, Girogos (2019): "Is green growth possible?", New Political Economy, Vol. 25, Issue 4, 469–486.

Hillekamp, Sven (2023): "Strategien der Klimaschutzbewegung". https://www.youtube.com/watch?v=lGCxl6Mgs6Q.

Hodgson, Camilla (10.10.2022): "Challenge against EU 'green' label for gas and nuclear energy steps up", Financial Times.

Holthaus, Eric (2020): *The future earth*. San Francisco: HarperOne.

Holzinger, Markus (2017): "Die Theorie funktionaler Differenzierung als integratives Programm einer Soziologie der Moderne?", Zeitschrift für Theoretische Soziologie. Band 6, Nr. 1, 2017, 44–73.

Honneth, Axel & Fraser, Nancy (2004): *Redistribution or Recognition*. London: Verso books.

Honneth, Axel & Rancière, Jacques (2021): *Anerkennung oder Unvernehmen?* Berlin: Suhrkamp Verlag.

Hooft van, Stan (2009): "Global justice: a cosmopolitan account", Ethics & Global Politics, December 2009.

hooks, bell (2000): *Feminist theory*. London: Pluto Press.

Horkheimer, Max (2013): *Critique of instrumental reason*. London: Verso.

Hornborg, Alf (2017): *Global magic*. London: Palgrave Macmillan.

HU (2023): Klimaschutzkonzept. https://humboldts17.de/media/pages/nachhaltigkeit-an-der-humboldt-universitaet/akteure/klimaschutzmanagement/f549d375cc-1677581010/klimaschutzkonzept_hu_2023.pdf.

Immordino-Yang, Mary Helen (2015): *Emotions, learning, and the Brain*. New York: W.W. Norton.

IPCC (2023): https://twitter.com/IPCC_CH/status/1666750514699071488.

IPCC (2022): "Climate change: a threat to human wellbeing and health of the planet", https://www.ipcc.ch/2022/02/28/pr-wgii-ar6.

IPCCSR15 / Spezial-Rapport (2018): IPCC (Intergovernmental Panel on Climate Change) 2018. Global warming of 1.5 °C. Special Report. IPCC with World

Meteorological Organisation (WMO), and United Nations Environmental Program (UNEP): Geneva, Switzerland, https://www.ipcc.ch/report/sr15.

IPCC/Land-Rapport (2019): "Land is a critical resource", https://www.ipcc.ch/2019/08/08/land-is-a-critical-resource_srccl/

Jacobs, Lawrence & Shapiro, Robert (1994): "Studying Substantive Democracy", PS: Political Science & Politics, 27(1), 9–17.

Jacobson, Mark Z. et al. (2019): "Impacts of Green New Deal Energy Plans on Grid Stability, Costs, Jobs, Health, and Climate in 143 Countries", One Earth 1, 449–463.

Jafry, Tahseen (2020): *Routledge Handbook of Climate Justice*. London: Routledge.

Jakubowska, Joanna (5.7.2021): "The New Climate and Energy Package. Is Europe Fit for 55?", euractiv.pl.

Johnstone, Keith (1987): *Impro*. London: Routledge.

Kalmus, Peter (6.4.2022): "Climate scientists are desperate: we're crying, begging and getting arrested", The Guardian.

Kelly, Marjoire; Howard, Ted (2019): *The Making of a Democratic Economy*. San Francisco: Berrett-Koehler Publishers.

Kemfert, Claudia (2020): *Mondays For Future*. Hamburg: Murmann publishers.

Ketterer, Hanna & Becker, Karina (2019): *Was stimmt nicht mit der Demokratie?* Berlin: Suhrkamp.

Kimmerer, Robin Wall (2022): "Mending Our Relationship with the Earth", in: Thunberg (2022), 415–421.

Kimmerer, Robin Wall (2015): *Braiding Sweetgrass: Indigenous Wisdom, Scientific Knowledge and the Teachings of Plants*. Minneapolis: Milkweed.

King, Michaela et al. (2020): "Dynamic ice loss from the Greenland Ice Sheet driven by sustained glacier retreat." Nature, Commun Earth Environ 1, 1.

Klimatpolitiska rådet (21.3.2019): "Kritisk rapport fick positivt mottagande". https://www.klimatpolitiskaradet.se/pressrummet/den-21-mars-klimatpolitiska-radets-rapport-2019.

Klein, Naomi (2019): *On fire*. London: Penguin.

Kottasová, Ivana; Gupta, Swati; Regan, Helen (9. 2020): "The one chance we have", CNN.

Kotzé, Louis & Kim, Rakhyun (2022): "Towards planetary nexus governance in the Anthropocene: An earth system law perspective", Global Policy, 13(S3), 86–97.

Kountouris, Yiannis & Williams, Eleri (2023): "Do protests influence environmental attitudes? Evidence from Extinction Rebellion", Environmental Research Communications, Volume 5, Number 1.

Kraftl, Peter (2020): *After childhood*. London: Routledge.
Latour, Bruno (2018): *Down to Earth*. Cambridge: Polity Press.
Latour, Bruno (2017): *Facing Gaia: Eight Lectures on the New Climatic Regime*. Cambridge: Polity Press.
Leavy, Patricia (2009): *Method meets art*. New York: Guilford.
Lenton, Timothy et al. (2023): "Quantifying the human cost of global warming", Nat Sustain (2023).
Lenton, Thimothy et al. (2019): "Climate tipping points — too risky to bet against", Nature, 27 November 2019
Levinas, Emmanuel (1969): *Totality and Infinity: An Essay on Exteriority*. Pittsburgh: Duquesne University Press.
Lind, Gustaf & Sandahl, Johanna (27.2.2023): "EU's Nature Restoration Law: make or break for Swedish forests?", Euractive.
Lindgren, Astrid (2018): *Ronja the Robber's Daughter*. Oxford: Oxford University Press.
Linnér, Björn-Ola & Wibeck, Victoria (2019): *Sustainability Transformations: Agents and Drivers across Societies*. Cambridge: Cambridge University Press.
Loick, Daniel & Thompson, Vanessa (2022): *Abolitionismus*. Berlin: Suhrkamp.
Lohmann, Georg (1991): *Indifferenz und Gesellschaft*. Frankfurt: Suhrkamp.
Lynas, Mark (2020): *Our final warning*. London: Fourth Estate.
MacDougall, Andrew; Swart, Neil; Knutti, Reto (2017): "The Uncertainty in the Transient Climate Response to Cumulative CO2 Emissions Arising from the Uncertainty in Physical Climate Parameters", J. Climate, 30, S. 813–827.
Machin Amanda (2023): "Democracy, Agony, and Rupture: A Critique of Climate Citizens' Assemblies", Politische Vierteljahresschrift 2023, Mar 7:1–20.
Mackintosh, Eliza (23.8.2019): "The Amazon is burning because the world eats so much meat", CNN.
Macy, Joanna (2014): *Coming back to life*. Gabriola: New Society.
Maier, Lucas (18.1.2023): "Sanitäter über Einsatz in Lützerath: Polizei ist sehr brutal vorgegangen", Frankfurter Rundschau.
Malm, Andreas (2021): *How to blow up a pipeline*. London: Verso books.
Malm, Andreas (2020): *Corona, Climate, Chronic Emergency*. London: Verso books.
Malm, Andreas (2017): *The progress of this storm*. London: Verso books.
Malm, Andreas (2016): *Fossil capital*. London: Verso books.
Mao, Frances (30.11.2018): "Climate change: Australian students skip school for mass protest", BBC.

Marchart, Oliver (2018): *Post-Foundational Theories of Democracy*. Edinburgh: Edinburgh University Press.
Margolin, Jamie (2020): *Youth to power*. Boston: DaCapo Lifelong.
Marusczyk, Ivo (13.11.2017): "Peruanischer Bauer verklagt RWE", Deutschlandfunk.
Mathiesen, Karl (16.4.2019): "Leading climate lawyer arrested after gluing herself to Shell headquarters", ClimateHomeNews.
Maxton, Graeme (2018): *Change!* Grünwald: Komplett-Media.
McAvoy, Mary & O'Connor, Peter (2022): *The Routledge Companion to Drama in Education*. London: Routledge.
McGeown, Calum & Barry, John (2023): "Agents of (un)sustainability: democratising universities for the planetary crisis", Frontiers in Sustainability, Volume 4.
McGrath, Matt (29.9.2022): "Over 1,700 environment activists killed in decade", BBC.
Meisner, Sanford (1987): *On Acting*. New York: Random House.
Menke, Christoph (2023): "The Historicity of Spirit, in a Materialistic Understanding", Deutsche Vierteljahrsschrift für Literaturwissenschaft und Geistesgeschichte 97, 175–182 (2023).
Menke, Christoph & Pollmann, Arnd (2017): *Philosophie der Menschenrechte zur Einführung*. Hamburg: Junius.
Merleau-Ponty, Maurice (1974): *Phenomenology of Perception*. London/New York: Routledge.
Meyer, Katrin (2017): *Theorien der Intersektionalität zur Einführung*. Hamburg: Junius.
Millward-Hopkins, Joel; Steinberger, Julia K.; Rao, Narasimha D.; Oswald, Yannick (2020): "Providing decent living with minimum energy: A global scenario", Global Environmental Change, Volume 65, 2020, 102168.
Milman, Oliver (18.10.2022): "'Buckle up': US backers of Just Stop Oil vow more Van Gogh-style protest", The Guardian.
Milman, Oliver (9.3.2022): "'This is a fossil fuel war': Ukraine's top climate scientist speaks out", The Guardian.
Milman, Oliver (12.2.2020): "Last decade was Earth's hottest on record as climate crisis accelerates", The Guardian.
Mirowski, Philip & Plehwe, Dieter (2015): *The road from Mont Pelerin*. Cambridge: Harvard University Press
Moberg et al. (7.12.2022): "De unga gör helt rätt när de stämmer staten", Aftonbladet.

Moellendorf, Darrel (2014): *The Moral Challenge of Dangerous Climate Change: Values, Poverty, and Policy*. Cambridge: Cambridge University Press.
Monbiot, George (7.5.2019): "Grünes Wachstum ist eine Illusion", Freitag.
Monbiot, George (2014): *Feral. Rewilding the Land, Sea and Human Life*. London: Penguin books.
Mowle, Thomas (2003): "Worldviews in Foreign Policy: Realism, Liberalism, and External Conflict", Political psychology 2003, Vol. 4, Issue 3, 561–592.
Müller, Tadzio (2020): "Climate justice", in: Burkhart et al. 2020, 114–128.
Nagel, Thomas (1989): *The View from Nowhere*. Oxford: Oxford University Press.
Nakabuye, Hilda (2019): "COP25 speech", https://www.youtube.com/watch?v=wgpYF9iVotg.
Nakamura, Kate (5.11.2021): "Malala Yousafzai, Vanessa Nakate & Greta Thunberg Speak Out at COP26", GlobalCitizen.
NEF (2010): "The Great Transition: Social justice and the core economy". Nef working paper 1, New Economics Foundation, London.
Neubauer, Luisa & Repenning, Alexander (2019): *Vom Ende der Klimakrise*. Stuttgart: Tropen.
Neumayer, Eric (2010): *Weak versus Strong Sustainability*. Cheltenham: Edward Elgar Publisher.
Newell, Peter & Simms, Andrew (2019): "Towards a fossil fuel non-proliferationtreaty", Climate Policy, Volume 20, 2020 – Issue 8, 1043–1054.
NewYorkTimes (9.7. 2020): "Landmark Supreme Court Ruling Affirms Native American Rights in Oklahoma".
NHS (2021): Alexander technique. https://www.nhs.uk/conditions/alexander-technique.
Nicholson, Kate (2.11.2021): "Greta Thunberg Caught Chanting 'You Can Shove Your Climate Crisis Up Your Arse' Outside COP26", Huffington Post.
Noë, Ava (2010): *Out of Our Heads*. New York: Hill & Wang.
Nowshin, Tonny (18.6.2020): "Grüner Rassismus", TAZ.
Nussbaum, Martha (1992): *Love's knowledge*. Oxford: Oxford University Press.
Oei, Pao-Yu et al. (11.12023): "Offener Brief: Ein Moratorium für die Räumung von Lützerath", https://www.scientistsforfuture.org.
Ogette, Tupoka (2018): *exit RACISM*. Münster: Unrast Verlag.
Oreskes, Naomi & Conway, Erik (2012): *Merchants of doubt*. Bloomsbury, London.
Osten, Suzanne (2009): *Babydrama: en konstnärlig forskningsrapport*. Stockholm: Dramatiska Institutet.

Osten, Suzanne (2000): "Tabu och barnteater". In: Barn-Teater-Drama. Stockholm: CBK.
Oxfam (16.1.2023): "Richest 1 % bag nearly twice as much wealth as the rest of the world put together over the past two years", https://www.oxfam.org/en/press-releases/richest-1-bag-nearly-twice-much-wealth-rest-world-put-together-over-past-two-years.
Oxfam (2.12.2015): "World's richest 10 % produce half of carbon emissions while poorest 3.5 billion account for just a tenth", https://www.oxfam.org/en/press-releases/worlds-richest-10-produce-half-carbon-emissions-while-poorest-35-billion-account.
Panksepp, Jaak (2004): *Affective neuroscience*. Oxford: Oxford University Press.
Pansardi, Pamela (2016): "Democracy, domination and the distribution of power: Substantive Political Equality as a Procedural Requirement", Revue internationale de philosophie, 275, 91–108.
Parfit, Derek (1986): *Reasons and Persons*. Oxford: Oxford University Press.
Parlament (2022): "Wie viel russische Rohstoffe werden wirklich über die Schweiz gehandelt?" 22.3483 Interpellation. https://www.parlament.ch/de/ratsbetrieb/suche-curia-vista/geschaeft?AffairId=20223483
Patel, Leigh (2015): *Decolonizing Educational Research*. New York: Routledge.
Pe'er, Guy et al. (2020): "Action needed for the EU Common Agricultural Policy to address sustainability challenges", People and Nature, 305–316.
Pelluchon, Corine (2019): *Nourishment: A Philosophy of the Political Body*. London: Bloomsbury Academic.
Perry, Keisha-Khan Y. (2016): "Geographies of Power: Black Women Mobilizing Intersectionality in Brazil", Meridians 1 June 2016; 14 (1): 94–120.
Pettifor, Ann (2019): *The Case for the Green New Deal*. London: Verso books.
Pettit, Philip (2015): *Just Freedom*. New York & London: W.W. Norton.
Pfaff, Isabel (25.2.2022): "Die Schweiz bleibt bei Sanktionen gegen Russland vage", Sueddeutsche Zeitung.
Phillips, Tom & Maisonnave, Fabiano (30.4.2020): "'Utter disaster': Manaus fills mass graves as Covid-19 hits the Amazon", The Guardian.
Pickett, Kate & Wilkinson, Richard (2010): *The spirit level*. London: Penguin.
Piketty, Thomas (2022): *A Brief History of Equality*. Cambridge: The Belknap Press.
Piketty, Thomas (2018): *Das Kapital im 21. Jahrhundert*. München: Beck.
Prentki, Tim & Abraham, Nicola (2021): *The Applied Theatre Reader*. London: Routledge.

Pullmann, Philip (2001): *His dark materials: Der goldene Kompass.* Carlsen, Hamburg.
Raffoul, Alexandre (2023): "Listen to the science! Which science? Regenerative research for times of planetary crises", Frontiers in Sustainability, Volume 4.
Raffoul, Alexandre et al. (8.9.2022): "Vuxna, förena er bakom skolstrejkande ungdomar", ETC.
Raffoul, Alexandre; Fopp, David; Elfversson, Emma; Avery, Helen (5.11.2021): "The climate crisis gives science a new role", The Conversation.
Rahmstorf, Stefan & Schnellhuber, Hans Joachim (2019): *Der Klimawandel.* München: Beck.
Rahmstorf, Stefan (18.9.2019): "Es sieht zwar nicht so aus, aber wir können die Klimakrise noch abwenden", Spiegel Wissenschaft.
Ranciére, Jacques (2002): *Das Unvernehmen.* Berlin: Suhrkamp Verlag.
Rankin, Jennifer (2.2.2022): "EU includes gas and nuclear in guidebook for <green> investments", The Guardian.
Raworth, Kate (2018): *Doughnut economics.* London: Random House.
Read, Rupert (2018): "This civilization is finished: So what is to be done?", Institute for Leadership and Sustainability.
Readfearn, Graham (14.5.2023): "Oceans have been absorbing the world's extra heat", The Guardian.
Reuters (22.9.2016): "REFILE-EU clears Vattenfall's sale of German lignite assets", Reuters.
Rich, Nathaniel (2019): *Losing earth.* New York: MacMillan.
Riley, Tess (10.7.2017): "Just 100 companies responsible for 71 % of global emissions, study says", The Guardian.
Ringberg, Amie et al. (19.11.2022): "Kyrkan borde ha råd att skydda sin skog", Aftonbladet.
River-Roberts, Emma (2023): Tweet, 17 May 2023.
Robertson, Brian (2016): *The Revolutionary Management System that Abolishes Hierarchy.* London: Penguin.
Robertson, Joe & Murphy, Joe (2017): *The jungle.* London: Faber&Faber.
Rockström, Johan et al. (2023): "Safe and just Earth system boundaries", Nature 619, 102–111.
Rockström, Johan (2020): "Vinter i P1", https://sverigesradio.se/avsnitt/1425542.
Rockström, Johan & Klum, Mattias (2012): *The human quest.* Stockholm: Langenskiöld.

Rockström, Johan & Steffen, W. & Noone, K. et al. (2009): "A safe operating space for humanity", Nature 461, 472–475.

Rodriguez, Cheryl (1998). "Activist Stories: Culture and Continuity in Black Women's Narratives of Grassroots Community Work", Frontiers: A Journal of Women Studies, 19(2), 94–112.

Rosa, Hartmut (2019): *Resonance*. Cambridge: Polity Press.

Rosenberg, Marshall B. (2015): *Nonviolent Communication: A Language of Life*. Encinitas: Puddle Dancer Press.

Röstlund, Lisa (2022): *Skogslandet*. Forum: Stockholm.

Ryser, Daniel (23.11.2019): "Herr Knutti, sind wir noch zu retten?", Republik.

Saad, Layla (2020): *Me and white supremacy*. London: Quercus.

Samenow, Jason & Patel, Kasha (18.3.2022): "It's 70 degrees warmer than normal in eastern Antarctica. Scientists are flabbergasted", Washington Post.

Samuelsson, Marcus (2017): *Lärandets ordning och reda: ledarskap i klassrummet*. Stockholm: Natur & Kultur.

Sen, Amartya (2001): *Development as Freedom*. Oxford: Oxford University Press.

Scharmer, Otto (2009): *Theory U: Leading from the Future as It Emerges*. San Francisco: Berrett-Koehler Publishers.

Schellnhuber, Hans Joachim (2017): https://www.fu-berlin.de/campusleben/campus/2017/171101-einstein-lecture/index.html.

Schellnhuber, Hans Joachim (2015): *Selbstverbrennung*. München: C.Bertelsmann.

Schmelzer, Matthias & Vansintjan, Aaron & Vetter, Andrea (2022): *The Future is Degrowth*. London: Verso Books.

Schmelzer, Matthias & Vetter, Andrea (2019): *Degrowth/Postwachstum zur Einführung*. Hamburg: Junius.

Schmid, Hans Bernhard (2013): "Shared Intentionality and the Origins of Human Communication", in: Salice, Alessandro (ed.), *Intentionality*. München: Philosophia.

Schmidt, Vivien A. (2013). "Democracy and Legitimacy in the European Union Revisited: Input, Output and Throughput", Political Studies, 61(1), 2–22.

Schüpf, Dennis (21.2.2023): "Lützerath bleibt!", Resilience.

Sengupta, Somini (6.8.2020): "Here is what extreme heat looks like", New York Times.

Sharp, Gene (2014): *Von der Diktatur zur Demokratie*. München: Beck.

Sharp, Gene (1973): *The Politics of Nonviolent Action*. Boston: Porter Sargent.

Shiva, Vandana (2020): *Reclaiming the Commons: Biodiversity, Traditional Knowledge, and the Rights of Mother Earth*. Santa Fe: Synergetic Press.

Simon, Frédéric (15.7.2020): "EU states spent more on fossil fuels than clean energy during the crisis, new data shows", euractiv.com.
Skelton, Alasdair et al. (2020): "10 myths about net zero targets and carbon offsetting, busted", https://www.climatechangenews.com/2020/12/11/10-myths-net-zero-targets-carbon-offsetting-busted.
Solnit, Rebecca & Young Lutunatabua, Thelma (2023): *Not too late*. Chicago: Haymarket Books.
Sommer, Christian et al. (2020): "Rapid glacier retreat and downwasting throughout the European Alps in the early 21st century", Nature Communications 11, 3209.
Sommer, Dion (2012): *A childhood psychology*. London: Bloomsbury Academic.
Sommer, Dion et al. (2010): *Child Perspectives and Children's Perspectives in Theory and Practice*. Berlin: Springer.
Spade, Dean (2021): *Mutual aid*. London: Verso Books.
Spring, Jake & Kelly, Bruno (9.7.2022): "Deforestation in Brazil's Amazon hits record for first half of 2022", Reuters.
Springer, Kimberley (2005): *Living for the revolution: Black feminist organizations, 1968–1980*. Durham: Duke University Press.
Steffen, Will (2019): "All you need to know about climate change". https://klimatstudenterna.se/klimatstudentpodden/?fbclid=IwAR38oaD8vJ1D9pzE7M8lWEBwgMaYm1mZdV2TZptmO0oq_RTJ4oHGY-YURhI
Steffen, Will et al. (2004): *Global change and the earth system*. Berlin/Heidelberg: Springer.
Sterling, Stephen & Huckle, John (1996): *Education for sustainability*. New York: Routdledge.
Stern, Daniel (1985): *The interpersonal world of the infant*. London: Karnac.
Stiegler, Bernard (2020): *Qu'appelle-t-on panser? La leçon de Greta Thunberg*. Paris: Les liens qui libèrent.
Stillwell, Matthew (2019): https://www.systems-change.net.
Stoddard, Isak et al. (2021): "Three Decades of Climate Mitigation: Why Haven't We Bent the Global Emissions Curve?", Annual Review of Environment and Resources, 46(1).
Strittmatter, Kai (26.4.2019): "Die Reifeprüfung", Sueddeutsche Zeitung.
SU (2020): https://www.su.se/forskning/profilomr%C3%A5den/klimat-hav-o ch-milj%C3%B6/varmaste-vintern-i-stockholm-p%C3%A5-mer-%C3%A4n-250-%C3%A5r-1.487862.
Sverigesradio (2022): https://sverigesradio.se/avsnitt/sa-kan-klimatarbetet-forandras-nar-politiken-kort-fast.

SVT (19.10.22): "SD-ledamot: Klimatkris saknar stöd i vetenskap", https://www.svt.se/nyheter/inrikes/sd-ledamot-klimatkris-saknar-stod-i-vetenskap.

SVT (12.9.2022): "SD-profilen i natt: Helg seger", https://www.svt.se/nyheter/inrikes/sd-profilen-i-natt-hel-g-seger.

Swenson, Kyle (23.4.2019): "More than a thousand arrested in 'Extinction Rebellion' protests against climate change", Washington Post.

SZ (28.12.2022): "So unterschiedlich reagieren Bayerns Unis auf Klima-Besetzungen", Sueddeutsche Zeitung.

Taylor, Charles (2018): The Ethics of Authenticity. Cambridge: Harvard University Press.

Taylor, Charles (2015): Hegel and Modern Society. Cambridge: Cambridge University Press.

Taylor, Matthew (27.12.2021): "Asad Rehman on climate justice", The Guardian.

Taylor, Matthew; Gayle, Damien; Brooks, Libby (17.4.2019): "Extinction Rebellion keep control of major London sites into a third day", The Guardian.

Teske, Sven et al. (2019): Achieving the Paris Climate Agreement Goals. Berlin: Springer.

Thanki, Nathan (2019): "A new chance for climate justice?" https://www.opendemocracy.net/en/opendemocracyuk/new-chance-climate-justice.

Theunissen, Michael (1984): The Other. Cambridge: MIT Press.

Theunissen, Michael (1970): Hegels Lehre vom absoluten Geist als theologisch-politischer Traktat. Berlin: De Gruyter.

Thompson, Evan (2010): Mind in Life: Biology, Phenomenology, and the Sciences of Mind. Cambridge: Harvard University Press.

Thunberg, Greta (2022): The climate book. London: Penguin.

Thunberg, Greta (2020/3): "Humanity has not yet failed", https://sverigesradio.se/avsnitt/1535269.

Thunberg, Greta et al. (2020/2): #FaceTheClimateEmergency. https://climateemergencyeu.org.

Thunberg, Greta et al. (10.1.2020): "At Davos we will tell world leaders to abandon the fossil fuel economy", The Guardian.

Thunberg, Greta (2019): No one is too small to make a difference. London: Penguin.

Tsui, Tori (2023): It's Not Just You. New York: Simon&Schuster.

Twidale, Susanna & Chestney, Nina (8.3.2022): "Global energy-related CO2 emissions rose to record high in 2021", Reuters.

UNEP (20.12.2022): "COP15 ends with landmark biodiversity agreement". https://www.unep.org/news-and-stories/story/cop15-ends-landmark-biodiversity-agreement.

UNESCO (2022): https://www.whec2022.org/EN/homepage/WHEC2022.
UNHCR (2022): "2022, Floods Response Plan, Pakistan", https://reporting.unhcr.org/pakistan-revised-2022-floods-response-plan.
UNICEF (6.10.2022): "Annalkande svältkatastrof på Afrikas horn". https://blog.unicef.se/2022/10/06/annalkande-svaltkatastrof-pa-afrikas-horn-agerar-vi-inte-nu-far-vi-betala-priset-for-var-passivitet-under-resten-av-vara-liv.
Urisman Otto, Alexandra (2022): "The Truth about Government Climate Targets", in: Thunberg (2022), 210–216.
Van den Berg, Bas (2021/1): "Regenerative Education for The Ecological University in Times of Socio-Ecological Crises – Educational Design Dispositions, Qualities, Opportunities & Barriers", IAFOR 2nd Barcelona Conference on Education.
Van den Berg, Bas et. al. (2021/2): "Navigating Personal and Systemic Barriers to Enact Regenerative Higher Education for a more Sustainable World". https://library.wur.nl/ojs/index.php/CircularWUR2022/article/view/18285.
Vetter, Andrea (2021): *Konviviale Technik*. Bielefeld: transcript.
Vogl, Joseph (2021): *Kapital und Ressentiment*. München: Beck.
Von Redecker, Eva (2020): *Revolution für das Leben*. Frankfurt: Fischer Verlag.
Voosen, Paul (25.3.2021): "U.S. needs solar geoengineering research program, National Academies says", https://www.science.org/content/article/us-needs-solar-geoengineering-research-program-national-academies-says.
Voss, Regina (13.4.2019): "Wie sich die Generation Greta das Reisen vorstellt", Deutschlandfunk.
Wallace-Wells, David (2019): *The uninhabitable earth*. London/New York: Penguin.
Warren, Mark R. (2014): "Transforming Public Education: The Need for an Educational Justice Movement", New England Journal of Public Policy: Vol. 26: Iss. 1, Article 11.
Welch, Craig (9.2021): "Arctic permafrost is thawing fast. That affects us all", National Geographic.
Wesche, Tilo (2023): *Die Rechte der Natur*. Berlin: Suhrkamp.
Westberg, Marcus (16.4.2021): "<Forests are not renewable>: the felling of Sweden's ancient trees", The Guardian.
Westhoek, Henk et al. (2014): "Food choices, health and environment: Effects of cutting Europe's meat and dairy intake", Global Environmental Change, Volume 26, 2014, 196–205.

Wetzel, Jakob (2.4.2019): "Mit Bußgeld gegen die Schülerstreiks", Sueddeutsche Zeitung.
WFP (24.5.2023): "Horn of Africa hunger crisis pushes millions to the brink", https://www.wfp.org/stories/horn-africa-hunger-crisis-pushes-millions-brink.
WHO (2019): "1 in 3 people globally do not have access to safe drinking water", https://www.who.int/news/item/18-06-2019-1-in-3-people-globally-do-not-have-access-to-safe-drinking-water-unicef-who.
WHO (25.3.2014): "7 million premature deaths annually linked to air pollution", https://www.who.int/news/item/25-03-2014-7-million-premature-deaths-annually-linked-to-air-pollution.
Winnicott, Donald (2005): *Playing and Reality*. London: Routledge.
WMO (31.7.2023): "July 2023 is set to be the hottest month on record", https://public.wmo.int/en/media/press-release/july-2023-set-be-hottest-month-record.
WMO (17.5.2023): "WMO Global Annual to Decadal Climate Update", https://mcusercontent.com/618614864060486033e4590d6/files/318fc86d-b700-3a55-72b6-19cf6a531ded/WMO_GADCU_2023_2027.pdf.
WMO (8.7.2020): "New climate predictions assess global temperatures in coming five years", https://public.wmo.int/en/media/press-release/new-climate-predictions-assess-global-temperatures-coming-five-years.
Wohlleben, Peter (2017): *The Hidden Life of Trees*. Glasgow: William Collins.
Wolff, Jonathan (22.11.2022): „Justice Across Ages: Treating Young and Old as Equals, by Juliana Uhuru Bidadanure", Mind, fzac057.
Xu, Chi et al. (2020): "Future of the human climate niche", PNAS Mai 26, 117 (21), S. 11350–11355.
Young, Iris Marion (2000): *Inclusion and Democracy*. Oxford: Oxford University Press.
Young, Iris Marion (1997): *Intersecting voices*. Princeton: Princeton University Press.
ZDI (2022): "Education for Sustainable Development", https://zdi-portal.de/en/blog/bne-lernen-fuer-eine-bessere-welt.